Measurement and Instrumentation Principles

To Jane, Nicola and Julia

Measurement and Instrumentation Principles

Alan S. Morris

OXFORD   AUCKLAND   BOSTON   JOHANNESBURG   MELBOURNE   NEW DELHI

Butterworth-Heinemann
Linacre House, Jordan Hill, Oxford OX2 8DP
225 Wildwood Avenue, Woburn, MA 01801-2041
A division of Reed Educational and Professional Publishing Ltd

A member of the Reed Elsevier plc group

First published 2001

© Alan S. Morris 2001

All rights reserved. No part of this publication
may be reproduced in any material form (including
photocopying or storing in any medium by electronic
means and whether or not transiently or incidentally
to some other use of this publication) without the
written permission of the copyright holder except
in accordance with the provisions of the Copyright,
Designs and Patents Act 1988 or under the terms of a
licence issued by the Copyright Licensing Agency Ltd,
90 Tottenham Court Road, London, England W1P 9HE.
Applications for the copyright holder's written permission
to reproduce any part of this publication should be addressed
to the publishers

British Library Cataloguing in Publication Data
A catalogue record for this book is available from the British Library

ISBN 0 7506 5081 8

Typeset in 10/12pt Times Roman by Laser Words, Madras, India
Printed and bound in Great Britain by MPG Books Ltd, Bodmin, Cornwall

Contents

Preface xvii
Acknowledgements xx

Part 1: Principles of Measurement 1

1 INTRODUCTION TO MEASUREMENT 3
   1.1 Measurement units 3
   1.2 Measurement system applications 6
   1.3 Elements of a measurement system 8
   1.4 Choosing appropriate measuring instruments 9

2 INSTRUMENT TYPES AND PERFORMANCE CHARACTERISTICS 12
   2.1 Review of instrument types 12
      2.1.1 Active and passive instruments 12
      2.1.2 Null-type and deflection-type instruments 13
      2.1.3 Analogue and digital instruments 14
      2.1.4 Indicating instruments and instruments with a signal output 15
      2.1.5 Smart and non-smart instruments 16
   2.2 Static characteristics of instruments 16
      2.2.1 Accuracy and inaccuracy (measurement uncertainty) 16
      2.2.2 Precision/repeatability/reproducibility 17
      2.2.3 Tolerance 17
      2.2.4 Range or span 18
      2.2.5 Linearity 19
      2.2.6 Sensitivity of measurement 19
      2.2.7 Threshold 20
      2.2.8 Resolution 20
      2.2.9 Sensitivity to disturbance 20
      2.2.10 Hysteresis effects 22
      2.2.11 Dead space 23
   2.3 Dynamic characteristics of instruments 23

		2.3.1	Zero order instrument	25

		2.3.1	Zero order instrument	25
		2.3.2	First order instrument	25
		2.3.3	Second order instrument	28
	2.4	Necessity for calibration		29
	2.5	Self-test questions		30

3 ERRORS DURING THE MEASUREMENT PROCESS — 32

	3.1	Introduction		32
	3.2	Sources of systematic error		33
		3.2.1	System disturbance due to measurement	33
		3.2.2	Errors due to environmental inputs	37
		3.2.3	Wear in instrument components	38
		3.2.4	Connecting leads	38
	3.3	Reduction of systematic errors		39
		3.3.1	Careful instrument design	39
		3.3.2	Method of opposing inputs	39
		3.3.3	High-gain feedback	39
		3.3.4	Calibration	41
		3.3.5	Manual correction of output reading	42
		3.3.6	Intelligent instruments	42
	3.4	Quantification of systematic errors		42
	3.5	Random errors		42
		3.5.1	Statistical analysis of measurements subject to random errors	43
		3.5.2	Graphical data analysis techniques – frequency distributions	46
	3.6	Aggregation of measurement system errors		56
		3.6.1	Combined effect of systematic and random errors	56
		3.6.2	Aggregation of errors from separate measurement system components	56
		3.6.3	Total error when combining multiple measurements	59
	3.7	Self-test questions		60
	References and further reading			63

4 CALIBRATION OF MEASURING SENSORS AND INSTRUMENTS — 64

	4.1	Principles of calibration	64
	4.2	Control of calibration environment	66
	4.3	Calibration chain and traceability	67
	4.4	Calibration records	71
	References and further reading		72

5 MEASUREMENT NOISE AND SIGNAL PROCESSING — 73

	5.1	Sources of measurement noise		73
		5.1.1	Inductive coupling	74
		5.1.2	Capacitive (electrostatic) coupling	74
		5.1.3	Noise due to multiple earths	74

Contents   vii

		5.1.4	Noise in the form of voltage transients	75
		5.1.5	Thermoelectric potentials	75
		5.1.6	Shot noise	76
		5.1.7	Electrochemical potentials	76
	5.2	Techniques for reducing measurement noise		76
		5.2.1	Location and design of signal wires	76
		5.2.2	Earthing	77
		5.2.3	Shielding	77
		5.2.4	Other techniques	77
	5.3	Introduction to signal processing		78
	5.4	Analogue signal filtering		78
		5.4.1	Passive analogue filters	81
		5.4.2	Active analogue filters	85
	5.5	Other analogue signal processing operations		86
		5.5.1	Signal amplification	87
		5.5.2	Signal attenuation	88
		5.5.3	Differential amplification	89
		5.5.4	Signal linearization	90
		5.5.5	Bias (zero drift) removal	91
		5.5.6	Signal integration	92
		5.5.7	Voltage follower (pre-amplifier)	92
		5.5.8	Voltage comparator	92
		5.5.9	Phase-sensitive detector	93
		5.5.10	Lock-in amplifier	94
		5.5.11	Signal addition	94
		5.5.12	Signal multiplication	95
	5.6	Digital signal processing		95
		5.6.1	Signal sampling	95
		5.6.2	Sample and hold circuit	97
		5.6.3	Analogue-to-digital converters	97
		5.6.4	Digital-to-analogue (D/A) conversion	99
		5.6.5	Digital filtering	100
		5.6.6	Autocorrelation	100
		5.6.7	Other digital signal processing operations	101
	References and further reading			101
6	**ELECTRICAL INDICATING AND TEST INSTRUMENTS**			**102**
	6.1	Digital meters		102
		6.1.1	Voltage-to-time conversion digital voltmeter	103
		6.1.2	Potentiometric digital voltmeter	103
		6.1.3	Dual-slope integration digital voltmeter	103
		6.1.4	Voltage-to-frequency conversion digital voltmeter	104
		6.1.5	Digital multimeter	104
	6.2	Analogue meters		104
		6.2.1	Moving-coil meters	105
		6.2.2	Moving-iron meter	106
		6.2.3	Electrodynamic meters	107

		6.2.4	Clamp-on meters	108
		6.2.5	Analogue multimeter	108
		6.2.6	Measuring high-frequency signals	109
		6.2.7	Thermocouple meter	110
		6.2.8	Electronic analogue voltmeters	111
		6.2.9	Calculation of meter outputs for non-standard waveforms	112
	6.3	Cathode ray oscilloscope		114
		6.3.1	Cathode ray tube	115
		6.3.2	Channel	116
		6.3.3	Single-ended input	117
		6.3.4	Differential input	117
		6.3.5	Timebase circuit	117
		6.3.6	Vertical sensitivity control	117
		6.3.7	Display position control	118
	6.4	Digital storage oscilloscopes		118
	References and further reading			118
7	**VARIABLE CONVERSION ELEMENTS**			**119**
	7.1	Bridge circuits		119
		7.1.1	Null-type, d.c. bridge (Wheatstone bridge)	120
		7.1.2	Deflection-type d.c. bridge	121
		7.1.3	Error analysis	128
		7.1.4	A.c. bridges	130
	7.2	Resistance measurement		134
		7.2.1	D.c. bridge circuit	135
		7.2.2	Voltmeter–ammeter method	135
		7.2.3	Resistance-substitution method	135
		7.2.4	Use of the digital voltmeter to measure resistance	136
		7.2.5	The ohmmeter	136
		7.2.6	Codes for resistor values	137
	7.3	Inductance measurement		138
	7.4	Capacitance measurement		138
		7.4.1	Alphanumeric codes for capacitor values	139
	7.5	Current measurement		140
	7.6	Frequency measurement		141
		7.6.1	Digital counter-timers	142
		7.6.2	Phase-locked loop	142
		7.6.3	Cathode ray oscilloscope	143
		7.6.4	The Wien bridge	144
	7.7	Phase measurement		145
		7.7.1	Electronic counter-timer	145
		7.7.2	X–Y plotter	145
		7.7.3	Oscilloscope	147
		7.7.4	Phase-sensitive detector	147
	7.8	Self-test questions		147
	References and further reading			150

8 SIGNAL TRANSMISSION — 151
- 8.1 Electrical transmission — 151
  - 8.1.1 Transmission as varying voltages — 151
  - 8.1.2 Current loop transmission — 152
  - 8.1.3 Transmission using an a.c. carrier — 153
- 8.2 Pneumatic transmission — 154
- 8.3 Fibre-optic transmission — 155
  - 8.3.1 Principles of fibre optics — 156
  - 8.3.2 Transmission characteristics — 158
  - 8.3.3 Multiplexing schemes — 160
- 8.4 Optical wireless telemetry — 160
- 8.5 Radio telemetry (radio wireless transmission) — 161
- 8.6 Digital transmission protocols — 163
- References and further reading — 164

9 DIGITAL COMPUTATION AND INTELLIGENT DEVICES — 165
- 9.1 Principles of digital computation — 165
  - 9.1.1 Elements of a computer — 165
  - 9.1.2 Computer operation — 168
  - 9.1.3 Interfacing — 174
  - 9.1.4 Practical considerations in adding computers to measurement systems — 176
- 9.2 Intelligent devices — 177
  - 9.2.1 Intelligent instruments — 177
  - 9.2.2 Smart sensors — 179
  - 9.2.3 Smart transmitters — 180
  - 9.2.4 Communication with intelligent devices — 183
  - 9.2.5 Computation in intelligent devices — 184
  - 9.2.6 Future trends in intelligent devices — 185
- 9.3 Self-test questions — 185
- References and further reading — 186

10 INSTRUMENTATION/COMPUTER NETWORKS — 187
- 10.1 Introduction — 187
- 10.2 Serial communication lines — 188
  - 10.2.1 Asynchronous transmission — 189
- 10.3 Parallel data bus — 190
- 10.4 Local area networks (LANs) — 192
  - 10.4.1 Star networks — 193
  - 10.4.2 Ring and bus networks — 194
- 10.5 Gateways — 195
- 10.6 HART — 195
- 10.7 Digital fieldbuses — 196
- 10.8 Communication protocols for very large systems — 198
  - 10.8.1 Protocol standardization — 198
- 10.9 Future development of networks — 199
- References and further reading — 199

11 DISPLAY, RECORDING AND PRESENTATION OF MEASUREMENT DATA — 200

- 11.1 Display of measurement signals — 200
  - 11.1.1 Electronic output displays — 200
  - 11.1.2 Computer monitor displays — 201
- 11.2 Recording of measurement data — 202
  - 11.2.1 Mechanical chart recorders — 202
  - 11.2.2 Ultra-violet recorders — 208
  - 11.2.3 Fibre-optic recorders (recording oscilloscopes) — 209
  - 11.2.4 Hybrid chart recorders — 209
  - 11.2.5 Magnetic tape recorders — 209
  - 11.2.6 Digital recorders — 210
  - 11.2.7 Storage oscilloscopes — 211
- 11.3 Presentation of data — 212
  - 11.3.1 Tabular data presentation — 212
  - 11.3.2 Graphical presentation of data — 213
- 11.4 Self-test questions — 222
- References and further reading — 223

12 MEASUREMENT RELIABILITY AND SAFETY SYSTEMS — 224

- 12.1 Reliability — 224
  - 12.1.1 Principles of reliability — 224
  - 12.1.2 Laws of reliability in complex systems — 228
  - 12.1.3 Improving measurement system reliability — 229
  - 12.1.4 Software reliability — 232
- 12.2 Safety systems — 236
  - 12.2.1 Introduction to safety systems — 236
  - 12.2.2 Operation of safety systems — 237
  - 12.2.3 Design of a safety system — 238
- 12.3 Self-test questions — 241
- References and further reading — 242

Part 2: Measurement Sensors and Instruments — 245

13 SENSOR TECHNOLOGIES — 247

- 13.1 Capacitive and resistive sensors — 247
- 13.2 Magnetic sensors — 247
- 13.3 Hall-effect sensors — 249
- 13.4 Piezoelectric transducers — 250
- 13.5 Strain gauges — 251
- 13.6 Piezoresistive sensors — 252
- 13.7 Optical sensors (air path) — 252
- 13.8 Optical sensors (fibre-optic) — 253
  - 13.8.1 Intrinsic sensors — 254
  - 13.8.2 Extrinsic sensors — 258
  - 13.8.3 Distributed sensors — 259

	13.9	Ultrasonic transducers	259
		13.9.1 Transmission speed	260
		13.9.2 Direction of travel of ultrasound waves	261
		13.9.3 Directionality of ultrasound waves	261
		13.9.4 Relationship between wavelength, frequency and directionality of ultrasound waves	262
		13.9.5 Attenuation of ultrasound waves	262
		13.9.6 Ultrasound as a range sensor	263
		13.9.7 Use of ultrasound in tracking 3D object motion	264
		13.9.8 Effect of noise in ultrasonic measurement systems	265
		13.9.9 Exploiting Doppler shift in ultrasound transmission	265
		13.9.10 Ultrasonic imaging	267
	13.10	Nuclear sensors	267
	13.11	Microsensors	268
	References and further reading		270
14	**TEMPERATURE MEASUREMENT**		**271**
	14.1	Principles of temperature measurement	271
	14.2	Thermoelectric effect sensors (thermocouples)	272
		14.2.1 Thermocouple tables	276
		14.2.2 Non-zero reference junction temperature	277
		14.2.3 Thermocouple types	279
		14.2.4 Thermocouple protection	280
		14.2.5 Thermocouple manufacture	281
		14.2.6 The thermopile	282
		14.2.7 Digital thermometer	282
		14.2.8 The continuous thermocouple	282
	14.3	Varying resistance devices	283
		14.3.1 Resistance thermometers (resistance temperature devices)	284
		14.3.2 Thermistors	285
	14.4	Semiconductor devices	286
	14.5	Radiation thermometers	287
		14.5.1 Optical pyrometers	289
		14.5.2 Radiation pyrometers	290
	14.6	Thermography (thermal imaging)	293
	14.7	Thermal expansion methods	294
		14.7.1 Liquid-in-glass thermometers	295
		14.7.2 Bimetallic thermometer	296
		14.7.3 Pressure thermometers	296
	14.8	Quartz thermometers	297
	14.9	Fibre-optic temperature sensors	297
	14.10	Acoustic thermometers	298
	14.11	Colour indicators	299
	14.12	Change of state of materials	299
	14.13	Intelligent temperature-measuring instruments	300
	14.14	Choice between temperature transducers	300

	14.15	Self-test questions	302
		References and further reading	303
15	**PRESSURE MEASUREMENT**		**304**
	15.1	Diaphragms	305
	15.2	Capacitive pressure sensor	306
	15.3	Fibre-optic pressure sensors	306
	15.4	Bellows	307
	15.5	Bourdon tube	308
	15.6	Manometers	310
	15.7	Resonant-wire devices	311
	15.8	Dead-weight gauge	312
	15.9	Special measurement devices for low pressures	312
	15.10	High-pressure measurement (greater than 7000 bar)	315
	15.11	Intelligent pressure transducers	316
	15.12	Selection of pressure sensors	316
16	**FLOW MEASUREMENT**		**319**
	16.1	Mass flow rate	319
		16.1.1 Conveyor-based methods	319
		16.1.2 Coriolis flowmeter	320
		16.1.3 Thermal mass flow measurement	320
		16.1.4 Joint measurement of volume flow rate and fluid density	321
	16.2	Volume flow rate	321
		16.2.1 Differential pressure (obstruction-type) meters	322
		16.2.2 Variable area flowmeters (Rotameters)	327
		16.2.3 Positive displacement flowmeters	328
		16.2.4 Turbine meters	329
		16.2.5 Electromagnetic flowmeters	330
		16.2.6 Vortex-shedding flowmeters	332
		16.2.7 Ultrasonic flowmeters	332
		16.2.8 Other types of flowmeter for measuring volume flow rate	336
	16.3	Intelligent flowmeters	338
	16.4	Choice between flowmeters for particular applications	338
		References and further reading	339
17	**LEVEL MEASUREMENT**		**340**
	17.1	Dipsticks	340
	17.2	Float systems	340
	17.3	Pressure-measuring devices (hydrostatic systems)	341
	17.4	Capacitive devices	343
	17.5	Ultrasonic level gauge	344
	17.6	Radar (microwave) methods	346

	17.7	Radiation methods	346
	17.8	Other techniques	348
		17.8.1 Vibrating level sensor	348
		17.8.2 Hot-wire elements/carbon resistor elements	348
		17.8.3 Laser methods	349
		17.8.4 Fibre-optic level sensors	349
		17.8.5 Thermography	349
	17.9	Intelligent level-measuring instruments	351
	17.10	Choice between different level sensors	351
		References and further reading	351
18	**MASS, FORCE AND TORQUE MEASUREMENT**		**352**
	18.1	Mass (weight) measurement	352
		18.1.1 Electronic load cell (electronic balance)	352
		18.1.2 Pneumatic/hydraulic load cells	354
		18.1.3 Intelligent load cells	355
		18.1.4 Mass-balance (weighing) instruments	356
		18.1.5 Spring balance	359
	18.2	Force measurement	359
		18.2.1 Use of accelerometers	360
		18.2.2 Vibrating wire sensor	360
	18.3	Torque measurement	361
		18.3.1 Reaction forces in shaft bearings	361
		18.3.2 Prony brake	361
		18.3.3 Measurement of induced strain	362
		18.3.4 Optical torque measurement	364
19	**TRANSLATIONAL MOTION TRANSDUCERS**		**365**
	19.1	Displacement	365
		19.1.1 The resistive potentiometer	365
		19.1.2 Linear variable differential transformer (LVDT)	368
		19.1.3 Variable capacitance transducers	370
		19.1.4 Variable inductance transducers	371
		19.1.5 Strain gauges	371
		19.1.6 Piezoelectric transducers	373
		19.1.7 Nozzle flapper	373
		19.1.8 Other methods of measuring small displacements	374
		19.1.9 Measurement of large displacements (range sensors)	378
		19.1.10 Proximity sensors	381
		19.1.11 Selection of translational measurement transducers	382
	19.2	Velocity	382
		19.2.1 Differentiation of displacement measurements	382
		19.2.2 Integration of the output of an accelerometer	383
		19.2.3 Conversion to rotational velocity	383
	19.3	Acceleration	383
		19.3.1 Selection of accelerometers	385

xiv Contents

	19.4	Vibration	386
		19.4.1 Nature of vibration	386
		19.4.2 Vibration measurement	386
	19.5	Shock	388

20 ROTATIONAL MOTION TRANSDUCERS 390
	20.1	Rotational displacement	390
		20.1.1 Circular and helical potentiometers	390
		20.1.2 Rotational differential transformer	391
		20.1.3 Incremental shaft encoders	392
		20.1.4 Coded-disc shaft encoders	394
		20.1.5 The resolver	398
		20.1.6 The synchro	399
		20.1.7 The induction potentiometer	402
		20.1.8 The rotary inductosyn	402
		20.1.9 Gyroscopes	402
		20.1.10 Choice between rotational displacement transducers	406
	20.2	Rotational velocity	407
		20.2.1 Digital tachometers	407
		20.2.2 Stroboscopic methods	410
		20.2.3 Analogue tachometers	411
		20.2.4 Mechanical flyball	413
		20.2.5 The rate gyroscope	415
		20.2.6 Fibre-optic gyroscope	416
		20.2.7 Differentiation of angular displacement measurements	417
		20.2.8 Integration of the output from an accelerometer	417
		20.2.9 Choice between rotational velocity transducers	417
	20.3	Measurement of rotational acceleration	417
	References and further reading		418

21 SUMMARY OF OTHER MEASUREMENTS 419
	21.1	Dimension measurement	419
		21.1.1 Rules and tapes	419
		21.1.2 Callipers	421
		21.1.3 Micrometers	422
		21.1.4 Gauge blocks (slip gauges) and length bars	423
		21.1.5 Height and depth measurement	425
	21.2	Angle measurement	426
	21.3	Flatness measurement	428
	21.4	Volume measurement	428
	21.5	Viscosity measurement	429
		21.5.1 Capillary and tube viscometers	430
		21.5.2 Falling body viscometer	431
		21.5.3 Rotational viscometers	431
	21.6	Moisture measurement	432
		21.6.1 Industrial moisture measurement techniques	432
		21.6.2 Laboratory techniques for moisture measurement	434

		21.6.3 Humidity measurement	435
21.7	Sound measurement		436
21.8	pH measurement		437
		21.8.1 The glass electrode	438
		21.8.2 Other methods of pH measurement	439
21.9	Gas sensing and analysis		439
		21.9.1 Catalytic (calorimetric) sensors	440
		21.9.2 Paper tape sensors	441
		21.9.3 Liquid electrolyte electrochemical cells	441
		21.9.4 Solid-state electrochemical cells (zirconia sensor)	442
		21.9.5 Catalytic gate FETs	442
		21.9.6 Semiconductor (metal oxide) sensors	442
		21.9.7 Organic sensors	442
		21.9.8 Piezoelectric devices	443
		21.9.9 Infra-red absorption	443
		21.9.10 Mass spectrometers	443
		21.9.11 Gas chromatography	443
References and further reading			444
APPENDIX 1	Imperial–metric–SI conversion tables		445
APPENDIX 2	Thévenin's theorem		452
APPENDIX 3	Thermocouple tables		458
APPENDIX 4	Solutions to self-test questions		464
INDEX			469

Preface

The foundations of this book lie in the highly successful text *Principles of Measurement and Instrumentation* by the same author. The first edition of this was published in 1988, and a second, revised and extended edition appeared in 1993. Since that time, a number of new developments have occurred in the field of measurement. In particular, there have been significant advances in smart sensors, intelligent instruments, microsensors, digital signal processing, digital recorders, digital fieldbuses and new methods of signal transmission. The rapid growth of digital components within measurement systems has also created a need to establish procedures for measuring and improving the reliability of the software that is used within such components. Formal standards governing instrument calibration procedures and measurement system performance have also extended beyond the traditional area of quality assurance systems (BS 5781, BS 5750 and more recently ISO 9000) into new areas such as environmental protection systems (BS 7750 and ISO 14000). Thus, an up-to-date book incorporating all of the latest developments in measurement is strongly needed. With so much new material to include, the opportunity has been taken to substantially revise the order and content of material presented previously in *Principles of Measurement and Instrumentation*, and several new chapters have been written to cover the many new developments in measurement and instrumentation that have occurred over the past few years. To emphasize the substantial revision that has taken place, a decision has been made to publish the book under a new title rather than as a third edition of the previous book. Hence, *Measurement and Instrumentation Principles* has been born.

The overall aim of the book is to present the topics of sensors and instrumentation, and their use within measurement systems, as an integrated and coherent subject. Measurement systems, and the instruments and sensors used within them, are of immense importance in a wide variety of domestic and industrial activities. The growth in the sophistication of instruments used in industry has been particularly significant as advanced automation schemes have been developed. Similar developments have also been evident in military and medical applications.

Unfortunately, the crucial part that measurement plays in all of these systems tends to get overlooked, and measurement is therefore rarely given the importance that it deserves. For example, much effort goes into designing sophisticated automatic control systems, but little regard is given to the accuracy and quality of the raw measurement data that such systems use as their inputs. This disregard of measurement system quality and performance means that such control systems will never achieve their full potential, as it is very difficult to increase their performance beyond the quality of the raw measurement data on which they depend.

Ideally, the principles of good measurement and instrumentation practice should be taught throughout the duration of engineering courses, starting at an elementary level and moving on to more advanced topics as the course progresses. With this in mind, the material contained in this book is designed both to support introductory courses in measurement and instrumentation, and also to provide in-depth coverage of advanced topics for higher-level courses. In addition, besides its role as a student course text, it is also anticipated that the book will be useful to practising engineers, both to update their knowledge of the latest developments in measurement theory and practice, and also to serve as a guide to the typical characteristics and capabilities of the range of sensors and instruments that are currently in use.

The text is divided into two parts. The principles and theory of measurement are covered first in Part 1 and then the ranges of instruments and sensors that are available for measuring various physical quantities are covered in Part 2. This order of coverage has been chosen so that the general characteristics of measuring instruments, and their behaviour in different operating environments, are well established before the reader is introduced to the procedures involved in choosing a measurement device for a particular application. This ensures that the reader will be properly equipped to appreciate and critically appraise the various merits and characteristics of different instruments when faced with the task of choosing a suitable instrument.

It should be noted that, whilst measurement theory inevitably involves some mathematics, the mathematical content of the book has deliberately been kept to the minimum necessary for the reader to be able to design and build measurement systems that perform to a level commensurate with the needs of the automatic control scheme or other system that they support. Where mathematical procedures are necessary, worked examples are provided as necessary throughout the book to illustrate the principles involved. Self-assessment questions are also provided in critical chapters to enable readers to test their level of understanding, with answers being provided in Appendix 4.

Part 1 is organized such that all of the elements in a typical measurement system are presented in a logical order, starting with the capture of a measurement signal by a sensor and then proceeding through the stages of signal processing, sensor output transducing, signal transmission and signal display or recording. Ancillary issues, such as calibration and measurement system reliability, are also covered. Discussion starts with a review of the different classes of instrument and sensor available, and the sort of applications in which these different types are typically used. This opening discussion includes analysis of the static and dynamic characteristics of instruments and exploration of how these affect instrument usage. A comprehensive discussion of measurement system errors then follows, with appropriate procedures for quantifying and reducing errors being presented. The importance of calibration procedures in all aspects of measurement systems, and particularly to satisfy the requirements of standards such as ISO 9000 and ISO 14000, is recognized by devoting a full chapter to the issues involved. This is followed by an analysis of measurement noise sources, and discussion on the various analogue and digital signal-processing procedures that are used to attenuate noise and improve the quality of signals. After coverage of the range of electrical indicating and test instruments that are used to monitor electrical measurement signals, a chapter is devoted to presenting the range of variable conversion elements (transducers) and techniques that are used to convert non-electrical sensor outputs into electrical signals, with particular emphasis on electrical bridge circuits. The problems of signal transmission are considered next, and various means of improving the quality of transmitted signals are presented. This is followed by an introduction to digital computation techniques, and then a description of their use within intelligent measurement devices. The methods used to combine a number of intelligent devices into a large measurement network, and the current status of development of digital fieldbuses, are also explained. Then, the final element in a measurement system, of displaying, recording and presenting measurement data, is covered. To conclude Part 1, the issues of measurement system reliability, and the effect of unreliability on plant safety systems, are discussed. This discussion also includes the subject of software reliability, since computational elements are now embedded in many measurement systems.

Part 2 commences in the opening chapter with a review of the various technologies used in measurement sensors. The chapters that follow then provide comprehensive coverage of the main types of sensor and instrument that exist for measuring all the physical quantities that a practising engineer is likely to meet in normal situations. However, whilst the coverage is as comprehensive as possible, the distinction is emphasized between (a) instruments that are current and in common use, (b) instruments that are current but not widely used except in special applications, for reasons of cost or limited capabilities, and (c) instruments that are largely obsolete as regards new industrial implementations, but are still encountered on older plant that was installed some years ago. As well as emphasizing this distinction, some guidance is given about how to go about choosing an instrument for a particular measurement application.

Acknowledgements

The author gratefully acknowledges permission by John Wiley and Sons Ltd to reproduce some material that was previously published in *Measurement and Calibration Requirements for Quality Assurance to ISO 9000* by A. S. Morris (published 1997). The material involved are Tables 1.1, 1.2 and 3.1, Figures 3.1, 4.2 and 4.3, parts of sections 2.1, 2.2, 2.3, 3.1, 3.2, 3.6, 4.3 and 4.4, and Appendix 1.

Part 1 Principles of Measurement

1

Introduction to measurement

Measurement techniques have been of immense importance ever since the start of human civilization, when measurements were first needed to regulate the transfer of goods in barter trade to ensure that exchanges were fair. The industrial revolution during the nineteenth century brought about a rapid development of new instruments and measurement techniques to satisfy the needs of industrialized production techniques. Since that time, there has been a large and rapid growth in new industrial technology. This has been particularly evident during the last part of the twentieth century, encouraged by developments in electronics in general and computers in particular. This, in turn, has required a parallel growth in new instruments and measurement techniques.

The massive growth in the application of computers to industrial process control and monitoring tasks has spawned a parallel growth in the requirement for instruments to measure, record and control process variables. As modern production techniques dictate working to tighter and tighter accuracy limits, and as economic forces limiting production costs become more severe, so the requirement for instruments to be both accurate and cheap becomes ever harder to satisfy. This latter problem is at the focal point of the research and development efforts of all instrument manufacturers. In the past few years, the most cost-effective means of improving instrument accuracy has been found in many cases to be the inclusion of digital computing power within instruments themselves. These intelligent instruments therefore feature prominently in current instrument manufacturers' catalogues.

1.1 Measurement units

The very first measurement units were those used in barter trade to quantify the amounts being exchanged and to establish clear rules about the relative values of different commodities. Such early systems of measurement were based on whatever was available as a measuring unit. For purposes of measuring length, the human torso was a convenient tool, and gave us units of the hand, the foot and the cubit. Although generally adequate for barter trade systems, such measurement units are of course imprecise, varying as they do from one person to the next. Therefore, there has been a progressive movement towards measurement units that are defined much more accurately.

4 Introduction to measurement

The first improved measurement unit was a unit of length (the metre) defined as 10^{-7} times the polar quadrant of the earth. A platinum bar made to this length was established as a standard of length in the early part of the nineteenth century. This was superseded by a superior quality standard bar in 1889, manufactured from a platinum–iridium alloy. Since that time, technological research has enabled further improvements to be made in the standard used for defining length. Firstly, in 1960, a standard metre was redefined in terms of 1.65076373×10^6 wavelengths of the radiation from krypton-86 in vacuum. More recently, in 1983, the metre was redefined yet again as the length of path travelled by light in an interval of 1/299 792 458 seconds. In a similar fashion, standard units for the measurement of other physical quantities have been defined and progressively improved over the years. The latest standards for defining the units used for measuring a range of physical variables are given in Table 1.1.

The early establishment of standards for the measurement of physical quantities proceeded in several countries at broadly parallel times, and in consequence, several sets of units emerged for measuring the same physical variable. For instance, length can be measured in yards, metres, or several other units. Apart from the major units of length, subdivisions of standard units exist such as feet, inches, centimetres and millimetres, with a fixed relationship between each fundamental unit and its subdivisions.

Table 1.1 Definitions of standard units

Physical quantity	Standard unit	Definition
Length	metre	The length of path travelled by light in an interval of 1/299 792 458 seconds
Mass	kilogram	The mass of a platinum–iridium cylinder kept in the International Bureau of Weights and Measures, Sèvres, Paris
Time	second	9.192631770×10^9 cycles of radiation from vaporized caesium-133 (an accuracy of 1 in 10^{12} or 1 second in 36 000 years)
Temperature	kelvin	The temperature difference between absolute zero and the triple point of water is defined as 273.16 kelvin
Current	ampere	One ampere is the current flowing through two infinitely long parallel conductors of negligible cross-section placed 1 metre apart in a vacuum and producing a force of 2×10^{-7} newtons per metre length of conductor
Luminous intensity	candela	One candela is the luminous intensity in a given direction from a source emitting monochromatic radiation at a frequency of 540 terahertz (Hz $\times 10^{12}$) and with a radiant density in that direction of 1.4641 mW/steradian. (1 steradian is the solid angle which, having its vertex at the centre of a sphere, cuts off an area of the sphere surface equal to that of a square with sides of length equal to the sphere radius)
Matter	mole	The number of atoms in a 0.012 kg mass of carbon-12

Table 1.2 Fundamental and derived SI units

(a) Fundamental units

Quantity	Standard unit	Symbol
Length	metre	m
Mass	kilogram	kg
Time	second	s
Electric current	ampere	A
Temperature	kelvin	K
Luminous intensity	candela	cd
Matter	mole	mol

(b) Supplementary fundamental units

Quantity	Standard unit	Symbol
Plane angle	radian	rad
Solid angle	steradian	sr

(c) Derived units

Quantity	Standard unit	Symbol	Derivation formula
Area	square metre	m^2	
Volume	cubic metre	m^3	
Velocity	metre per second	m/s	
Acceleration	metre per second squared	m/s^2	
Angular velocity	radian per second	rad/s	
Angular acceleration	radian per second squared	rad/s^2	
Density	kilogram per cubic metre	kg/m^3	
Specific volume	cubic metre per kilogram	m^3/kg	
Mass flow rate	kilogram per second	kg/s	
Volume flow rate	cubic metre per second	m^3/s	
Force	newton	N	$kg\,m/s^2$
Pressure	newton per square metre	N/m^2	
Torque	newton metre	N m	
Momentum	kilogram metre per second	kg m/s	
Moment of inertia	kilogram metre squared	$kg\,m^2$	
Kinematic viscosity	square metre per second	m^2/s	
Dynamic viscosity	newton second per square metre	$N\,s/m^2$	
Work, energy, heat	joule	J	N m
Specific energy	joule per cubic metre	J/m^3	
Power	watt	W	J/s
Thermal conductivity	watt per metre kelvin	W/m K	
Electric charge	coulomb	C	A s
Voltage, e.m.f., pot. diff.	volt	V	W/A
Electric field strength	volt per metre	V/m	
Electric resistance	ohm	Ω	V/A
Electric capacitance	farad	F	A s/V
Electric inductance	henry	H	V s/A
Electric conductance	siemen	S	A/V
Resistivity	ohm metre	Ωm	
Permittivity	farad per metre	F/m	
Permeability	henry per metre	H/m	
Current density	ampere per square metre	A/m^2	

(*continued overleaf*)

Table 1.2 (*continued*)

(c) Derived units

Quantity	Standard unit	Symbol	Derivation formula
Magnetic flux	weber	Wb	V s
Magnetic flux density	tesla	T	Wb/m^2
Magnetic field strength	ampere per metre	A/m	
Frequency	hertz	Hz	s^{-1}
Luminous flux	lumen	lm	cd sr
Luminance	candela per square metre	cd/m^2	
Illumination	lux	lx	lm/m^2
Molar volume	cubic metre per mole	m^3/mol	
Molarity	mole per kilogram	mol/kg	
Molar energy	joule per mole	J/mol	

Yards, feet and inches belong to the Imperial System of units, which is characterized by having varying and cumbersome multiplication factors relating fundamental units to subdivisions such as 1760 (miles to yards), 3 (yards to feet) and 12 (feet to inches). The metric system is an alternative set of units, which includes for instance the unit of the metre and its centimetre and millimetre subdivisions for measuring length. All multiples and subdivisions of basic metric units are related to the base by factors of ten and such units are therefore much easier to use than Imperial units. However, in the case of derived units such as velocity, the number of alternative ways in which these can be expressed in the metric system can lead to confusion.

As a result of this, an internationally agreed set of standard units (SI units or Systèmes Internationales d'Unités) has been defined, and strong efforts are being made to encourage the adoption of this system throughout the world. In support of this effort, the SI system of units will be used exclusively in this book. However, it should be noted that the Imperial system is still widely used, particularly in America and Britain. The European Union has just deferred planned legislation to ban the use of Imperial units in Europe in the near future, and the latest proposal is to introduce such legislation to take effect from the year 2010.

The full range of fundamental SI measuring units and the further set of units derived from them are given in Table 1.2. Conversion tables relating common Imperial and metric units to their equivalent SI units can also be found in Appendix 1.

1.2 Measurement system applications

Today, the techniques of measurement are of immense importance in most facets of human civilization. Present-day applications of measuring instruments can be classified into three major areas. The first of these is their use in regulating trade, applying instruments that measure physical quantities such as length, volume and mass in terms of standard units. The particular instruments and transducers employed in such applications are included in the general description of instruments presented in Part 2 of this book.

The second application area of measuring instruments is in monitoring functions. These provide information that enables human beings to take some prescribed action accordingly. The gardener uses a thermometer to determine whether he should turn the heat on in his greenhouse or open the windows if it is too hot. Regular study of a barometer allows us to decide whether we should take our umbrellas if we are planning to go out for a few hours. Whilst there are thus many uses of instrumentation in our normal domestic lives, the majority of monitoring functions exist to provide the information necessary to allow a human being to control some industrial operation or process. In a chemical process for instance, the progress of chemical reactions is indicated by the measurement of temperatures and pressures at various points, and such measurements allow the operator to take correct decisions regarding the electrical supply to heaters, cooling water flows, valve positions etc. One other important use of monitoring instruments is in calibrating the instruments used in the automatic process control systems described below.

Use as part of automatic feedback control systems forms the third application area of measurement systems. Figure 1.1 shows a functional block diagram of a simple temperature control system in which the temperature T_a of a room is maintained at a reference value T_d. The value of the controlled variable T_a, as determined by a temperature-measuring device, is compared with the reference value T_d, and the difference e is applied as an error signal to the heater. The heater then modifies the room temperature until $T_a = T_d$. The characteristics of the measuring instruments used in any feedback control system are of fundamental importance to the quality of control achieved. The accuracy and resolution with which an output variable of a process is controlled can never be better than the accuracy and resolution of the measuring instruments used. This is a very important principle, but one that is often inadequately discussed in many texts on automatic control systems. Such texts explore the theoretical aspects of control system design in considerable depth, but fail to give sufficient emphasis to the fact that all gain and phase margin performance calculations etc. are entirely dependent on the quality of the process measurements obtained.

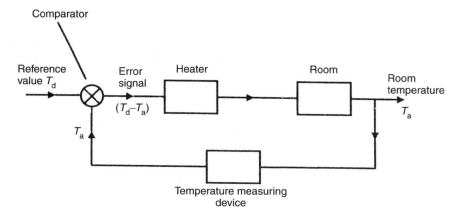

Fig. 1.1 Elements of a simple closed-loop control system.

1.3 Elements of a measurement system

A *measuring system* exists to provide information about the physical value of some variable being measured. In simple cases, the system can consist of only a single unit that gives an output reading or signal according to the magnitude of the unknown variable applied to it. However, in more complex measurement situations, a measuring system consists of several separate elements as shown in Figure 1.2. These components might be contained within one or more boxes, and the boxes holding individual measurement elements might be either close together or physically separate. The term *measuring instrument* is commonly used to describe a measurement system, whether it contains only one or many elements, and this term will be widely used throughout this text.

The first element in any measuring system is the primary *sensor*: this gives an output that is a function of the measurand (the input applied to it). For most but not all sensors, this function is at least approximately linear. Some examples of primary sensors are a liquid-in-glass thermometer, a thermocouple and a strain gauge. In the case of the mercury-in-glass thermometer, the output reading is given in terms of the level of the mercury, and so this particular primary sensor is also a complete measurement system in itself. However, in general, the primary sensor is only part of a measurement system. The types of primary sensors available for measuring a wide range of physical quantities are presented in Part 2 of this book.

Variable conversion elements are needed where the output variable of a primary transducer is in an inconvenient form and has to be converted to a more convenient form. For instance, the displacement-measuring strain gauge has an output in the form of a varying resistance. The resistance change cannot be easily measured and so it is converted to a change in voltage by a *bridge circuit*, which is a typical example of a variable conversion element. In some cases, the primary sensor and variable conversion element are combined, and the combination is known as a *transducer*.*

Signal processing elements exist to improve the quality of the output of a measurement system in some way. A very common type of signal processing element is the electronic amplifier, which amplifies the output of the primary transducer or variable conversion element, thus improving the sensitivity and resolution of measurement. This element of a measuring system is particularly important where the primary transducer has a low output. For example, thermocouples have a typical output of only a few millivolts. Other types of signal processing element are those that filter out induced noise and remove mean levels etc. In some devices, signal processing is incorporated into a transducer, which is then known as a *transmitter*.*

In addition to these three components just mentioned, some measurement systems have one or two other components, firstly to transmit the signal to some remote point and secondly to display or record the signal if it is not fed automatically into a feedback control system. Signal transmission is needed when the observation or application point of the output of a measurement system is some distance away from the site of the primary transducer. Sometimes, this separation is made solely for purposes of convenience, but more often it follows from the physical inaccessibility or environmental unsuitability of the site of the primary transducer for mounting the signal

---

* In some cases, the word 'sensor' is used generically to refer to both transducers and transmitters.

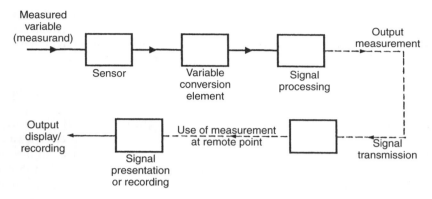

Fig. 1.2 Elements of a measuring instrument.

presentation/recording unit. The signal transmission element has traditionally consisted of single or multi-cored cable, which is often screened to minimize signal corruption by induced electrical noise. However, fibre-optic cables are being used in ever increasing numbers in modern installations, in part because of their low transmission loss and imperviousness to the effects of electrical and magnetic fields.

The final optional element in a measurement system is the point where the measured signal is utilized. In some cases, this element is omitted altogether because the measurement is used as part of an automatic control scheme, and the transmitted signal is fed directly into the control system. In other cases, this element in the measurement system takes the form either of a signal presentation unit or of a signal-recording unit. These take many forms according to the requirements of the particular measurement application, and the range of possible units is discussed more fully in Chapter 11.

1.4 Choosing appropriate measuring instruments

The starting point in choosing the most suitable instrument to use for measurement of a particular quantity in a manufacturing plant or other system is the specification of the instrument characteristics required, especially parameters like the desired measurement accuracy, resolution, sensitivity and dynamic performance (see next chapter for definitions of these). It is also essential to know the environmental conditions that the instrument will be subjected to, as some conditions will immediately either eliminate the possibility of using certain types of instrument or else will create a requirement for expensive protection of the instrument. It should also be noted that protection reduces the performance of some instruments, especially in terms of their dynamic characteristics (for example, sheaths protecting thermocouples and resistance thermometers reduce their speed of response). Provision of this type of information usually requires the expert knowledge of personnel who are intimately acquainted with the operation of the manufacturing plant or system in question. Then, a skilled instrument engineer, having knowledge of all the instruments that are available for measuring the quantity in question, will be able to evaluate the possible list of instruments in terms of their accuracy, cost and suitability for the environmental conditions and thus choose the most appropriate instrument. As far as possible, measurement systems and instruments should be chosen that are as insensitive as possible to the operating environment, although this requirement is often difficult to meet because of cost and other performance considerations. The extent to which the measured system will be disturbed during the measuring process is another important factor in instrument choice. For example, significant pressure loss can be caused to the measured system in some techniques of flow measurement.

Published literature is of considerable help in the choice of a suitable instrument for a particular measurement situation. Many books are available that give valuable assistance in the necessary evaluation by providing lists and data about all the instruments available for measuring a range of physical quantities (e.g. Part 2 of this text). However, new techniques and instruments are being developed all the time, and therefore a good instrumentation engineer must keep abreast of the latest developments by reading the appropriate technical journals regularly.

The instrument characteristics discussed in the next chapter are the features that form the technical basis for a comparison between the relative merits of different instruments. Generally, the better the characteristics, the higher the cost. However, in comparing the cost and relative suitability of different instruments for a particular measurement situation, considerations of durability, maintainability and constancy of performance are also very important because the instrument chosen will often have to be capable of operating for long periods without performance degradation and a requirement for costly maintenance. In consequence of this, the initial cost of an instrument often has a low weighting in the evaluation exercise.

Cost is very strongly correlated with the performance of an instrument, as measured by its static characteristics. Increasing the accuracy or resolution of an instrument, for example, can only be done at a penalty of increasing its manufacturing cost. Instrument choice therefore proceeds by specifying the minimum characteristics required by a measurement situation and then searching manufacturers' catalogues to find an instrument whose characteristics match those required. To select an instrument with characteristics superior to those required would only mean paying more than necessary for a level of performance greater than that needed.

As well as purchase cost, other important factors in the assessment exercise are instrument durability and the maintenance requirements. Assuming that one had £10 000 to spend, one would not spend £8000 on a new motor car whose projected life was five years if a car of equivalent specification with a projected life of ten years was available for £10 000. Likewise, durability is an important consideration in the choice of instruments. The projected life of instruments often depends on the conditions in which the instrument will have to operate. Maintenance requirements must also be taken into account, as they also have cost implications.

As a general rule, a good assessment criterion is obtained if the total purchase cost and estimated maintenance costs of an instrument over its life are divided by the period of its expected life. The figure obtained is thus a cost per year. However, this rule becomes modified where instruments are being installed on a process whose life is expected to be limited, perhaps in the manufacture of a particular model of car. Then, the total costs can only be divided by the period of time that an instrument is expected to be used for, unless an alternative use for the instrument is envisaged at the end of this period.

To summarize therefore, instrument choice is a compromise between performance characteristics, ruggedness and durability, maintenance requirements and purchase cost. To carry out such an evaluation properly, the instrument engineer must have a wide knowledge of the range of instruments available for measuring particular physical quantities, and he/she must also have a deep understanding of how instrument characteristics are affected by particular measurement situations and operating conditions.

2

Instrument types and performance characteristics

2.1 Review of instrument types

Instruments can be subdivided into separate classes according to several criteria. These subclassifications are useful in broadly establishing several attributes of particular instruments such as accuracy, cost, and general applicability to different applications.

2.1.1 Active and passive instruments

Instruments are divided into active or passive ones according to whether the instrument output is entirely produced by the quantity being measured or whether the quantity being measured simply modulates the magnitude of some external power source. This is illustrated by examples.

An example of a passive instrument is the pressure-measuring device shown in Figure 2.1. The pressure of the fluid is translated into a movement of a pointer against a scale. The energy expended in moving the pointer is derived entirely from the change in pressure measured: there are no other energy inputs to the system.

An example of an active instrument is a float-type petrol tank level indicator as sketched in Figure 2.2. Here, the change in petrol level moves a potentiometer arm, and the output signal consists of a proportion of the external voltage source applied across the two ends of the potentiometer. The energy in the output signal comes from the external power source: the primary transducer float system is merely modulating the value of the voltage from this external power source.

In active instruments, the external power source is usually in electrical form, but in some cases, it can be other forms of energy such as a pneumatic or hydraulic one.

One very important difference between active and passive instruments is the level of measurement resolution that can be obtained. With the simple pressure gauge shown, the amount of movement made by the pointer for a particular pressure change is closely defined by the nature of the instrument. Whilst it is possible to increase measurement resolution by making the pointer longer, such that the pointer tip moves through a longer arc, the scope for such improvement is clearly restricted by the practical limit of how long the pointer can conveniently be. In an active instrument, however, adjustment of the magnitude of the external energy input allows much greater control over

Measurement and Instrumentation Principles 13

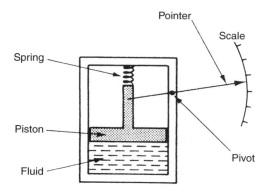

Fig. 2.1 Passive pressure gauge.

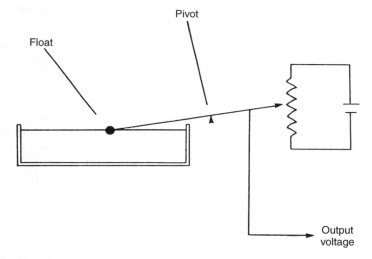

Fig. 2.2 Petrol-tank level indicator.

measurement resolution. Whilst the scope for improving measurement resolution is much greater incidentally, it is not infinite because of limitations placed on the magnitude of the external energy input, in consideration of heating effects and for safety reasons.

In terms of cost, passive instruments are normally of a more simple construction than active ones and are therefore cheaper to manufacture. Therefore, choice between active and passive instruments for a particular application involves carefully balancing the measurement resolution requirements against cost.

2.1.2 Null-type and deflection-type instruments

The pressure gauge just mentioned is a good example of a deflection type of instrument, where the value of the quantity being measured is displayed in terms of the amount of

14 Instrument types and performance characteristics

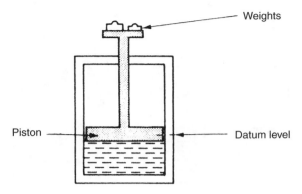

Fig. 2.3 Deadweight pressure gauge.

movement of a pointer. An alternative type of pressure gauge is the deadweight gauge shown in Figure 2.3, which is a null-type instrument. Here, weights are put on top of the piston until the downward force balances the fluid pressure. Weights are added until the piston reaches a datum level, known as the null point. Pressure measurement is made in terms of the value of the weights needed to reach this null position.

The accuracy of these two instruments depends on different things. For the first one it depends on the linearity and calibration of the spring, whilst for the second it relies on the calibration of the weights. As calibration of weights is much easier than careful choice and calibration of a linear-characteristic spring, this means that the second type of instrument will normally be the more accurate. This is in accordance with the general rule that null-type instruments are more accurate than deflection types.

In terms of usage, the deflection type instrument is clearly more convenient. It is far simpler to read the position of a pointer against a scale than to add and subtract weights until a null point is reached. A deflection-type instrument is therefore the one that would normally be used in the workplace. However, for calibration duties, the null-type instrument is preferable because of its superior accuracy. The extra effort required to use such an instrument is perfectly acceptable in this case because of the infrequent nature of calibration operations.

2.1.3 Analogue and digital instruments

An analogue instrument gives an output that varies continuously as the quantity being measured changes. The output can have an infinite number of values within the range that the instrument is designed to measure. The deflection-type of pressure gauge described earlier in this chapter (Figure 2.1) is a good example of an analogue instrument. As the input value changes, the pointer moves with a smooth continuous motion. Whilst the pointer can therefore be in an infinite number of positions within its range of movement, the number of different positions that the eye can discriminate between is strictly limited, this discrimination being dependent upon how large the scale is and how finely it is divided.

A digital instrument has an output that varies in discrete steps and so can only have a finite number of values. The rev counter sketched in Figure 2.4 is an example of

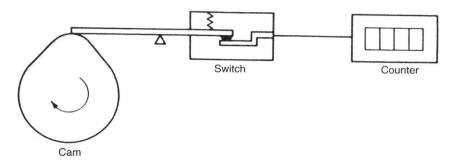

Fig. 2.4 Rev counter.

a digital instrument. A cam is attached to the revolving body whose motion is being measured, and on each revolution the cam opens and closes a switch. The switching operations are counted by an electronic counter. This system can only count whole revolutions and cannot discriminate any motion that is less than a full revolution.

The distinction between analogue and digital instruments has become particularly important with the rapid growth in the application of microcomputers to automatic control systems. Any digital computer system, of which the microcomputer is but one example, performs its computations in digital form. An instrument whose output is in digital form is therefore particularly advantageous in such applications, as it can be interfaced directly to the control computer. Analogue instruments must be interfaced to the microcomputer by an analogue-to-digital (A/D) converter, which converts the analogue output signal from the instrument into an equivalent digital quantity that can be read into the computer. This conversion has several disadvantages. Firstly, the A/D converter adds a significant cost to the system. Secondly, a finite time is involved in the process of converting an analogue signal to a digital quantity, and this time can be critical in the control of fast processes where the accuracy of control depends on the speed of the controlling computer. Degrading the speed of operation of the control computer by imposing a requirement for A/D conversion thus impairs the accuracy by which the process is controlled.

2.1.4 Indicating instruments and instruments with a signal output

The final way in which instruments can be divided is between those that merely give an audio or visual indication of the magnitude of the physical quantity measured and those that give an output in the form of a measurement signal whose magnitude is proportional to the measured quantity.

The class of indicating instruments normally includes all null-type instruments and most passive ones. Indicators can also be further divided into those that have an analogue output and those that have a digital display. A common analogue indicator is the liquid-in-glass thermometer. Another common indicating device, which exists in both analogue and digital forms, is the bathroom scale. The older mechanical form of this is an analogue type of instrument that gives an output consisting of a rotating pointer moving against a scale (or sometimes a rotating scale moving against a pointer). More recent electronic forms of bathroom scale have a digital output consisting of numbers presented on an electronic display. One major drawback with indicating devices is that human intervention is required to read and record a measurement. This process is particularly prone to error in the case of analogue output displays, although digital displays are not very prone to error unless the human reader is careless.

Instruments that have a signal-type output are commonly used as part of automatic control systems. In other circumstances, they can also be found in measurement systems where the output measurement signal is recorded in some way for later use. This subject is covered in later chapters. Usually, the measurement signal involved is an electrical voltage, but it can take other forms in some systems such as an electrical current, an optical signal or a pneumatic signal.

2.1.5 Smart and non-smart instruments

The advent of the microprocessor has created a new division in instruments between those that do incorporate a microprocessor (smart) and those that don't. Smart devices are considered in detail in Chapter 9.

2.2 Static characteristics of instruments

If we have a thermometer in a room and its reading shows a temperature of 20°C, then it does not really matter whether the true temperature of the room is 19.5°C or 20.5°C. Such small variations around 20°C are too small to affect whether we feel warm enough or not. Our bodies cannot discriminate between such close levels of temperature and therefore a thermometer with an inaccuracy of ±0.5°C is perfectly adequate. If we had to measure the temperature of certain chemical processes, however, a variation of 0.5°C might have a significant effect on the rate of reaction or even the products of a process. A measurement inaccuracy much less than ±0.5°C is therefore clearly required.

Accuracy of measurement is thus one consideration in the choice of instrument for a particular application. Other parameters such as sensitivity, linearity and the reaction to ambient temperature changes are further considerations. These attributes are collectively known as the static characteristics of instruments, and are given in the data sheet for a particular instrument. It is important to note that the values quoted for instrument characteristics in such a data sheet only apply when the instrument is used under specified standard calibration conditions. Due allowance must be made for variations in the characteristics when the instrument is used in other conditions.

The various static characteristics are defined in the following paragraphs.

2.2.1 Accuracy and inaccuracy (measurement uncertainty)

The *accuracy* of an instrument is a measure of how close the output reading of the instrument is to the correct value. In practice, it is more usual to quote the *inaccuracy* figure rather than the accuracy figure for an instrument. Inaccuracy is the extent to which a reading might be wrong, and is often quoted as a percentage of the full-scale (f.s.) reading of an instrument. If, for example, a pressure gauge of range 0–10 bar has a quoted inaccuracy of $\pm 1.0\%$ f.s. ($\pm 1\%$ of full-scale reading), then the maximum error to be expected in any reading is 0.1 bar. This means that when the instrument is reading 1.0 bar, the possible error is 10% of this value. For this reason, it is an important system design rule that instruments are chosen such that their range is appropriate to the spread of values being measured, in order that the best possible accuracy is maintained in instrument readings. Thus, if we were measuring pressures with expected values between 0 and 1 bar, we would not use an instrument with a range of 0–10 bar. The term *measurement uncertainty* is frequently used in place of inaccuracy.

2.2.2 Precision/repeatability/reproducibility

Precision is a term that describes an instrument's degree of freedom from random errors. If a large number of readings are taken of the same quantity by a high precision instrument, then the spread of readings will be very small. Precision is often, though incorrectly, confused with accuracy. High precision does not imply anything about measurement accuracy. A high precision instrument may have a low accuracy. Low accuracy measurements from a high precision instrument are normally caused by a bias in the measurements, which is removable by recalibration.

The terms repeatability and reproducibility mean approximately the same but are applied in different contexts as given below. *Repeatability* describes the closeness of output readings when the same input is applied repetitively over a short period of time, with the same measurement conditions, same instrument and observer, same location and same conditions of use maintained throughout. *Reproducibility* describes the closeness of output readings for the same input when there are changes in the method of measurement, observer, measuring instrument, location, conditions of use and time of measurement. Both terms thus describe the spread of output readings for the same input. This spread is referred to as repeatability if the measurement conditions are constant and as reproducibility if the measurement conditions vary.

The degree of repeatability or reproducibility in measurements from an instrument is an alternative way of expressing its precision. Figure 2.5 illustrates this more clearly. The figure shows the results of tests on three industrial robots that were programmed to place components at a particular point on a table. The target point was at the centre of the concentric circles shown, and the black dots represent the points where each robot actually deposited components at each attempt. Both the accuracy and precision of Robot 1 are shown to be low in this trial. Robot 2 consistently puts the component down at approximately the same place but this is the wrong point. Therefore, it has high precision but low accuracy. Finally, Robot 3 has both high precision and high accuracy, because it consistently places the component at the correct target position.

2.2.3 Tolerance

Tolerance is a term that is closely related to accuracy and defines the maximum error that is to be expected in some value. Whilst it is not, strictly speaking, a static

18 Instrument types and performance characteristics

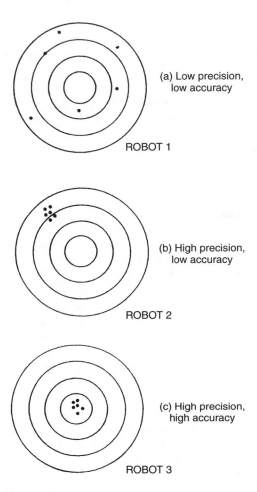

Fig. 2.5 Comparison of accuracy and precision.

characteristic of measuring instruments, it is mentioned here because the accuracy of some instruments is sometimes quoted as a tolerance figure. When used correctly, tolerance describes the maximum deviation of a manufactured component from some specified value. For instance, crankshafts are machined with a diameter tolerance quoted as so many microns (10^{-6} m), and electric circuit components such as resistors have tolerances of perhaps 5%. One resistor chosen at random from a batch having a nominal value 1000 W and tolerance 5% might have an actual value anywhere between 950 W and 1050 W.

2.2.4 Range or span

The *range* or *span* of an instrument defines the minimum and maximum values of a quantity that the instrument is designed to measure.

2.2.5 Linearity

It is normally desirable that the output reading of an instrument is linearly proportional to the quantity being measured. The Xs marked on Figure 2.6 show a plot of the typical output readings of an instrument when a sequence of input quantities are applied to it. Normal procedure is to draw a good fit straight line through the Xs, as shown in Figure 2.6. (Whilst this can often be done with reasonable accuracy by eye, it is always preferable to apply a mathematical least-squares line-fitting technique, as described in Chapter 11.) The non-linearity is then defined as the maximum deviation of any of the output readings marked X from this straight line. Non-linearity is usually expressed as a percentage of full-scale reading.

2.2.6 Sensitivity of measurement

The sensitivity of measurement is a measure of the change in instrument output that occurs when the quantity being measured changes by a given amount. Thus, sensitivity is the ratio:

$$\frac{\text{scale deflection}}{\text{value of measurand producing deflection}}$$

The sensitivity of measurement is therefore the slope of the straight line drawn on Figure 2.6. If, for example, a pressure of 2 bar produces a deflection of 10 degrees in a pressure transducer, the sensitivity of the instrument is 5 degrees/bar (assuming that the deflection is zero with zero pressure applied).

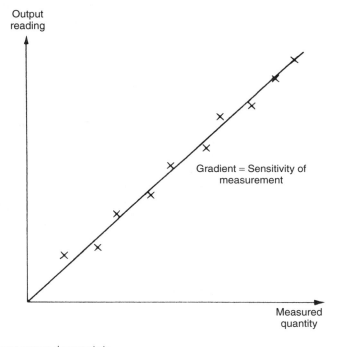

Fig. 2.6 Instrument output characteristic.

Example 2.1
The following resistance values of a platinum resistance thermometer were measured at a range of temperatures. Determine the measurement sensitivity of the instrument in ohms/°C.

Resistance (Ω)	Temperature (°C)
307	200
314	230
321	260
328	290

Solution
If these values are plotted on a graph, the straight-line relationship between resistance change and temperature change is obvious.

For a change in temperature of 30°C, the change in resistance is 7 Ω. Hence the measurement sensitivity $= 7/30 = 0.233\ \Omega/°C$.

2.2.7 Threshold

If the input to an instrument is gradually increased from zero, the input will have to reach a certain minimum level before the change in the instrument output reading is of a large enough magnitude to be detectable. This minimum level of input is known as the *threshold* of the instrument. Manufacturers vary in the way that they specify threshold for instruments. Some quote absolute values, whereas others quote threshold as a percentage of full-scale readings. As an illustration, a car speedometer typically has a threshold of about 15 km/h. This means that, if the vehicle starts from rest and accelerates, no output reading is observed on the speedometer until the speed reaches 15 km/h.

2.2.8 Resolution

When an instrument is showing a particular output reading, there is a lower limit on the magnitude of the change in the input measured quantity that produces an observable change in the instrument output. Like threshold, *resolution* is sometimes specified as an absolute value and sometimes as a percentage of f.s. deflection. One of the major factors influencing the resolution of an instrument is how finely its output scale is divided into subdivisions. Using a car speedometer as an example again, this has subdivisions of typically 20 km/h. This means that when the needle is between the scale markings, we cannot estimate speed more accurately than to the nearest 5 km/h. This figure of 5 km/h thus represents the resolution of the instrument.

2.2.9 Sensitivity to disturbance

All calibrations and specifications of an instrument are only valid under controlled conditions of temperature, pressure etc. These standard ambient conditions are usually defined in the instrument specification. As variations occur in the ambient temperature etc., certain static instrument characteristics change, and the *sensitivity to disturbance* is a measure of the magnitude of this change. Such environmental changes affect instruments in two main ways, known as *zero drift* and *sensitivity drift*. Zero drift is sometimes known by the alternative term, *bias*.

Zero drift or *bias* describes the effect where the zero reading of an instrument is modified by a change in ambient conditions. This causes a constant error that exists over the full range of measurement of the instrument. The mechanical form of bathroom scale is a common example of an instrument that is prone to bias. It is quite usual to find that there is a reading of perhaps 1 kg with no one stood on the scale. If someone of known weight 70 kg were to get on the scale, the reading would be 71 kg, and if someone of known weight 100 kg were to get on the scale, the reading would be 101 kg. Zero drift is normally removable by calibration. In the case of the bathroom scale just described, a thumbwheel is usually provided that can be turned until the reading is zero with the scales unloaded, thus removing the bias.

Zero drift is also commonly found in instruments like voltmeters that are affected by ambient temperature changes. Typical units by which such zero drift is measured are volts/°C. This is often called the *zero drift coefficient* related to temperature changes. If the characteristic of an instrument is sensitive to several environmental parameters, then it will have several zero drift coefficients, one for each environmental parameter. A typical change in the output characteristic of a pressure gauge subject to zero drift is shown in Figure 2.7(a).

Sensitivity drift (also known as *scale factor drift*) defines the amount by which an instrument's sensitivity of measurement varies as ambient conditions change. It is quantified by sensitivity drift coefficients that define how much drift there is for a unit change in each environmental parameter that the instrument characteristics are sensitive to. Many components within an instrument are affected by environmental fluctuations, such as temperature changes: for instance, the modulus of elasticity of a spring is temperature dependent. Figure 2.7(b) shows what effect sensitivity drift can have on the output characteristic of an instrument. Sensitivity drift is measured in units of the form (angular degree/bar)/°C. If an instrument suffers both zero drift and sensitivity drift at the same time, then the typical modification of the output characteristic is shown in Figure 2.7(c).

Example 2.2
A spring balance is calibrated in an environment at a temperature of 20°C and has the following deflection/load characteristic.

Load (kg)	0	1	2	3
Deflection (mm)	0	20	40	60

It is then used in an environment at a temperature of 30°C and the following deflection/load characteristic is measured.

Load (kg):	0	1	2	3
Deflection (mm)	5	27	49	71

Determine the zero drift and sensitivity drift per °C change in ambient temperature.

22 Instrument types and performance characteristics

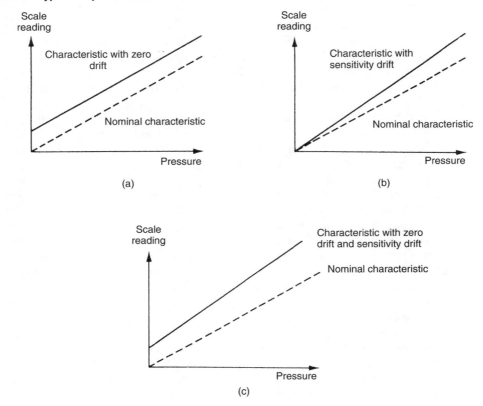

Fig. 2.7 Effects of disturbance: (a) zero drift; (b) sensitivity drift; (c) zero drift plus sensitivity drift.

Solution
At 20°C, deflection/load characteristic is a straight line. Sensitivity = 20 mm/kg.
At 30°C, deflection/load characteristic is still a straight line. Sensitivity = 22 mm/kg.
Bias (zero drift) = 5 mm (the no-load deflection)
Sensitivity drift = 2 mm/kg
Zero drift/°C = 5/10 = 0.5 mm/°C
Sensitivity drift/°C = 2/10 = 0.2 (mm per kg)/°C

2.2.10 Hysteresis effects

Figure 2.8 illustrates the output characteristic of an instrument that exhibits *hysteresis*. If the input measured quantity to the instrument is steadily increased from a negative value, the output reading varies in the manner shown in curve (a). If the input variable is then steadily decreased, the output varies in the manner shown in curve (b). The non-coincidence between these loading and unloading curves is known as *hysteresis*. Two quantities are defined, *maximum input hysteresis* and *maximum output hysteresis*, as shown in Figure 2.8. These are normally expressed as a percentage of the full-scale input or output reading respectively.

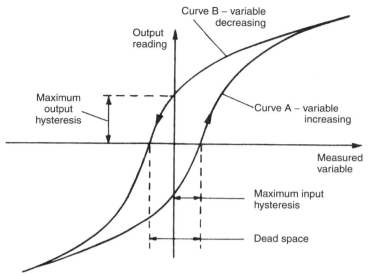

Fig. 2.8 Instrument characteristic with hysteresis.

Hysteresis is most commonly found in instruments that contain springs, such as the passive pressure gauge (Figure 2.1) and the Prony brake (used for measuring torque). It is also evident when friction forces in a system have different magnitudes depending on the direction of movement, such as in the pendulum-scale mass-measuring device. Devices like the mechanical flyball (a device for measuring rotational velocity) suffer hysteresis from both of the above sources because they have friction in moving parts and also contain a spring. Hysteresis can also occur in instruments that contain electrical windings formed round an iron core, due to magnetic hysteresis in the iron. This occurs in devices like the variable inductance displacement transducer, the LVDT and the rotary differential transformer.

2.2.11 Dead space

Dead space is defined as the range of different input values over which there is no change in output value. Any instrument that exhibits hysteresis also displays dead space, as marked on Figure 2.8. Some instruments that do not suffer from any significant hysteresis can still exhibit a dead space in their output characteristics, however. Backlash in gears is a typical cause of dead space, and results in the sort of instrument output characteristic shown in Figure 2.9. Backlash is commonly experienced in gearsets used to convert between translational and rotational motion (which is a common technique used to measure translational velocity).

2.3 Dynamic characteristics of instruments

The static characteristics of measuring instruments are concerned only with the steady-state reading that the instrument settles down to, such as the accuracy of the reading etc.

24 Instrument types and performance characteristics

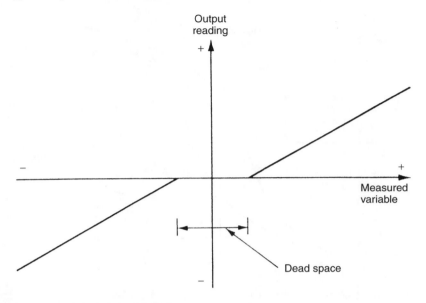

Fig. 2.9 Instrument characteristic with dead space.

The dynamic characteristics of a measuring instrument describe its behaviour between the time a measured quantity changes value and the time when the instrument output attains a steady value in response. As with static characteristics, any values for dynamic characteristics quoted in instrument data sheets only apply when the instrument is used under specified environmental conditions. Outside these calibration conditions, some variation in the dynamic parameters can be expected.

In any linear, time-invariant measuring system, the following general relation can be written between input and output for time $(t) > 0$:

$$a_n \frac{d^n q_0}{dt^n} + a_{n-1} \frac{d^{n-1} q_0}{dt^{n-1}} + \cdots + a_1 \frac{dq_0}{dt} + a_0 q_0$$
$$= b_m \frac{d^m q_i}{dt^m} + b_{m-1} \frac{d^{m-1} q_i}{dt^{m-1}} + \cdots + b_1 \frac{dq_i}{dt} + b_0 q_i \quad (2.1)$$

where q_i is the measured quantity, q_0 is the output reading and $a_0 \ldots a_n$, $b_0 \ldots b_m$ are constants.

The reader whose mathematical background is such that the above equation appears daunting should not worry unduly, as only certain special, simplified cases of it are applicable in normal measurement situations. The major point of importance is to have a practical appreciation of the manner in which various different types of instrument respond when the measurand applied to them varies.

If we limit consideration to that of step changes in the measured quantity only, then equation (2.1) reduces to:

$$a_n \frac{d^n q_0}{dt^n} + a_{n-1} \frac{d^{n-1} q_0}{dt^{n-1}} + \cdots + a_1 \frac{dq_0}{dt} + a_0 q_0 = b_0 q_i \quad (2.2)$$

2.3.1 Zero order instrument

Further simplification can be made by taking certain special cases of equation (2.2), which collectively apply to nearly all measurement systems.

If all the coefficients $a_1 \ldots a_n$ other than a_0 in equation (2.2) are assumed zero, then:

$$a_0 q_0 = b_0 q_i \quad \text{or} \quad q_0 = b_0 q_i / a_0 = K q_i \tag{2.3}$$

where K is a constant known as the instrument sensitivity as defined earlier.

Any instrument that behaves according to equation (2.3) is said to be of zero order type. Following a step change in the measured quantity at time t, the instrument output moves immediately to a new value at the same time instant t, as shown in Figure 2.10. A potentiometer, which measures motion, is a good example of such an instrument, where the output voltage changes instantaneously as the slider is displaced along the potentiometer track.

2.3.2 First order instrument

If all the coefficients $a_2 \ldots a_n$ except for a_0 and a_1 are assumed zero in equation (2.2) then:

$$a_1 \frac{dq_0}{dt} + a_0 q_0 = b_0 q_i \tag{2.4}$$

Any instrument that behaves according to equation (2.4) is known as a first order instrument. If d/dt is replaced by the D operator in equation (2.4), we get:

$$a_1 D q_0 + a_0 q_0 = b_0 q_i \quad \text{and rearranging this then gives} \quad q_0 = \frac{(b_0/a_0) q_i}{[1 + (a_1/a_0) D]} \tag{2.5}$$

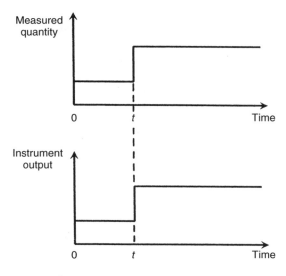

Fig. 2.10 Zero order instrument characteristic.

26 Instrument types and performance characteristics

Defining $K = b_0/a_0$ as the static sensitivity and $\tau = a_1/a_0$ as the time constant of the system, equation (2.5) becomes:

$$q_0 = \frac{Kq_i}{1 + \tau D} \tag{2.6}$$

If equation (2.6) is solved analytically, the output quantity q_0 in response to a step change in q_i at time t varies with time in the manner shown in Figure 2.11. The time constant τ of the step response is the time taken for the output quantity q_0 to reach 63% of its final value.

The liquid-in-glass thermometer (see Chapter 14) is a good example of a first order instrument. It is well known that, if a thermometer at room temperature is plunged into boiling water, the output e.m.f. does not rise instantaneously to a level indicating 100°C, but instead approaches a reading indicating 100°C in a manner similar to that shown in Figure 2.11.

A large number of other instruments also belong to this first order class: this is of particular importance in control systems where it is necessary to take account of the time lag that occurs between a measured quantity changing in value and the measuring instrument indicating the change. Fortunately, the time constant of many first order instruments is small relative to the dynamics of the process being measured, and so no serious problems are created.

Example 2.3
A balloon is equipped with temperature and altitude measuring instruments and has radio equipment that can transmit the output readings of these instruments back to ground. The balloon is initially anchored to the ground with the instrument output readings in steady state. The altitude-measuring instrument is approximately zero order and the temperature transducer first order with a time constant of 15 seconds. The

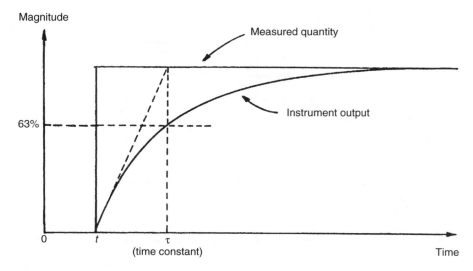

Fig. 2.11 First order instrument characteristic.

temperature on the ground, T_0, is 10°C and the temperature T_x at an altitude of x metres is given by the relation: $T_x = T_0 - 0.01x$

(a) If the balloon is released at time zero, and thereafter rises upwards at a velocity of 5 metres/second, draw a table showing the temperature and altitude measurements reported at intervals of 10 seconds over the first 50 seconds of travel. Show also in the table the error in each temperature reading.
(b) What temperature does the balloon report at an altitude of 5000 metres?

Solution
In order to answer this question, it is assumed that the solution of a first order differential equation has been presented to the reader in a mathematics course. If the reader is not so equipped, the following solution will be difficult to follow.

Let the temperature reported by the balloon at some general time t be T_r. Then T_x is related to T_r by the relation:

$$T_r = \frac{T_x}{1 + \tau D} = \frac{T_0 - 0.01x}{1 + \tau D} = \frac{10 - 0.01x}{1 + 15D}$$

It is given that $x = 5t$, thus: $T_r = \frac{10 - 0.05t}{1 + 15D}$
The transient or complementary function part of the solution ($T_x = 0$) is given by: $T_{r_{cf}} = Ce^{-t/15}$
The particular integral part of the solution is given by: $T_{r_{pi}} = 10 - 0.05(t - 15)$
Thus, the whole solution is given by: $T_r = T_{r_{cf}} + T_{r_{pi}} = Ce^{-t/15} + 10 - 0.05(t - 15)$
Applying initial conditions: At $t = 0$, $T_r = 10$, i.e. $10 = Ce^{-0} + 10 - 0.05(-15)$
Thus $C = -0.75$ and therefore: $T_r = 10 - 0.75e^{-t/15} - 0.05(t - 15)$
Using the above expression to calculate T_r for various values of t, the following table can be constructed:

Time	Altitude	Temperature reading	Temperature error
0	0	10	0
10	50	9.86	0.36
20	100	9.55	0.55
30	150	9.15	0.65
40	200	8.70	0.70
50	250	8.22	0.72

(b) At 5000 m, $t = 1000$ seconds. Calculating T_r from the above expression:

$$T_r = 10 - 0.75e^{-1000/15} - 0.05(1000 - 15)$$

The exponential term approximates to zero and so T_r can be written as:

$$T_r \approx 10 - 0.05(985) = -39.25°C$$

This result might have been inferred from the table above where it can be seen that the error is converging towards a value of 0.75. For large values of t, the transducer reading lags the true temperature value by a period of time equal to the time constant of

15 seconds. In this time, the balloon travels a distance of 75 metres and the temperature falls by 0.75°. Thus for large values of t, the output reading is always 0.75° less than it should be.

2.3.3 Second order instrument

If all coefficients $a_3 \ldots a_n$ other than a_0, a_1 and a_2 in equation (2.2) are assumed zero, then we get:

$$a_2 \frac{d^2 q_0}{dt^2} + a_1 \frac{dq_0}{dt} + a_0 q_0 = b_0 q_i \qquad (2.7)$$

Applying the D operator again: $a_2 D^2 q_0 + a_1 D q_0 + a_0 q_0 = b_0 q_i$, and rearranging:

$$q_0 = \frac{b_0 q_i}{a_0 + a_1 D + a_2 D^2} \qquad (2.8)$$

It is convenient to re-express the variables a_0, a_1, a_2 and b_0 in equation (2.8) in terms of three parameters K (static sensitivity), ω (undamped natural frequency) and ξ (damping ratio), where:

$$K = b_0/a_0; \quad \omega = a_0/a_2; \quad \xi = a_1/2a_0 a_2$$

Re-expressing equation (2.8) in terms of K, ω and ξ we get:

$$\frac{q_0}{q_i} = \frac{K}{D^2/\omega^2 + 2\xi D/\omega + 1} \qquad (2.9)$$

This is the standard equation for a second order system and any instrument whose response can be described by it is known as a second order instrument. If equation (2.9) is solved analytically, the shape of the step response obtained depends on the value of the damping ratio parameter ξ. The output responses of a second order instrument for various values of ξ following a step change in the value of the measured quantity at time t are shown in Figure 2.12. For case (A) where $\xi = 0$, there is no damping and the instrument output exhibits constant amplitude oscillations when disturbed by any change in the physical quantity measured. For light damping of $\xi = 0.2$, represented by case (B), the response to a step change in input is still oscillatory but the oscillations gradually die down. Further increase in the value of ξ reduces oscillations and overshoot still more, as shown by curves (C) and (D), and finally the response becomes very overdamped as shown by curve (E) where the output reading creeps up slowly towards the correct reading. Clearly, the extreme response curves (A) and (E) are grossly unsuitable for any measuring instrument. If an instrument were to be only ever subjected to step inputs, then the design strategy would be to aim towards a damping ratio of 0.707, which gives the critically damped response (C). Unfortunately, most of the physical quantities that instruments are required to measure do not change in the mathematically convenient form of steps, but rather in the form of ramps of varying slopes. As the form of the input variable changes, so the best value for ξ varies, and choice of ξ becomes one of compromise between those values that are best for each type of input variable behaviour anticipated. Commercial second order instruments, of which the accelerometer is a common example, are generally designed to have a damping ratio (ξ) somewhere in the range of 0.6–0.8.

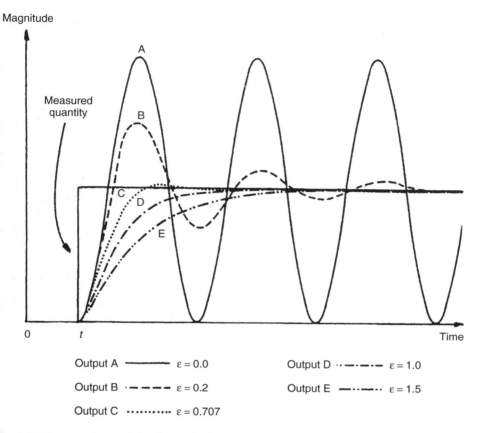

Fig. 2.12 Response characteristics of second order instruments.

2.4 Necessity for calibration

The foregoing discussion has described the static and dynamic characteristics of measuring instruments in some detail. However, an important qualification that has been omitted from this discussion is that an instrument only conforms to stated static and dynamic patterns of behaviour after it has been calibrated. It can normally be assumed that a new instrument will have been calibrated when it is obtained from an instrument manufacturer, and will therefore initially behave according to the characteristics stated in the specifications. During use, however, its behaviour will gradually diverge from the stated specification for a variety of reasons. Such reasons include mechanical wear, and the effects of dirt, dust, fumes and chemicals in the operating environment. The rate of divergence from standard specifications varies according to the type of instrument, the frequency of usage and the severity of the operating conditions. However, there will come a time, determined by practical knowledge, when the characteristics of the instrument will have drifted from the standard specification by an unacceptable amount. When this situation is reached, it is necessary to recalibrate the instrument to the standard specifications. Such recalibration is performed by adjusting the instrument

Instrument types and performance characteristics

at each point in its output range until its output readings are the same as those of a second standard instrument to which the same inputs are applied. This second instrument is one kept solely for calibration purposes whose specifications are accurately known. Calibration procedures are discussed more fully in Chapter 4.

2.5 Self-test questions

2.1 Explain what is meant by:
   (a) active instruments
   (b) passive instruments.
   Give examples of each and discuss the relative merits of these two classes of instruments.

2.2 Discuss the advantages and disadvantages of null and deflection types of measuring instrument. What are null types of instrument mainly used for and why?

2.3 Briefly define and explain all the static characteristics of measuring instruments.

2.4 Explain the difference between accuracy and precision in an instrument.

2.5 A tungsten/5% rhenium–tungsten/26% rhenium thermocouple has an output e.m.f. as shown in the following table when its hot (measuring) junction is at the temperatures shown. Determine the sensitivity of measurement for the thermocouple in mV/°C.

mV	4.37	8.74	13.11	17.48
°C	250	500	750	1000

2.6 Define sensitivity drift and zero drift. What factors can cause sensitivity drift and zero drift in instrument characteristics?

2.7 (a) An instrument is calibrated in an environment at a temperature of 20°C and the following output readings y are obtained for various input values x:

y	13.1	26.2	39.3	52.4	65.5	78.6
x	5	10	15	20	25	30

Determine the measurement sensitivity, expressed as the ratio y/x.

(b) When the instrument is subsequently used in an environment at a temperature of 50°C, the input/output characteristic changes to the following:

y	14.7	29.4	44.1	58.8	73.5	88.2
x	5	10	15	20	25	30

Determine the new measurement sensitivity. Hence determine the sensitivity drift due to the change in ambient temperature of 30°C.

2.8 A load cell is calibrated in an environment at a temperature of 21°C and has the following deflection/load characteristic:

Load (kg)	0	50	100	150	200
Deflection (mm)	0.0	1.0	2.0	3.0	4.0

When used in an environment at 35°C, its characteristic changes to the following:

Load (kg)	0	50	100	150	200
Deflection (mm)	0.2	1.3	2.4	3.5	4.6

(a) Determine the sensitivity at 21°C and 35°C.
(b) Calculate the total zero drift and sensitivity drift at 35°C.
(c) Hence determine the zero drift and sensitivity drift coefficients (in units of μm/°C and (μm per kg)/(°C)).

2.9 An unmanned submarine is equipped with temperature and depth measuring instruments and has radio equipment that can transmit the output readings of these instruments back to the surface. The submarine is initially floating on the surface of the sea with the instrument output readings in steady state. The depth-measuring instrument is approximately zero order and the temperature transducer first order with a time constant of 50 seconds. The water temperature on the sea surface, T_0, is 20°C and the temperature T_x at a depth of x metres is given by the relation:

$$T_x = T_0 - 0.01x$$

(a) If the submarine starts diving at time zero, and thereafter goes down at a velocity of 0.5 metres/second, draw a table showing the temperature and depth measurements reported at intervals of 100 seconds over the first 500 seconds of travel. Show also in the table the error in each temperature reading.
(b) What temperature does the submarine report at a depth of 1000 metres?

2.10 Write down the general differential equation describing the dynamic response of a second order measuring instrument and state the expressions relating the static sensitivity, undamped natural frequency and damping ratio to the parameters in this differential equation. Sketch the instrument response for the cases of heavy damping, critical damping and light damping, and state which of these is the usual target when a second order instrument is being designed.

3

Errors during the measurement process

3.1 Introduction

Errors in measurement systems can be divided into those that arise during the measurement process and those that arise due to later corruption of the measurement signal by induced noise during transfer of the signal from the point of measurement to some other point. This chapter considers only the first of these, with discussion on induced noise being deferred to Chapter 5.

It is extremely important in any measurement system to reduce errors to the minimum possible level and then to quantify the maximum remaining error that may exist in any instrument output reading. However, in many cases, there is a further complication that the final output from a measurement system is calculated by combining together two or more measurements of separate physical variables. In this case, special consideration must also be given to determining how the calculated error levels in each separate measurement should be combined together to give the best estimate of the most likely error magnitude in the calculated output quantity. This subject is considered in section 3.6.

The starting point in the quest to reduce the incidence of errors arising during the measurement process is to carry out a detailed analysis of all error sources in the system. Each of these error sources can then be considered in turn, looking for ways of eliminating or at least reducing the magnitude of errors. Errors arising during the measurement process can be divided into two groups, known as systematic errors and random errors.

Systematic errors describe errors in the output readings of a measurement system that are consistently on one side of the correct reading, i.e. either all the errors are positive or they are all negative. Two major sources of systematic errors are system disturbance during measurement and the effect of environmental changes (modifying inputs), as discussed in sections 3.2.1 and 3.2.2. Other sources of systematic error include bent meter needles, the use of uncalibrated instruments, drift in instrument characteristics and poor cabling practices. Even when systematic errors due to the above factors have been reduced or eliminated, some errors remain that are inherent in the manufacture of an instrument. These are quantified by the accuracy figure quoted in the published specifications contained in the instrument data sheet.

Random errors are perturbations of the measurement either side of the true value caused by random and unpredictable effects, such that positive errors and negative errors occur in approximately equal numbers for a series of measurements made of the same quantity. Such perturbations are mainly small, but large perturbations occur from time to time, again unpredictably. Random errors often arise when measurements are taken by human observation of an analogue meter, especially where this involves interpolation between scale points. Electrical noise can also be a source of random errors. To a large extent, random errors can be overcome by taking the same measurement a number of times and extracting a value by averaging or other statistical techniques, as discussed in section 3.5. However, any quantification of the measurement value and statement of error bounds remains a statistical quantity. Because of the nature of random errors and the fact that large perturbations in the measured quantity occur from time to time, the best that we can do is to express measurements in probabilistic terms: we may be able to assign a 95% or even 99% confidence level that the measurement is a certain value within error bounds of, say, $\pm 1\%$, but we can never attach a 100% probability to measurement values that are subject to random errors.

Finally, a word must be said about the distinction between systematic and random errors. Error sources in the measurement system must be examined carefully to determine what type of error is present, systematic or random, and to apply the appropriate treatment. In the case of manual data measurements, a human observer may make a different observation at each attempt, but it is often reasonable to assume that the errors are random and that the mean of these readings is likely to be close to the correct value. However, this is only true as long as the human observer is not introducing a parallax-induced systematic error as well by persistently reading the position of a needle against the scale of an analogue meter from one side rather than from directly above. In that case, correction would have to be made for this systematic error (bias) in the measurements before statistical techniques were applied to reduce the effect of random errors.

3.2 Sources of systematic error

Systematic errors in the output of many instruments are due to factors inherent in the manufacture of the instrument arising out of tolerances in the components of the instrument. They can also arise due to wear in instrument components over a period of time. In other cases, systematic errors are introduced either by the effect of environmental disturbances or through the disturbance of the measured system by the act of measurement. These various sources of systematic error, and ways in which the magnitude of the errors can be reduced, are discussed below.

3.2.1 System disturbance due to measurement

Disturbance of the measured system by the act of measurement is a common source of systematic error. If we were to start with a beaker of hot water and wished to measure its temperature with a mercury-in-glass thermometer, then we would take the thermometer, which would initially be at room temperature, and plunge it into the water. In so doing, we would be introducing a relatively cold mass (the thermometer) into the hot water and a heat transfer would take place between the water and the thermometer. This heat transfer would lower the temperature of the water. Whilst the reduction in temperature in this case would be so small as to be undetectable by the limited measurement resolution of such a thermometer, the effect is finite and clearly establishes the principle that, in nearly all measurement situations, the process of measurement disturbs the system and alters the values of the physical quantities being measured.

A particularly important example of this occurs with the orifice plate. This is placed into a fluid-carrying pipe to measure the flow rate, which is a function of the pressure that is measured either side of the orifice plate. This measurement procedure causes a permanent pressure loss in the flowing fluid. The disturbance of the measured system can often be very significant.

Thus, as a general rule, the process of measurement always disturbs the system being measured. The magnitude of the disturbance varies from one measurement system to the next and is affected particularly by the type of instrument used for measurement. Ways of minimizing disturbance of measured systems is an important consideration in instrument design. However, an accurate understanding of the mechanisms of system disturbance is a prerequisite for this.

Measurements in electric circuits

In analysing system disturbance during measurements in electric circuits, Thévenin's theorem (see Appendix 3) is often of great assistance. For instance, consider the circuit shown in Figure 3.1(a) in which the voltage across resistor R_5 is to be measured by a voltmeter with resistance R_m. Here, R_m acts as a shunt resistance across R_5, decreasing the resistance between points AB and so disturbing the circuit. Therefore, the voltage E_m measured by the meter is not the value of the voltage E_0 that existed prior to measurement. The extent of the disturbance can be assessed by calculating the open-circuit voltage E_0 and comparing it with E_m.

Thévenin's theorem allows the circuit of Figure 3.1(a) comprising two voltage sources and five resistors to be replaced by an equivalent circuit containing a single resistance and one voltage source, as shown in Figure 3.1(b). For the purpose of defining the equivalent single resistance of a circuit by Thévenin's theorem, all voltage sources are represented just by their internal resistance, which can be approximated to zero, as shown in Figure 3.1(c). Analysis proceeds by calculating the equivalent resistances of sections of the circuit and building these up until the required equivalent resistance of the whole of the circuit is obtained. Starting at C and D, the circuit to the left of C and D consists of a series pair of resistances (R_1 and R_2) in parallel with R_3, and the equivalent resistance can be written as:

$$\frac{1}{R_{CD}} = \frac{1}{R_1 + R_2} + \frac{1}{R_3} \quad \text{or} \quad R_{CD} = \frac{(R_1 + R_2)R_3}{R_1 + R_2 + R_3}$$

Moving now to A and B, the circuit to the left consists of a pair of series resistances (R_{CD} and R_4) in parallel with R_5. The equivalent circuit resistance R_{AB} can thus be

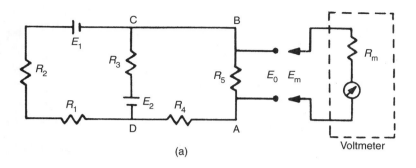

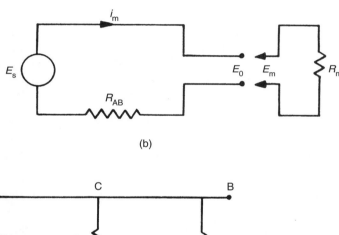

Fig. 3.1 Analysis of circuit loading: (a) a circuit in which the voltage across R_5 is to be measured; (b) equivalent circuit by Thévenin's theorem; (c) the circuit used to find the equivalent single resistance R_{AB}.

written as:

$$\frac{1}{R_{AB}} = \frac{1}{R_{CD} + R_4} + \frac{1}{R_5} \quad \text{or} \quad R_{AB} = \frac{(R_4 + R_{CD})R_5}{R_4 + R_{CD} + R_5}$$

Substituting for R_{CD} using the expression derived previously, we obtain:

$$R_{AB} = \frac{\left[\dfrac{(R_1 + R_2)R_3}{R_1 + R_2 + R_3} + R_4\right] R_5}{\dfrac{(R_1 + R_2)R_3}{R_1 + R_2 + R_3} + R_4 + R_5} \quad (3.1)$$

36 Errors during the measurement process

Defining I as the current flowing in the circuit when the measuring instrument is connected to it, we can write:

$$I = \frac{E_0}{R_{AB} + R_m},$$

and the voltage measured by the meter is then given by:

$$E_m = \frac{R_m E_0}{R_{AB} + R_m}.$$

In the absence of the measuring instrument and its resistance R_m, the voltage across AB would be the equivalent circuit voltage source whose value is E_0. The effect of measurement is therefore to reduce the voltage across AB by the ratio given by:

$$\frac{E_m}{E_0} = \frac{R_m}{R_{AB} + R_m} \tag{3.2}$$

It is thus obvious that as R_m gets larger, the ratio E_m/E_0 gets closer to unity, showing that the design strategy should be to make R_m as high as possible to minimize disturbance of the measured system. (Note that we did not calculate the value of E_0, since this is not required in quantifying the effect of R_m.)

Example 3.1
Suppose that the components of the circuit shown in Figure 3.1(a) have the following values:

$$R_1 = 400\,\Omega; R_2 = 600\,\Omega; R_3 = 1000\,\Omega, R_4 = 500\,\Omega; R_5 = 1000\,\Omega$$

The voltage across AB is measured by a voltmeter whose internal resistance is 9500 Ω. What is the measurement error caused by the resistance of the measuring instrument?

Solution
Proceeding by applying Thévenin's theorem to find an equivalent circuit to that of Figure 3.1(a) of the form shown in Figure 3.1(b), and substituting the given component values into the equation for R_{AB} (3.1), we obtain:

$$R_{AB} = \frac{[(1000^2/2000) + 500]1000}{(1000^2/2000) + 500 + 1000} = \frac{1000^2}{2000} = 500\,\Omega$$

From equation (3.2), we have:

$$\frac{E_m}{E_0} = \frac{R_m}{R_{AB} + R_m}$$

The measurement error is given by $(E_0 - E_m)$:

$$E_0 - E_m = E_0\left(1 - \frac{R_m}{R_{AB} + R_m}\right)$$

Substituting in values:

$$E_0 - E_m = E_0\left(1 - \frac{9500}{10\,000}\right) = 0.95 E_0$$

Thus, the error in the measured value is 5%.

At this point, it is interesting to note the constraints that exist when practical attempts are made to achieve a high internal resistance in the design of a moving-coil voltmeter. Such an instrument consists of a coil carrying a pointer mounted in a fixed magnetic field. As current flows through the coil, the interaction between the field generated and the fixed field causes the pointer it carries to turn in proportion to the applied current (see Chapter 6 for further details). The simplest way of increasing the input impedance (the resistance) of the meter is either to increase the number of turns in the coil or to construct the same number of coil turns with a higher-resistance material. However, either of these solutions decreases the current flowing in the coil, giving less magnetic torque and thus decreasing the measurement sensitivity of the instrument (i.e. for a given applied voltage, we get less deflection of the pointer). This problem can be overcome by changing the spring constant of the restraining springs of the instrument, such that less torque is required to turn the pointer by a given amount. However, this reduces the ruggedness of the instrument and also demands better pivot design to reduce friction. This highlights a very important but tiresome principle in instrument design: any attempt to improve the performance of an instrument in one respect generally decreases the performance in some other aspect. This is an inescapable fact of life with passive instruments such as the type of voltmeter mentioned, and is often the reason for the use of alternative active instruments such as digital voltmeters, where the inclusion of auxiliary power greatly improves performance.

Bridge circuits for measuring resistance values are a further example of the need for careful design of the measurement system. The impedance of the instrument measuring the bridge output voltage must be very large in comparison with the component resistances in the bridge circuit. Otherwise, the measuring instrument will load the circuit and draw current from it. This is discussed more fully in Chapter 7.

3.2.2 Errors due to environmental inputs

An environmental input is defined as an apparently real input to a measurement system that is actually caused by a change in the environmental conditions surrounding the measurement system. The fact that the static and dynamic characteristics specified for measuring instruments are only valid for particular environmental conditions (e.g. of temperature and pressure) has already been discussed at considerable length in Chapter 2. These specified conditions must be reproduced as closely as possible during calibration exercises because, away from the specified calibration conditions, the characteristics of measuring instruments vary to some extent and cause measurement errors. The magnitude of this environment-induced variation is quantified by the two constants known as sensitivity drift and zero drift, both of which are generally included in the published specifications for an instrument. Such variations of environmental conditions away from the calibration conditions are sometimes described as modifying inputs to the measurement system because they modify the output of the system. When such modifying inputs are present, it is often difficult to determine how much of the output change in a measurement system is due to a change in the measured variable and how much is due to a change in environmental conditions. This is illustrated by the following example. Suppose we are given a small closed box and told that it may contain either a mouse or a rat. We are also told that the box weighs 0.1 kg when empty. If we put the box onto bathroom scales and observe a reading of 1.0 kg, this does not immediately tell us what is in the box because the reading may be due to one of three things:

(a) a 0.9 kg rat in the box (real input)
(b) an empty box with a 0.9 kg bias on the scales due to a temperature change (environmental input)
(c) a 0.4 kg mouse in the box together with a 0.5 kg bias (real + environmental inputs).

Thus, the magnitude of any environmental input must be measured before the value of the measured quantity (the real input) can be determined from the output reading of an instrument.

In any general measurement situation, it is very difficult to avoid environmental inputs, because it is either impractical or impossible to control the environmental conditions surrounding the measurement system. System designers are therefore charged with the task of either reducing the susceptibility of measuring instruments to environmental inputs or, alternatively, quantifying the effect of environmental inputs and correcting for them in the instrument output reading. The techniques used to deal with environmental inputs and minimize their effect on the final output measurement follow a number of routes as discussed below.

3.2.3 Wear in instrument components

Systematic errors can frequently develop over a period of time because of wear in instrument components. Recalibration often provides a full solution to this problem.

3.2.4 Connecting leads

In connecting together the components of a measurement system, a common source of error is the failure to take proper account of the resistance of connecting leads (or pipes in the case of pneumatically or hydraulically actuated measurement systems). For instance, in typical applications of a resistance thermometer, it is common to find that the thermometer is separated from other parts of the measurement system by perhaps 100 metres. The resistance of such a length of 20 gauge copper wire is 7 Ω, and there is a further complication that such wire has a temperature coefficient of 1 mΩ/°C.

Therefore, careful consideration needs to be given to the choice of connecting leads. Not only should they be of adequate cross-section so that their resistance is minimized, but they should be adequately screened if they are thought likely to be subject to electrical or magnetic fields that could otherwise cause induced noise. Where screening is thought essential, then the routing of cables also needs careful planning. In one application in the author's personal experience involving instrumentation of an electric-arc steel making furnace, screened signal-carrying cables between transducers on the arc furnace and a control room at the side of the furnace were initially corrupted by high amplitude 50 Hz noise. However, by changing the route of the cables between the transducers and the control room, the magnitude of this induced noise was reduced by a factor of about ten.

3.3 Reduction of systematic errors

The prerequisite for the reduction of systematic errors is a complete analysis of the measurement system that identifies all sources of error. Simple faults within a system, such as bent meter needles and poor cabling practices, can usually be readily and cheaply rectified once they have been identified. However, other error sources require more detailed analysis and treatment. Various approaches to error reduction are considered below.

3.3.1 Careful instrument design

Careful instrument design is the most useful weapon in the battle against environmental inputs, by reducing the sensitivity of an instrument to environmental inputs to as low a level as possible. For instance, in the design of strain gauges, the element should be constructed from a material whose resistance has a very low temperature coefficient (i.e. the variation of the resistance with temperature is very small). However, errors due to the way in which an instrument is designed are not always easy to correct, and a choice often has to be made between the high cost of redesign and the alternative of accepting the reduced measurement accuracy if redesign is not undertaken.

3.3.2 Method of opposing inputs

The method of opposing inputs compensates for the effect of an environmental input in a measurement system by introducing an equal and opposite environmental input that cancels it out. One example of how this technique is applied is in the type of millivoltmeter shown in Figure 3.2. This consists of a coil suspended in a fixed magnetic field produced by a permanent magnet. When an unknown voltage is applied to the coil, the magnetic field due to the current interacts with the fixed field and causes the coil (and a pointer attached to the coil) to turn. If the coil resistance R_{coil} is sensitive to temperature, then any environmental input to the system in the form of a temperature change will alter the value of the coil current for a given applied voltage and so alter the pointer output reading. Compensation for this is made by introducing a compensating resistance R_{comp} into the circuit, where R_{comp} has a temperature coefficient that is equal in magnitude but opposite in sign to that of the coil. Thus, in response to an increase in temperature, R_{coil} increases but R_{comp} decreases, and so the total resistance remains approximately the same.

3.3.3 High-gain feedback

The benefit of adding high-gain feedback to many measurement systems is illustrated by considering the case of the voltage-measuring instrument whose block diagram is shown in Figure 3.3. In this system, the unknown voltage E_i is applied to a motor of torque constant K_m, and the induced torque turns a pointer against the restraining action of a spring with spring constant K_s. The effect of environmental inputs on the

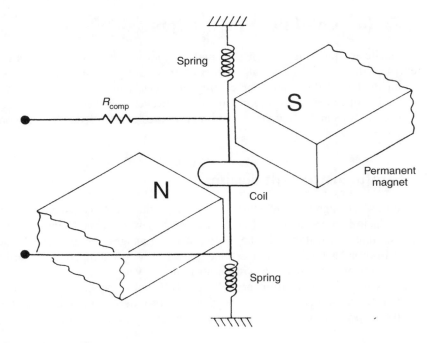

Fig. 3.2 Millivoltmeter.

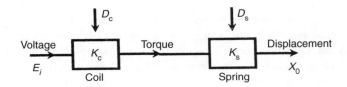

Fig. 3.3 Block diagram for voltage-measuring instrument.

motor and spring constants is represented by variables D_m and D_s. In the absence of environmental inputs, the displacement of the pointer X_0 is given by: $X_0 = K_m K_s E_i$. However, in the presence of environmental inputs, both K_m and K_s change, and the relationship between X_0 and E_i can be affected greatly. Therefore, it becomes difficult or impossible to calculate E_i from the measured value of X_0. Consider now what happens if the system is converted into a high-gain, closed-loop one, as shown in Figure 3.4, by adding an amplifier of gain constant K_a and a feedback device with gain constant K_f. Assume also that the effect of environmental inputs on the values of K_a and K_f are represented by D_a and D_f. The feedback device feeds back a voltage E_0 proportional to the pointer displacement X_0. This is compared with the unknown voltage E_i by a comparator and the error is amplified. Writing down the equations of the system, we have:

$$E_0 = K_f X_0; \quad X_0 = (E_i - E_0) K_a K_m K_s = (E_i - K_f X_0) K_a K_m K_s$$

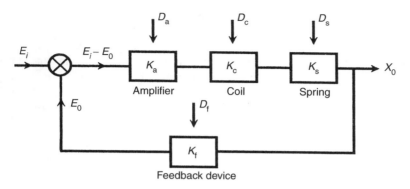

Fig. 3.4 Block diagram of voltage-measuring instrument with high-gain feedback.

Thus:
$$E_i K_a K_m K_s = (1 + K_f K_a K_m K_s) X_0$$

i.e.
$$X_0 = \frac{K_a K_m K_s}{1 + K_f K_a K_m K_s} E_i \qquad (3.3)$$

Because K_a is very large (it is a high-gain amplifier), $K_f . K_a . K_m . K_s \gg 1$, and equation (3.3) reduces to:

$$X_0 = E_i / K_f$$

This is a highly important result because we have reduced the relationship between X_0 and E_i to one that involves only K_f. The sensitivity of the gain constants K_a, K_m and K_s to the environmental inputs D_a, D_m and D_s has thereby been rendered irrelevant, and we only have to be concerned with one environmental input D_f. Conveniently, it is usually easy to design a feedback device that is insensitive to environmental inputs: this is much easier than trying to make a motor or spring insensitive. Thus, high-gain feedback techniques are often a very effective way of reducing a measurement system's sensitivity to environmental inputs. However, one potential problem that must be mentioned is that there is a possibility that high-gain feedback will cause instability in the system. Therefore, any application of this method must include careful stability analysis of the system.

3.3.4 Calibration

Instrument calibration is a very important consideration in measurement systems and calibration procedures are considered in detail in Chapter 4. All instruments suffer drift in their characteristics, and the rate at which this happens depends on many factors, such as the environmental conditions in which instruments are used and the frequency of their use. Thus, errors due to instruments being out of calibration can usually be rectified by increasing the frequency of recalibration.

3.3.5 Manual correction of output reading

In the case of errors that are due either to system disturbance during the act of measurement or due to environmental changes, a good measurement technician can substantially reduce errors at the output of a measurement system by calculating the effect of such systematic errors and making appropriate correction to the instrument readings. This is not necessarily an easy task, and requires all disturbances in the measurement system to be quantified. This procedure is carried out automatically by intelligent instruments.

3.3.6 Intelligent instruments

Intelligent instruments contain extra sensors that measure the value of environmental inputs and automatically compensate the value of the output reading. They have the ability to deal very effectively with systematic errors in measurement systems, and errors can be attenuated to very low levels in many cases. A more detailed analysis of intelligent instruments can be found in Chapter 9.

3.4 Quantification of systematic errors

Once all practical steps have been taken to eliminate or reduce the magnitude of systematic errors, the final action required is to estimate the maximum remaining error that may exist in a measurement due to systematic errors. Unfortunately, it is not always possible to quantify exact values of a systematic error, particularly if measurements are subject to unpredictable environmental conditions. The usual course of action is to assume mid-point environmental conditions and specify the maximum measurement error as $\pm x\%$ of the output reading to allow for the maximum expected deviation in environmental conditions away from this mid-point. Data sheets supplied by instrument manufacturers usually quantify systematic errors in this way, and such figures take account of all systematic errors that may be present in output readings from the instrument.

3.5 Random errors

Random errors in measurements are caused by unpredictable variations in the measurement system. They are usually observed as small perturbations of the measurement either side of the correct value, i.e. positive errors and negative errors occur in approximately equal numbers for a series of measurements made of the same constant quantity. Therefore, random errors can largely be eliminated by calculating the average of a number of repeated measurements, provided that the measured quantity remains constant during the process of taking the repeated measurements. This averaging process of repeated measurements can be done automatically by intelligent instruments, as discussed in Chapter 9. The degree of confidence in the calculated mean/median values can be quantified by calculating the standard deviation or variance of the data, these being parameters that describe how the measurements are distributed about the mean value/median. All of these terms are explained more fully in section 3.5.1.

3.5.1 Statistical analysis of measurements subject to random errors

Mean and median values

The average value of a set of measurements of a constant quantity can be expressed as either the mean value or the median value. As the number of measurements increases, the difference between the mean value and median values becomes very small. However, for any set of n measurements $x_1, x_2 \cdots x_n$ of a constant quantity, the most likely true value is the *mean* given by:

$$x_{\text{mean}} = \frac{x_1 + x_2 + \cdots x_n}{n} \tag{3.4}$$

This is valid for all data sets where the measurement errors are distributed equally about the zero error value, i.e. where the positive errors are balanced in quantity and magnitude by the negative errors.

The *median* is an approximation to the mean that can be written down without having to sum the measurements. The median is the middle value when the measurements in the data set are written down in ascending order of magnitude. For a set of n measurements $x_1, x_2 \cdots x_n$ of a constant quantity, written down in ascending order of magnitude, the median value is given by:

$$x_{\text{median}} = x_{n+1}/2 \tag{3.5}$$

Thus, for a set of 9 measurements $x_1, x_2 \cdots x_9$ arranged in order of magnitude, the median value is x_5. For an even number of measurements, the median value is midway between the two centre values, i.e. for 10 measurements $x_1 \cdots x_{10}$, the median value is given by: $(x_5 + x_6)/2$.

Suppose that the length of a steel bar is measured by a number of different observers and the following set of 11 measurements are recorded (units mm). We will call this measurement set A.

398 420 394 416 404 408 400 420 396 413 430   (Measurement set A)

Using (3.4) and (3.5), mean = 409.0 and median = 408. Suppose now that the measurements are taken again using a better measuring rule, and with the observers taking more care, to produce the following measurement set B:

409 406 402 407 405 404 407 404 407 407 408   (Measurement set B)

For these measurements, mean = 406.0 and median = 407. Which of the two measurement sets A and B, and the corresponding mean and median values, should we have most confidence in? Intuitively, we can regard measurement set B as being more reliable since the measurements are much closer together. In set A, the spread between the smallest (396) and largest (430) value is 34, whilst in set B, the spread is only 6.

44 Errors during the measurement process

- *Thus, the smaller the spread of the measurements, the more confidence we have in the mean or median value calculated.*

Let us now see what happens if we increase the number of measurements by extending measurement set B to 23 measurements. We will call this measurement set C.

$$409\ 406\ 402\ 407\ 405\ 404\ 407\ 404\ 407\ 407\ 408\ 406\ 410\ 406\ 405\ 408$$
$$406\ 409\ 406\ 405\ 409\ 406\ 407 \qquad \text{(Measurement set C)}$$

Now, mean = 406.5 and median = 406.

- *This confirms our earlier statement that the median value tends towards the mean value as the number of measurements increases.*

Standard deviation and variance

Expressing the spread of measurements simply as the range between the largest and smallest value is not in fact a very good way of examining how the measurement values are distributed about the mean value. A much better way of expressing the distribution is to calculate the variance or standard deviation of the measurements. The starting point for calculating these parameters is to calculate the deviation (error) d_i of each measurement x_i from the mean value x_{mean}:

$$d_i = x_i - x_{\text{mean}} \qquad (3.6)$$

The *variance* (V) is then given by:*

$$V = \frac{d_1^2 + d_2^2 \cdots d_n^2}{n - 1} \qquad (3.7)$$

The *standard deviation* (σ) is simply the square root of the variance. Thus*:

$$\sigma = \sqrt{V} = \sqrt{\frac{d_1^2 + d_2^2 \cdots d_n^2}{n - 1}} \qquad (3.8)$$

Example 3.2
Calculate σ and V for measurement sets A, B and C above.

---

*Mathematically minded readers may have observed that the expressions for V and σ differ from the formal mathematical definitions, which have (n) instead of ($n - 1$) in the denominator. This difference arises because the mathematical definition is for an infinite data set, whereas, in the case of measurements, we are concerned only with finite data sets. For a finite set of measurements (x_i) $i = 1, n$, the mean x_m will differ from the true mean μ of the infinite data set that the finite set is part of. If somehow we knew the true mean μ of a set of measurements, then the deviations d_i could be calculated as the deviation of each data value from the true mean and it would then be correct to calculate V and σ using (n) instead of ($n - 1$). However, in normal situations, using ($n - 1$) in the denominator of equations (3.7) and (3.8) produces a value that is statistically closer to the correct value.

Measurement and Instrumentation Principles 45

Solution
First, draw a table of measurements and deviations for set A (mean = 409 as calculated earlier):

Measurement	398	420	394	416	404	408	400	420	396	413	430
Deviation from mean	−11	+11	−15	+7	−5	−1	−9	+11	−13	+4	+21
(deviations)^2	121	121	225	49	25	1	81	121	169	16	441

$\sum(\text{deviations}^2) = 1370$; n = number of measurements = 11.
Then, from (3.7), $V = \sum(\text{deviations}^2)/n - 1; = 1370/10 = 137; \sigma = \sqrt{V} = 11.7$.
The measurements and deviations for set B are (mean = 406 as calculated earlier):

Measurement	409	406	402	407	405	404	407	404	407	407	408
Deviation from mean	+3	0	−4	+1	−1	−2	+1	−2	+1	+1	+2
(deviations)^2	9	0	16	1	1	4	1	4	1	1	4

From this data, using (3.7) and (3.8), $V = 4.2$ and $\sigma = 2.05$.
The measurements and deviations for set C are (mean = 406.5 as calculated earlier):

Measurement	409	406	402	407	405	404	407	404
Deviation from mean	+2.5	−0.5	−4.5	+0.5	−1.5	−2.5	+0.5	−2.5
(deviations)^2	6.25	0.25	20.25	0.25	2.25	6.25	0.25	6.25

Measurement	407	407	408	406	410	406	405	408
Deviation from mean	+0.5	+0.5	+1.5	−0.5	+3.5	−0.5	−1.5	+1.5
(deviations)^2	0.25	0.25	2.25	0.25	12.25	0.25	2.25	2.25

Measurement	406	409	406	405	409	406	407
Deviation from mean	−0.5	+2.5	−0.5	−1.5	+2.5	−0.5	+0.5
(deviations)^2	0.25	6.25	0.25	2.25	6.25	0.25	0.25

From this data, using (3.7) and (3.8), $V = 3.53$ and $\sigma = 1.88$.

Note that the smaller values of V and σ for measurement set B compared with A correspond with the respective size of the spread in the range between maximum and minimum values for the two sets.

- Thus, as V and σ decrease for a measurement set, we are able to express greater confidence that the calculated mean or median value is close to the true value, i.e. that the averaging process has reduced the random error value close to zero.
- Comparing V and σ for measurement sets B and C, V and σ get smaller as the number of measurements increases, confirming that confidence in the mean value increases as the number of measurements increases.

46 Errors during the measurement process

We have observed so far that random errors can be reduced by taking the average (mean or median) of a number of measurements. However, although the mean or median value is close to the true value, it would only become exactly equal to the true value if we could average an infinite number of measurements. As we can only make a finite number of measurements in a practical situation, the average value will still have some error. This error can be quantified as the *standard error of the mean*, which will be discussed in detail a little later. However, before that, the subject of graphical analysis of random measurement errors needs to be covered.

3.5.2 Graphical data analysis techniques – frequency distributions

Graphical techniques are a very useful way of analysing the way in which random measurement errors are distributed. The simplest way of doing this is to draw a *histogram*, in which bands of equal width across the range of measurement values are defined and the number of measurements within each band is counted. Figure 3.5 shows a histogram for set C of the length measurement data given in section 3.5.1, in which the bands chosen are 2 mm wide. For instance, there are 11 measurements in the range between 405.5 and 407.5 and so the height of the histogram for this range is 11 units. Also, there are 5 measurements in the range from 407.5 to 409.5 and so the height of the histogram over this range is 5 units. The rest of the histogram is completed in a similar fashion. (N.B. The scaling of the bands was deliberately chosen so that no measurements fell on the boundary between different bands and caused ambiguity about which band to put them in.) Such a histogram has the characteristic shape shown by truly random data, with symmetry about the mean value of the measurements.

As it is the actual value of measurement error that is usually of most concern, it is often more useful to draw a histogram of the deviations of the measurements

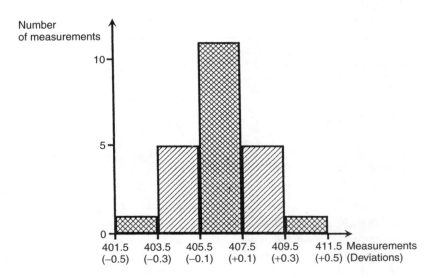

Fig. 3.5 Histogram of measurements and deviations.

from the mean value rather than to draw a histogram of the measurements themselves. The starting point for this is to calculate the deviation of each measurement away from the calculated mean value. Then a *histogram of deviations* can be drawn by defining deviation bands of equal width and counting the number of deviation values in each band. This histogram has exactly the same shape as the histogram of the raw measurements except that the scaling of the horizontal axis has to be redefined in terms of the deviation values (these units are shown in brackets on Figure 3.5).

Let us now explore what happens to the histogram of deviations as the number of measurements increases. As the number of measurements increases, smaller bands can be defined for the histogram, which retains its basic shape but then consists of a larger number of smaller steps on each side of the peak. In the limit, as the number of measurements approaches infinity, the histogram becomes a smooth curve known as a *frequency distribution curve* as shown in Figure 3.6. The ordinate of this curve is the frequency of occurrence of each deviation value, $F(D)$, and the abscissa is the magnitude of deviation, D.

The symmetry of Figures 3.5 and 3.6 about the zero deviation value is very useful for showing graphically that the measurement data only has random errors. Although these figures cannot easily be used to quantify the magnitude and distribution of the errors, very similar graphical techniques do achieve this. If the height of the frequency distribution curve is normalized such that the area under it is unity, then the curve in this form is known as a *probability curve*, and the height $F(D)$ at any particular deviation magnitude D is known as the *probability density function* (p.d.f.). The condition that

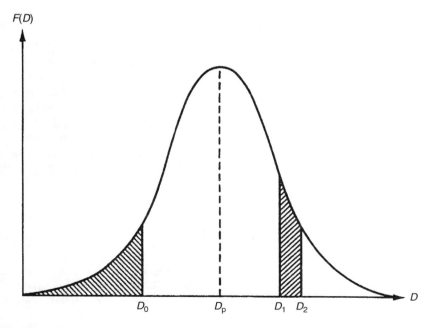

Fig. 3.6 Frequency distribution curve of deviations.

the area under the curve is unity can be expressed mathematically as:

$$\int_{-\infty}^{\infty} F(D)\,dD = 1$$

The probability that the error in any one particular measurement lies between two levels D_1 and D_2 can be calculated by measuring the area under the curve contained between two vertical lines drawn through D_1 and D_2, as shown by the right-hand hatched area in Figure 3.6. This can be expressed mathematically as:

$$P(D_1 \leq D \leq D_2) = \int_{D_1}^{D_2} F(D)\,dD \tag{3.9}$$

Of particular importance for assessing the maximum error likely in any one measurement is the *cumulative distribution function* (c.d.f.). This is defined as the probability of observing a value less than or equal to D_0, and is expressed mathematically as:

$$P(D \leq D_0) = \int_{-\infty}^{D_0} F(D)\,dD \tag{3.10}$$

Thus, the c.d.f. is the area under the curve to the left of a vertical line drawn through D_0, as shown by the left-hand hatched area on Figure 3.6.

The deviation magnitude D_p corresponding with the peak of the frequency distribution curve (Figure 3.6) is the value of deviation that has the greatest probability. If the errors are entirely random in nature, then the value of D_p will equal zero. Any non-zero value of D_p indicates systematic errors in the data, in the form of a bias that is often removable by recalibration.

Gaussian distribution

Measurement sets that only contain random errors usually conform to a distribution with a particular shape that is called *Gaussian*, although this conformance must always be tested (see the later section headed 'Goodness of fit'). The shape of a Gaussian curve is such that the frequency of small deviations from the mean value is much greater than the frequency of large deviations. This coincides with the usual expectation in measurements subject to random errors that the number of measurements with a small error is much larger than the number of measurements with a large error. Alternative names for the Gaussian distribution are the *Normal distribution* or *Bell-shaped distribution*. A Gaussian curve is formally defined as a normalized frequency distribution that is symmetrical about the line of zero error and in which the frequency and magnitude of quantities are related by the expression:

$$F(x) = \frac{1}{\sigma\sqrt{2\pi}} e^{[-(x-m)^2/2\sigma^2]} \tag{3.11}$$

where m is the mean value of the data set x and the other quantities are as defined before. Equation (3.11) is particularly useful for analysing a Gaussian set of measurements and predicting how many measurements lie within some particular defined range. If the measurement deviations D are calculated for all measurements such that $D = x - m$, then the curve of deviation frequency $F(D)$ plotted against deviation magnitude D is a Gaussian curve known as the *error frequency distribution curve*. The mathematical relationship between $F(D)$ and D can then be derived by modifying equation (3.11) to give:

$$F(D) = \frac{1}{\sigma\sqrt{2\pi}} e^{[-D^2/2\sigma^2]} \tag{3.12}$$

The shape of a Gaussian curve is strongly influenced by the value of σ, with the width of the curve decreasing as σ becomes smaller. As a smaller σ corresponds with the typical deviations of the measurements from the mean value becoming smaller, this confirms the earlier observation that the mean value of a set of measurements gets closer to the true value as σ decreases.

If the standard deviation is used as a unit of error, the Gaussian curve can be used to determine the probability that the deviation in any particular measurement in a Gaussian data set is greater than a certain value. By substituting the expression for $F(D)$ in (3.12) into the probability equation (3.9), the probability that the error lies in a band between error levels D_1 and D_2 can be expressed as:

$$P(D_1 \leq D \leq D_2) = \int_{D_1}^{D_2} \frac{1}{\sigma\sqrt{2\pi}} e^{(-D^2/2\sigma^2)} \, dD \tag{3.13}$$

Solution of this expression is simplified by the substitution:

$$z = D/\sigma \tag{3.14}$$

The effect of this is to change the error distribution curve into a new Gaussian distribution that has a standard deviation of one ($\sigma = 1$) and a mean of zero. This new form, shown in Figure 3.7, is known as a *standard Gaussian curve*, and the dependent

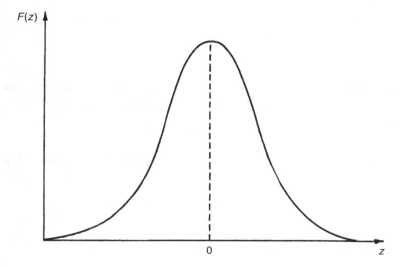

Fig. 3.7 Standard Gaussian curve ($F(z)$ versus z).

variable is now z instead of D. Equation (3.13) can now be re-expressed as:

$$P(D_1 \leq D \leq D_2) = P(z_1 \leq z \leq z_2) = \int_{z_1}^{z_2} \frac{1}{\sigma\sqrt{2\pi}} e^{(-z^2/2)} \, dz \qquad (3.15)$$

Unfortunately, neither equation (3.13) nor (3.15) can be solved analytically using tables of standard integrals, and numerical integration provides the only method of solution. However, in practice, the tedium of numerical integration can be avoided when analysing data because the standard form of equation (3.15), and its independence from the particular values of the mean and standard deviation of the data, means that standard Gaussian tables that tabulate $F(z)$ for various values of z can be used.

Standard Gaussian tables

A standard Gaussian table, such as that shown in Table 3.1, tabulates $F(z)$ for various values of z, where $F(z)$ is given by:

$$F(z) = \int_{-\infty}^{z} \frac{1}{\sigma\sqrt{2\pi}} e^{(-z^2/2)} \, dz \qquad (3.16)$$

Thus, $F(z)$ gives the proportion of data values that are less than or equal to z. This proportion is the area under the curve of $F(z)$ against z that is to the left of z. Therefore, the expression given in (3.15) has to be evaluated as $[F(z_2) - F(z_1)]$. Study of Table 3.1 shows that $F(z) = 0.5$ for $z = 0$. This confirms that, as expected, the number of data values ≤ 0 is 50% of the total. This must be so if the data only has random errors. It will also be observed that Table 3.1, in common with most published standard Gaussian tables, only gives $F(z)$ for positive values of z. For negative values of z, we can make use of the following relationship because the frequency distribution curve is normalized:

$$F(-z) = 1 - F(z) \qquad (3.17)$$

($F(-z)$ is the area under the curve to the left of $(-z)$, i.e. it represents the proportion of data values $\leq -z$.)

Example 3.3
How many measurements in a data set subject to random errors lie outside deviation boundaries of $+\sigma$ and $-\sigma$, i.e. how many measurements have a deviation greater than $|\sigma|$?

Solution
The required number is represented by the sum of the two shaded areas in Figure 3.8. This can be expressed mathematically as:

$$P(E < -\sigma \text{ or } E > +\sigma) = P(E < -\sigma) + P(E > +\sigma)$$

For $E = -\sigma$, $z = -1.0$ (from equation 3.12).
   Using Table 3.1:

$$P(E < -\sigma) = F(-1) = 1 - F(1) = 1 - 0.8413 = 0.1587$$

Table 3.1 Standard Gaussian table

z	0.00	0.01	0.02	0.03	0.04	0.05	0.06	0.07	0.08	0.09
					$F(z)$					
0.0	0.5000	0.5040	0.5080	0.5120	0.5160	0.5199	0.5239	0.5279	0.5319	0.5359
0.1	0.5398	0.5438	0.5478	0.5517	0.5557	0.5596	0.5636	0.5675	0.5714	0.5753
0.2	0.5793	0.5832	0.5871	0.5910	0.5948	0.5987	0.6026	0.6064	0.6103	0.6141
0.3	0.6179	0.6217	0.6255	0.6293	0.6331	0.6368	0.6406	0.6443	0.6480	0.6517
0.4	0.6554	0.6591	0.6628	0.6664	0.6700	0.6736	0.6772	0.6808	0.6844	0.6879
0.5	0.6915	0.6950	0.6985	0.7019	0.7054	0.7088	0.7123	0.7157	0.7190	0.7224
0.6	0.7257	0.7291	0.7324	0.7357	0.7389	0.7422	0.7454	0.7486	0.7517	0.7549
0.7	0.7580	0.7611	0.7642	0.7673	0.7703	0.7734	0.7764	0.7793	0.7823	0.7852
0.8	0.7881	0.7910	0.7939	0.7967	0.7995	0.8023	0.8051	0.8078	0.8106	0.8133
0.9	0.8159	0.8186	0.8212	0.8238	0.8264	0.8289	0.8315	0.8340	0.8365	0.8389
1.0	0.8413	0.8438	0.8461	0.8485	0.8508	0.8531	0.8554	0.8577	0.8599	0.8621
1.1	0.8643	0.8665	0.8686	0.8708	0.8729	0.8749	0.8770	0.8790	0.8810	0.8830
1.2	0.8849	0.8869	0.8888	0.8906	0.8925	0.8943	0.8962	0.8980	0.8997	0.9015
1.3	0.9032	0.9049	0.9066	0.9082	0.9099	0.9115	0.9131	0.9147	0.9162	0.9177
1.4	0.9192	0.9207	0.9222	0.9236	0.9251	0.9265	0.9279	0.9292	0.9306	0.9319
1.5	0.9332	0.9345	0.9357	0.9370	0.9382	0.9394	0.9406	0.9418	0.9429	0.9441
1.6	0.9452	0.9463	0.9474	0.9484	0.9495	0.9505	0.9515	0.9525	0.9535	0.9545
1.7	0.9554	0.9564	0.9573	0.9582	0.9591	0.9599	0.9608	0.9616	0.9625	0.9633
1.8	0.9641	0.9648	0.9656	0.9664	0.9671	0.9678	0.9686	0.9693	0.9699	0.9706
1.9	0.9713	0.9719	0.9726	0.9732	0.9738	0.9744	0.9750	0.9756	0.9761	0.9767
2.0	0.9772	0.9778	0.9783	0.9788	0.9793	0.9798	0.9803	0.9808	0.9812	0.9817
2.1	0.9821	0.9826	0.9830	0.9834	0.9838	0.9842	0.9846	0.9850	0.9854	0.9857
2.2	0.9861	0.9864	0.9868	0.9871	0.9875	0.9878	0.9881	0.9884	0.9887	0.9890
2.3	0.9893	0.9896	0.9898	0.9901	0.9904	0.9906	0.9909	0.9911	0.9913	0.9916
2.4	0.9918	0.9920	0.9922	0.9924	0.9926	0.9928	0.9930	0.9932	0.9934	0.9936
2.5	0.9938	0.9940	0.9941	0.9943	0.9945	0.9946	0.9948	0.9949	0.9951	0.9952
2.6	0.9953	0.9955	0.9956	0.9957	0.9959	0.9960	0.9961	0.9962	0.9963	0.9964
2.7	0.9965	0.9966	0.9967	0.9968	0.9969	0.9970	0.9971	0.9972	0.9973	0.9974
2.8	0.9974	0.9975	0.9976	0.9977	0.9977	0.9978	0.9979	0.9979	0.9980	0.9981
2.9	0.9981	0.9982	0.9982	0.9983	0.9984	0.9984	0.9985	0.9985	0.9986	0.9986
3.0	0.9986	0.9987	0.9987	0.9988	0.9988	0.9989	0.9989	0.9989	0.9990	0.9990
3.1	0.9990	0.9991	0.9991	0.9991	0.9992	0.9992	0.9992	0.9992	0.9993	0.9993
3.2	0.9993	0.9993	0.9994	0.9994	0.9994	0.9994	0.9994	0.9995	0.9995	0.9995
3.3	0.9995	0.9995	0.9995	0.9996	0.9996	0.9996	0.9996	0.9996	0.9996	0.9996
3.4	0.9997	0.9997	0.9997	0.9997	0.9997	0.9997	0.9997	0.9997	0.9997	0.9998
3.5	0.9998	0.9998	0.9998	0.9998	0.9998	0.9998	0.9998	0.9998	0.9998	0.9998
3.6	0.9998	0.9998	0.9998	0.9999	0.9999	0.9999	0.9999	0.9999	0.9999	0.9999

Similarly, for $E = +\sigma$, $z = +1.0$, Table 3.1 gives:

$$P(E > +\sigma) = 1 - P(E < +\sigma) = 1 - F(1) = 1 - 0.8413 = 0.1587.$$

(This last step is valid because the frequency distribution curve is normalized such that the total area under it is unity.)
Thus

$$P[E < -\sigma] + P[E > +\sigma] = 0.1587 + 0.1587 = 0.3174 \sim 32\%$$

i.e. 32% of the measurements lie outside the $\pm\sigma$ boundaries, then 68% of the measurements lie inside.

The above analysis shows that, for Gaussian-distributed data values, 68% of the measurements have deviations that lie within the bounds of $\pm\sigma$. Similar analysis shows

Errors during the measurement process

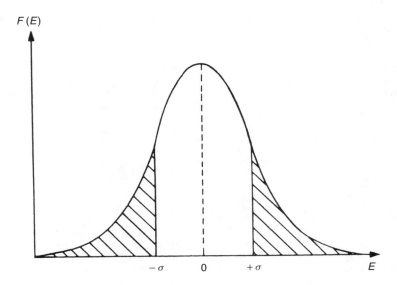

Fig. 3.8 $\pm\sigma$ boundaries.

that boundaries of $\pm 2\sigma$ contain 95.4% of data points, and extending the boundaries to $\pm 3\sigma$ encompasses 99.7% of data points. The probability of any data point lying outside particular deviation boundaries can therefore be expressed by the following table.

Deviation boundaries	% of data points within boundary	Probability of any particular data point being outside boundary
$\pm\sigma$	68.0	32.0%
$\pm 2\sigma$	95.4	4.6%
$\pm 3\sigma$	99.7	0.3%

Standard error of the mean

The foregoing analysis has examined the way in which measurements with random errors are distributed about the mean value. However, we have already observed that some error remains between the mean value of a set of measurements and the true value, i.e. averaging a number of measurements will only yield the true value if the number of measurements is infinite. If several subsets are taken from an infinite data population, then, by the central limit theorem, the means of the subsets will be distributed about the mean of the infinite data set. The error between the mean of a finite data set and the true measurement value (mean of the infinite data set) is defined as the *standard error of the mean*, α. This is calculated as:

$$\alpha = \sigma/\sqrt{n} \qquad (3.18)$$

α tends towards zero as the number of measurements in the data set expands towards infinity. The measurement value obtained from a set of n measurements, $x_1, x_2, \cdots x_n$,

can then be expressed as:
$$x = x_{\text{mean}} \pm \alpha$$

For the data set C of length measurements used earlier, $n = 23$, $\sigma = 1.88$ and $\alpha = 0.39$. The length can therefore be expressed as 406.5 ± 0.4 (68% confidence limit). However, it is more usual to express measurements with 95% confidence limits ($\pm 2\sigma$ boundaries). In this case, $2\sigma = 3.76$, $2\alpha = 0.78$ and the length can be expressed as 406.5 ± 0.8 (95% confidence limits).

Estimation of random error in a single measurement

In many situations where measurements are subject to random errors, it is not practical to take repeated measurements and find the average value. Also, the averaging process becomes invalid if the measured quantity does not remain at a constant value, as is usually the case when process variables are being measured. Thus, if only one measurement can be made, some means of estimating the likely magnitude of error in it is required. The normal approach to this is to calculate the error within 95% confidence limits, i.e. to calculate the value of the deviation D such that 95% of the area under the probability curve lies within limits of $\pm D$. These limits correspond to a deviation of $\pm 1.96\sigma$. Thus, it is necessary to maintain the measured quantity at a constant value whilst a number of measurements are taken in order to create a reference measurement set from which σ can be calculated. Subsequently, the maximum likely deviation in a single measurement can be expressed as: Deviation $= \pm 1.96\sigma$. However, this only expresses the maximum likely deviation of the measurement from the calculated mean of the reference measurement set, which is not the true value as observed earlier. Thus the calculated value for the standard error of the mean has to be added to the likely maximum deviation value. Thus, the maximum likely error in a single measurement can be expressed as:

$$\text{Error} = \pm(1.96\sigma + \alpha) \tag{3.19}$$

Example 3.4
Suppose that a standard mass is measured 30 times with the same instrument to create a reference data set, and the calculated values of σ and α are $\sigma = 0.43$ and $\alpha = 0.08$. If the instrument is then used to measure an unknown mass and the reading is 105.6 kg, how should the mass value be expressed?

Solution
Using (3.19), $1.96\sigma + \alpha = 0.92$. The mass value should therefore be expressed as: 105.6 ± 0.9 kg.

Before leaving this matter, it must be emphasized that the maximum error specified for a measurement is only specified for the confidence limits defined. Thus, if the maximum error is specified as $\pm 1\%$ with 95% confidence limits, this means that there is still 1 chance in 20 that the error will exceed $\pm 1\%$.

Distribution of manufacturing tolerances

Many aspects of manufacturing processes are subject to random variations caused by factors that are similar to those that cause random errors in measurements. In most cases, these random variations in manufacturing, which are known as *tolerances*, fit a

Errors during the measurement process

Gaussian distribution, and the previous analysis of random measurement errors can be applied to analyse the distribution of these variations in manufacturing parameters.

Example 3.5
An integrated circuit chip contains 10^5 transistors. The transistors have a mean current gain of 20 and a standard deviation of 2. Calculate the following:

(a) the number of transistors with a current gain between 19.8 and 20.2
(b) the number of transistors with a current gain greater than 17.

Solution
(a) The proportion of transistors where $19.8 <$ gain < 20.2 is:

$$P[X < 20] - P[X < 19.8] = P[z < 0.2] - P[z < -0.2] \quad \text{(for } z = (X - \mu)/\sigma\text{)}$$

For $X = 20.2$; $z = 0.1$ and for $X = 19.8$; $z = -0.1$
From tables, $P[z < 0.1] = 0.5398$ and thus $P[z < -0.1] = 1 - P[z < 0.1] = 1 - 0.5398 = 0.4602$
Hence, $P[z < 0.1] - P[z < -0.1] = 0.5398 - 0.4602 = 0.0796$
Thus $0.0796 \times 10^5 = 7960$ transistors have a current gain in the range from 19.8 to 20.2.

(b) The number of transistors with gain >17 is given by:

$$P[x > 17] = 1 - P[x < 17] = 1 - P[z < -1.5] = P[z < +1.5] = 0.9332$$

Thus, 93.32%, i.e. 93 320 transistors have a gain >17.

Goodness of fit to a Gaussian distribution

All of the analysis of random deviations presented so far only applies when the data being analysed belongs to a Gaussian distribution. Hence, the degree to which a set of data fits a Gaussian distribution should always be tested before any analysis is carried out. This test can be carried out in one of three ways:

(a) *Simple test*: The simplest way to test for Gaussian distribution of data is to plot a histogram and look for a 'Bell-shape' of the form shown earlier in Figure 3.5. Deciding whether or not the histogram confirms a Gaussian distribution is a matter of judgement. For a Gaussian distribution, there must always be approximate symmetry about the line through the centre of the histogram, the highest point of the histogram must always coincide with this line of symmetry, and the histogram must get progressively smaller either side of this point. However, because the histogram can only be drawn with a finite set of measurements, some deviation from the perfect shape of the histogram as described above is to be expected even if the data really is Gaussian.
(b) *Using a normal probability plot*: A normal probability plot involves dividing the data values into a number of ranges and plotting the cumulative probability of summed data frequencies against the data values on special graph paper.* This line should be a straight line if the data distribution is Gaussian. However, careful judgement is required since only a finite number of data values can be used and

---
* This is available from specialist stationery suppliers.

therefore the line drawn will not be entirely straight even if the distribution is Gaussian. Considerable experience is needed to judge whether the line is straight enough to indicate a Gaussian distribution. This will be easier to understand if the data in measurement set C is used as an example. Using the same five ranges as used to draw the histogram, the following table is first drawn:

Range:	401.5 to 403.5	403.5 to 405.5	405.5 to 407.5	407.5 to 409.5	409.5 to 411.5
Number of data items in range	1	5	11	5	1
Cumulative number of data items	1	6	17	22	23
Cumulative number of data items as %	4.3	26.1	73.9	95.7	100.0

The normal probability plot drawn from the above table is shown in Figure 3.9. This is sufficiently straight to indicate that the data in measurement set C is Gaussian.

(c) *Chi-squared test*: A further test that can be applied is based on the chi-squared (χ^2) distribution. This is beyond the scope of this book but full details can be found in Caulcott (1973).

Rogue data points

In a set of measurements subject to random error, measurements with a very large error sometimes occur at random and unpredictable times, where the magnitude of the error is much larger than could reasonably be attributed to the expected random variations in measurement value. Sources of such abnormal error include sudden transient voltage surges on the mains power supply and incorrect recording of data (e.g. writing down 146.1 when the actual measured value was 164.1). It is accepted practice in such

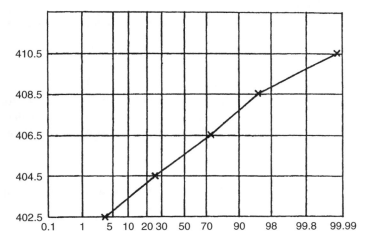

Fig. 3.9 Normal probability plot.

cases to discard these rogue measurements, and a threshold level of a $\pm 3\sigma$ deviation is often used to determine what should be discarded. It is extremely rare for measurement errors to exceed $\pm 3\sigma$ limits when only normal random effects are affecting the measured value.

Special case when the number of measurements is small

When the number of measurements of a quantity is particularly small and statistical analysis of the distribution of error values is required, problems can arise when using standard Gaussian tables in terms of z as defined in equation (3.16) because the mean of only a small number of measurements may deviate significantly from the true measurement value. In response to this, an alternative distribution function called the Student-t distribution can be used which gives a more accurate prediction of the error distribution when the number of samples is small. This is discussed more fully in Miller (1990).

3.6 Aggregation of measurement system errors

Errors in measurement systems often arise from two or more different sources, and these must be aggregated in the correct way in order to obtain a prediction of the total likely error in output readings from the measurement system. Two different forms of aggregation are required. Firstly, a single measurement component may have both systematic and random errors and, secondly, a measurement system may consist of several measurement components that each have separate errors.

3.6.1 Combined effect of systematic and random errors

If a measurement is affected by both systematic and random errors that are quantified as $\pm x$ (systematic errors) and $\pm y$ (random errors), some means of expressing the combined effect of both types of error is needed. One way of expressing the combined error would be to sum the two separate components of error, i.e. to say that the total possible error is $e = \pm(x + y)$. However, a more usual course of action is to express the likely maximum error as follows:

$$e = \sqrt{(x^2 + y^2)} \qquad (3.20)$$

It can be shown (ANSI/ASME, 1985) that this is the best expression for the error statistically, since it takes account of the reasonable assumption that the systematic and random errors are independent and so are unlikely to both be at their maximum or minimum value at the same time.

3.6.2 Aggregation of errors from separate measurement system components

A measurement system often consists of several separate components, each of which is subject to errors. Therefore, what remains to be investigated is how the errors associated with each measurement system component combine together, so that a total error calculation can be made for the complete measurement system. All four mathematical operations of addition, subtraction, multiplication and division may be performed on measurements derived from different instruments/transducers in a measurement system. Appropriate techniques for the various situations that arise are covered below.

Error in a sum
If the two outputs y and z of separate measurement system components are to be added together, we can write the sum as $S = y + z$. If the maximum errors in y and z are $\pm ay$ and $\pm bz$ respectively, we can express the maximum and minimum possible values of S as:

$$S_{\max} = (y + ay) + (z + bz); \quad S_{\min} = (y - ay) + (z - bz); \quad \text{or } S = y + z \pm (ay + bz)$$

This relationship for S is not convenient because in this form the error term cannot be expressed as a fraction or percentage of the calculated value for S. Fortunately, statistical analysis can be applied (see Topping, 1962) that expresses S in an alternative form such that the most probable maximum error in S is represented by a quantity e, where e is calculated in terms of the *absolute* errors as:

$$e = \sqrt{(ay)^2 + (bz)^2} \tag{3.21}$$

Thus $S = (y + z) \pm e$. This can be expressed in the alternative form:

$$S = (y + z)(1 \pm f) \quad \text{where } f = e/(y + z) \tag{3.22}$$

It should be noted that equations (3.21) and (3.22) are only valid provided that the measurements are uncorrelated (i.e. each measurement is entirely independent of the others).

Example 3.6
A circuit requirement for a resistance of 550 Ω is satisfied by connecting together two resistors of nominal values 220 Ω and 330 Ω in series. If each resistor has a tolerance of $\pm 2\%$, the error in the sum calculated according to equations (3.21) and (3.22) is given by:

$$e = \sqrt{(0.02 \times 220)^2 + (0.02 \times 330)^2} = 7.93; \quad f = 7.93/50 = 0.0144$$

Thus the total resistance S can be expressed as

$$S = 550\,\Omega \pm 7.93\,\Omega \text{ or } S = 550\,(1 \pm 0.0144)\,\Omega, \quad \text{i.e. } S = 550\,\Omega \pm 1.4\%$$

Error in a difference
If the two outputs y and z of separate measurement systems are to be subtracted from one another, and the possible errors are $\pm ay$ and $\pm bz$, then the difference S can be expressed (using statistical analysis as for calculating the error in a sum and assuming that the measurements are uncorrelated) as:

$$S = (y - z) \pm e \quad \text{or} \quad S = (y - z)(1 \pm f)$$

where e is calculated as above (equation 3.21), and $f = e/(y - z)$

58 Errors during the measurement process

Example 3.7
A fluid flow rate is calculated from the difference in pressure measured on both sides of an orifice plate. If the pressure measurements are 10.0 bar and 9.5 bar and the error in the pressure measuring instruments is specified as ±0.1%, then values for e and f can be calculated as:

$$e = \sqrt{(0.001 \times 10)^2 + (0.001 \times 9.5)^2} = 0.0138; \quad f = 0.0138/0.5 = 0.0276$$

- This example illustrates very poignantly the relatively large error that can arise when calculations are made based on the difference between two measurements.

Error in a product

If the outputs y and z of two measurement system components are multiplied together, the product can be written as $P = yz$. If the possible error in y is $\pm ay$ and in z is $\pm bz$, then the maximum and minimum values possible in P can be written as:

$$P_{\max} = (y + ay)(z + bz) = yz + ayz + byz + aybz;$$
$$P_{\min} = (y - ay)(z - bz) = yz - ayz - byz + aybz$$

For typical measurement system components with output errors of up to one or two per cent in magnitude, both a and b are very much less than one in magnitude and thus terms in $aybz$ are negligible compared with other terms. Therefore, we have $P_{\max} = yz(1 + a + b); P_{\min} = yz(1 - a - b)$. Thus the maximum error in the product P is $\pm(a + b)$. Whilst this expresses the maximum possible error in P, it tends to overestimate the likely maximum error since it is very unlikely that the errors in y and z will both be at the maximum or minimum value at the same time. A statistically better estimate of the likely maximum error e in the product P, provided that the measurements are uncorrelated, is given by Topping (1962):

$$e = \sqrt{a^2 + b^2} \tag{3.23}$$

Note that in the case of multiplicative errors, e is calculated in terms of the *fractional* errors in y and z (as opposed to the *absolute* error values used in calculating additive errors).

Example 3.8
If the power in a circuit is calculated from measurements of voltage and current in which the calculated maximum errors are respectively ±1% and ±2%, then the maximum likely error in the calculated power value, calculated using (3.23) is $\pm\sqrt{0.01^2 + 0.02^2} = \pm 0.022$ or ±2.2%.

Error in a quotient

If the output measurement y of one system component with possible error $\pm ay$ is divided by the output measurement z of another system component with possible error $\pm bz$, then the maximum and minimum possible values for the quotient can be written as:

$$Q_{max} = \frac{y+ay}{z-bz} = \frac{(y+ay)(z+bz)}{(z-bz)(z+bz)} = \frac{yz+ayz+byz+aybz}{z^2-b^2z^2};$$

$$Q_{min} = \frac{y-ay}{z+bz} = \frac{(y-ay)(z-bz)}{(z+bz)(z-bz)} = \frac{yz-ayz-byz+aybz}{z^2-b^2z^2}$$

For $a \ll 1$ and $b \ll 1$, terms in ab and b^2 are negligible compared with the other terms. Hence:

$$Q_{max} = \frac{yz(1+a+b)}{z^2}; \quad Q_{min} = \frac{yz(1-a-b)}{z^2}; \quad \text{i.e. } Q = \frac{y}{z} \pm \frac{y}{z}(a+b)$$

Thus the maximum error in the quotient is $\pm(a+b)$. However, using the same argument as made above for the product of measurements, a statistically better estimate (see Topping, 1962) of the likely maximum error in the quotient Q, provided that the measurements are uncorrelated, is that given in (3.23).

Example 3.9
If the density of a substance is calculated from measurements of its mass and volume where the respective errors are $\pm 2\%$ and $\pm 3\%$, then the maximum likely error in the density value using (3.23) is $\pm\sqrt{0.02^2 + 0.003^2} = \pm 0.036$ or $\pm 3.6\%$.

3.6.3 Total error when combining multiple measurements

The final case to be covered is where the final measurement is calculated from several measurements that are combined together in a way that involves more than one type of arithmetic operation. For example, the density of a rectangular-sided solid block of material can be calculated from measurements of its mass divided by the product of measurements of its length, height and width. The errors involved in each stage of arithmetic are cumulative, and so the total measurement error can be calculated by adding together the two error values associated with the two multiplication stages involved in calculating the volume and then calculating the error in the final arithmetic operation when the mass is divided by the volume.

Example 3.10
A rectangular-sided block has edges of lengths a, b and c, and its mass is m. If the values and possible errors in quantities a, b, c and m are as shown below, calculate the value of density and the possible error in this value.

$$a = 100 \text{ mm} \pm 1\%, b = 200 \text{ mm} \pm 1\%, c = 300 \text{ mm} \pm 1\%, m = 20 \text{ kg} \pm 0.5\%.$$

Solution
Value of $ab = 0.02 \text{ m}^2 \pm 2\%$ (possible error $= 1\% + 1\% = 2\%$)
Value of $(ab)c = 0.006 \text{ m}^3 \pm 3\%$ (possible error $= 2\% + 1\% = 3\%$)
Value of $\frac{m}{abc} = \frac{20}{0.006} = 3330 \text{ kg/m}^3 \pm 3.5\%$ (possible error $= 3\% + 0.5\% = 3.5\%$)

3.7 Self-test questions

3.1 Explain the difference between systematic and random errors. What are the typical sources of these two types of error?

3.2 In what ways can the act of measurement cause a disturbance in the system being measured?

3.3 Suppose that the components in the circuit shown in Figure 3.1(a) have the following values:

$$R_1 = 330\,\Omega; \quad R_2 = 1000\,\Omega; \quad R_3 = 1200\,\Omega; \quad R_4 = 220\,\Omega; \quad R_5 = 270\,\Omega.$$

If the instrument measuring the output voltage across AB has a resistance of $5000\,\Omega$, what is the measurement error caused by the loading effect of this instrument?

3.4 Instruments are normally calibrated and their characteristics defined for particular standard ambient conditions. What procedures are normally taken to avoid measurement errors when using instruments that are subjected to changing ambient conditions?

3.5 The voltage across a resistance R_5 in the circuit of Figure 3.10 is to be measured by a voltmeter connected across it.
   (a) If the voltmeter has an internal resistance (R_m) of $4750\,\Omega$, what is the measurement error?
   (b) What value would the voltmeter internal resistance need to be in order to reduce the measurement error to 1%?

3.6 In the circuit shown in Figure 3.11, the current flowing between A and B is measured by an ammeter whose internal resistance is $100\,\Omega$. What is the measurement error caused by the resistance of the measuring instrument?

3.7 What steps can be taken to reduce the effect of environmental inputs in measurement systems?

3.8 The output of a potentiometer is measured by a voltmeter having a resistance R_m, as shown in Figure 3.12. R_t is the resistance of the total length X_t of the potentiometer and R_i is the resistance between the wiper and common point C for a general wiper position X_i. Show that the measurement error due to the resistance

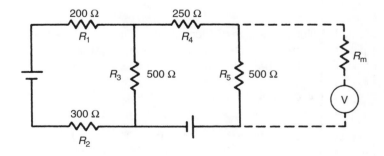

Fig. 3.10 Circuit for question 3.5.

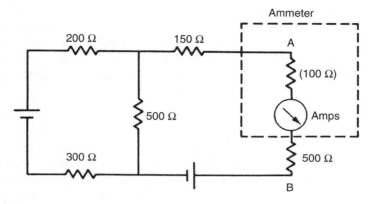

Fig. 3.11 Circuit for question 3.6.

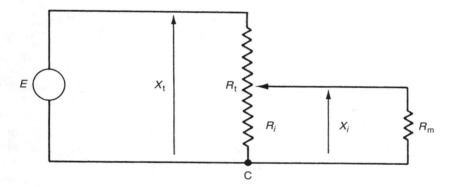

Fig. 3.12 Circuit for question 3.8.

R_m of the measuring instrument is given by:

$$\text{Error} = E \frac{R_i^2 (R_t - R_i)}{R_t (R_i R_t + R_m R_t - R_i^2)}$$

Hence show that the maximum error occurs when X_i is approximately equal to $2X_t/3$. (Hint – differentiate the error expression with respect to R_i and set to 0. Note that maximum error does not occur exactly at $X_i = 2X_t/3$, but this value is very close to the position where the maximum error occurs.)

3.9 In a survey of 15 owners of a certain model of car, the following figures for average petrol consumption were reported.

25.5 30.3 31.1 29.6 32.4 39.4 28.9 30.0 33.3 31.4 29.5 30.5 31.7 33.0 29.2

Calculate the mean value, the median value and the standard deviation of the data set.

3.10 (a) What do you understand by the term *probability density function*?
   (b) Write down an expression for a Gaussian probability density function of given mean value μ and standard deviation σ and show how you would obtain the best estimate of these two quantities from a sample of population n.
   (c) The following ten measurements are made of the output voltage from a high-gain amplifier that is contaminated due to noise fluctuations:

$$1.53, \ 1.57, \ 1.54, \ 1.54, \ 1.50, \ 1.51, \ 1.55, \ 1.54, \ 1.56, \ 1.53$$

Determine the mean value and standard deviation. Hence estimate the accuracy to which the mean value is determined from these ten measurements. If one thousand measurements were taken, instead of ten, but σ remained the same, by how much would the accuracy of the calculated mean value be improved?

3.11 The following measurements were taken with an analogue meter of the current flowing in a circuit (the circuit was in steady state and therefore, although the measurements varied due to random errors, the current flowing was actually constant):

$$21.5 \, \text{mA}, \ 22.1 \, \text{mA}, \ 21.3 \, \text{mA}, \ 21.7 \, \text{mA}, \ 22.0 \, \text{mA}, \ 22.2 \, \text{mA}, \ 21.8 \, \text{mA},$$

$$21.4 \, \text{mA}, \ 21.9 \, \text{mA}, \ 22.1 \, \text{mA}$$

Calculate the mean value, the deviations from the mean and the standard deviation.

3.12 The measurements in a data set are subject to random errors but it is known that the data set fits a Gaussian distribution. Use standard Gaussian tables to determine the percentage of measurements that lie within the boundaries of $\pm 1.5\sigma$, where σ is the standard deviation of the measurements.

3.13 The thickness of a set of gaskets varies because of random manufacturing disturbances but the thickness values measured belong to a Gaussian distribution. If the mean thickness is 3 mm and the standard deviation is 0.25, calculate the percentage of gaskets that have a thickness greater than 2.5 mm.

3.14 A 3 volt d.c. power source required for a circuit is obtained by connecting together two 1.5 V batteries in series. If the error in the voltage output of each battery is specified as ±1%, calculate the likely maximum possible error in the 3 volt power source that they make up.

3.15 In order to calculate the heat loss through the wall of a building, it is necessary to know the temperature difference between the inside and outside walls. If temperatures of 5°C and 20°C are measured on each side of the wall by mercury-in-glass thermometers with a range of 0°C to +50°C and a quoted inaccuracy figure of ±1% of full-scale reading, calculate the likely maximum possible error in the calculated figure for the temperature difference.

3.16 The power dissipated in a car headlight is calculated by measuring the d.c. voltage drop across it and the current flowing through it ($P = V \times I$). If the possible errors in the measured voltage and current values are ±1% and ±2% respectively, calculate the likely maximum possible error in the power value deduced.

3.17 The resistance of a carbon resistor is measured by applying a d.c. voltage across it and measuring the current flowing ($R = V/I$). If the voltage and current values are measured as 10 ± 0.1 V and 214 ± 5 mA respectively, express the value of the carbon resistor.

3.18 The density (d) of a liquid is calculated by measuring its depth (c) in a calibrated rectangular tank and then emptying it into a mass measuring system. The length and width of the tank are (a) and (b) respectively and thus the density is given by:

$$d = m/(a \times b \times c)$$

where m is the measured mass of the liquid emptied out.

If the possible errors in the measurements of a, b, c and m are 1%, 1%, 2% and 0.5% respectively, determine the likely maximum possible error in the calculated value of the density (d).

3.19 The volume flow rate of a liquid is calculated by allowing the liquid to flow into a cylindrical tank (stood on its flat end) and measuring the height of the liquid surface before and after the liquid has flowed for 10 minutes. The volume collected after 10 minutes is given by:

$$\text{Volume} = (h_2 - h_1)\pi(d/2)^2$$

where h_1 and h_2 are the starting and finishing surface heights and d is the measured diameter of the tank.

(a) If $h_1 = 2$ m, $h_2 = 3$ m and $d = 2$ m, calculate the volume flow rate in m³/min.
(b) If the possible error in each measurement h_1, h_2 and d is $\pm 1\%$, determine the likely maximum possible error in the calculated value of volume flow rate.

References and further reading

ANSI/ASME standards (1985) ASME performance test codes, supplement on instruments and apparatus, part 1: measurement uncertainty, American Society of Mechanical Engineers, New York.

Bennington, P.R. and Robinson, D.K. (1992) *Data Reduction and Error Analysis for the Physical Sciences*, McGraw-Hill.

Caulcott, E. (1973) *Significance Tests*, Routledge and Kegan Paul.

Miller, I.R., Freung, J.E. and Johnson, R. (1990) *Probability and Statistics for Engineers*, Prentice-Hall.

Topping, J. (1962) *Errors of Observation and Their Treatment*, Chapman and Hall.

4

Calibration of measuring sensors and instruments

4.1 Principles of calibration

Calibration consists of comparing the output of the instrument or sensor under test against the output of an instrument of known accuracy when the same input (the measured quantity) is applied to both instruments. This procedure is carried out for a range of inputs covering the whole measurement range of the instrument or sensor. Calibration ensures that the measuring accuracy of all instruments and sensors used in a measurement system is known over the whole measurement range, provided that the calibrated instruments and sensors are used in environmental conditions that are the same as those under which they were calibrated. For use of instruments and sensors under different environmental conditions, appropriate correction has to be made for the ensuing modifying inputs, as described in Chapter 3. Whether applied to instruments or sensors, calibration procedures are identical, and hence only the term instrument will be used for the rest of this chapter, with the understanding that whatever is said for instruments applies equally well to single measurement sensors.

Instruments used as a standard in calibration procedures are usually chosen to be of greater inherent accuracy than the process instruments that they are used to calibrate. Because such instruments are only used for calibration purposes, greater accuracy can often be achieved by specifying a type of instrument that would be unsuitable for normal process measurements. For instance, ruggedness is not a requirement, and freedom from this constraint opens up a much wider range of possible instruments. In practice, high-accuracy, null-type instruments are very commonly used for calibration duties, because the need for a human operator is not a problem in these circumstances.

Instrument calibration has to be repeated at prescribed intervals because the characteristics of any instrument change over a period. Changes in instrument characteristics are brought about by such factors as mechanical wear, and the effects of dirt, dust, fumes, chemicals and temperature changes in the operating environment. To a great extent, the magnitude of the drift in characteristics depends on the amount of use an instrument receives and hence on the amount of wear and the length of time that it is subjected to the operating environment. However, some drift also occurs even in storage, as a result of ageing effects in components within the instrument.

Determination of the frequency at which instruments should be calibrated is dependent upon several factors that require specialist knowledge. If an instrument is required to measure some quantity and an inaccuracy of $\pm 2\%$ is acceptable, then a certain amount of performance degradation can be allowed if its inaccuracy immediately after recalibration is $\pm 1\%$. What is important is that the pattern of performance degradation be quantified, such that the instrument can be recalibrated before its accuracy has reduced to the limit defined by the application.

Susceptibility to the various factors that can cause changes in instrument characteristics varies according to the type of instrument involved. Possession of an in-depth knowledge of the mechanical construction and other features involved in the instrument is necessary in order to be able to quantify the effect of these quantities on the accuracy and other characteristics of an instrument. The type of instrument, its frequency of use and the prevailing environmental conditions all strongly influence the calibration frequency necessary, and because so many factors are involved, it is difficult or even impossible to determine the required frequency of instrument recalibration from theoretical considerations. Instead, practical experimentation has to be applied to determine the rate of such changes. Once the maximum permissible measurement error has been defined, knowledge of the rate at which the characteristics of an instrument change allows a time interval to be calculated that represents the moment in time when an instrument will have reached the bounds of its acceptable performance level. The instrument must be recalibrated either at this time or earlier. This measurement error level that an instrument reaches just before recalibration is the error bound that must be quoted in the documented specifications for the instrument.

A proper course of action must be defined that describes the procedures to be followed when an instrument is found to be out of calibration, i.e. when its output is different to that of the calibration instrument when the same input is applied. The required action depends very much upon the nature of the discrepancy and the type of instrument involved. In many cases, deviations in the form of a simple output bias can be corrected by a small adjustment to the instrument (following which the adjustment screws must be sealed to prevent tampering). In other cases, the output scale of the instrument may have to be redrawn, or scaling factors altered where the instrument output is part of some automatic control or inspection system. In extreme cases, where the calibration procedure shows up signs of instrument damage, it may be necessary to send the instrument for repair or even scrap it.

Whatever system and frequency of calibration is established, it is important to review this from time to time to ensure that the system remains effective and efficient. It may happen that a cheaper (but equally effective) method of calibration becomes available with the passage of time, and such an alternative system must clearly be adopted in the interests of cost efficiency. However, the main item under scrutiny in this review is normally whether the calibration interval is still appropriate. Records of the calibration history of the instrument will be the primary basis on which this review is made. It may happen that an instrument starts to go out of calibration more quickly after a period of time, either because of ageing factors within the instrument or because of changes in the operating environment. The conditions or mode of usage of the instrument may also be subject to change. As the environmental and usage conditions of an instrument may change beneficially as well as adversely, there is the possibility that the recommended calibration interval may decrease as well as increase.

4.2 Control of calibration environment

Any instrument that is used as a standard in calibration procedures must be kept solely for calibration duties and must never be used for other purposes. Most particularly, it must not be regarded as a spare instrument that can be used for process measurements if the instrument normally used for that purpose breaks down. Proper provision for process instrument failures must be made by keeping a spare set of process instruments. Standard calibration instruments must be totally separate.

To ensure that these conditions are met, the calibration function must be managed and executed in a professional manner. This will normally mean setting aside a particular place within the instrumentation department of a company where all calibration operations take place and where all instruments used for calibration are kept. As far as possible this should take the form of a separate room, rather than a sectioned-off area in a room used for other purposes as well. This will enable better environmental control to be applied in the calibration area and will also offer better protection against unauthorized handling or use of the calibration instruments. The level of environmental control required during calibration should be considered carefully with due regard to what level of accuracy is required in the calibration procedure, but should not be overspecified as this will lead to unnecessary expense. Full air conditioning is not normally required for calibration at this level, as it is very expensive, but sensible precautions should be taken to guard the area from extremes of heat or cold, and also good standards of cleanliness should be maintained. Useful guidance on the operation of standards facilities can be found elsewhere (British Standards Society, 1979).

Whilst it is desirable that all calibration functions are performed in this carefully controlled environment, it is not always practical to achieve this. Sometimes, it is not convenient or possible to remove instruments from process plant, and in these cases, it is standard practice to calibrate them *in situ*. In these circumstances, appropriate corrections must be made for the deviation in the calibration environmental conditions away from those specified. This practice does not obviate the need to protect calibration instruments and maintain them in constant conditions in a calibration laboratory at all times other than when they are involved in such calibration duties on plant.

As far as management of calibration procedures is concerned, it is important that the performance of all calibration operations is assigned as the clear responsibility of just one person. That person should have total control over the calibration function, and be able to limit access to the calibration laboratory to designated, approved personnel only. Only by giving this appointed person total control over the calibration function can the function be expected to operate efficiently and effectively. Lack of such definite management can only lead to unintentional neglect of the calibration system, resulting in the use of equipment in an out-of-date state of calibration and subsequent loss of traceability to reference standards. Professional management is essential so that the customer can be assured that an efficient calibration system is in operation and that the accuracy of measurements is guaranteed.

Calibration procedures that relate in any way to measurements that are used for quality control functions are controlled by the international standard ISO 9000 (this subsumes the old British quality standard BS 5750). One of the clauses in ISO 9000 requires that all persons using calibration equipment be adequately trained. The manager in charge of the calibration function is clearly responsible for ensuring that this condition is met. Training must be adequate and targeted at the particular needs of the calibration systems involved. People must understand what they need to know and especially why they must have this information. Successful completion of training courses should be marked by the award of qualification certificates. These attest to the proficiency of personnel involved in calibration duties and are a convenient way of demonstrating that the ISO 9000 training requirement has been satisfied.

4.3 Calibration chain and traceability

The calibration facilities provided within the instrumentation department of a company provide the first link in the calibration chain. Instruments used for calibration at this level are known as *working standards*. As such working standard instruments are kept by the instrumentation department of a company solely for calibration duties, and for no other purpose, then it can be assumed that they will maintain their accuracy over a reasonable period of time because use-related deterioration in accuracy is largely eliminated. However, over the longer term, the characteristics of even such standard instruments will drift, mainly due to ageing effects in components within them. Therefore, over this longer term, a programme must be instituted for calibrating working standard instruments at appropriate intervals of time against instruments of yet higher accuracy. The instrument used for calibrating working standard instruments is known as a *secondary reference standard*. This must obviously be a very well-engineered instrument that gives high accuracy and is stabilized against drift in its performance with time. This implies that it will be an expensive instrument to buy. It also requires that the environmental conditions in which it is used be carefully controlled in respect of ambient temperature, humidity etc.

When the working standard instrument has been calibrated by an authorized standards laboratory, a calibration certificate will be issued. This will contain at least the following information:

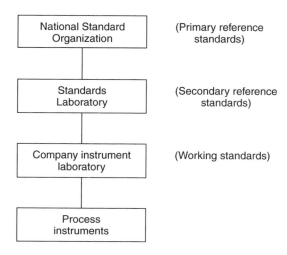

Fig. 4.1 Instrument calibration chain.

- the identification of the equipment calibrated
- the calibration results obtained
- the measurement uncertainty
- any use limitations on the equipment calibrated
- the date of calibration
- the authority under which the certificate is issued.

The establishment of a company Standards Laboratory to provide a calibration facility of the required quality is economically viable only in the case of very large companies where large numbers of instruments need to be calibrated across several factories. In the case of small to medium size companies, the cost of buying and maintaining such equipment is not justified. Instead, they would normally use the calibration service provided by various companies that specialize in offering a Standards Laboratory. What these specialist calibration companies effectively do is to share out the high cost of providing this highly accurate but infrequently used calibration service over a large number of companies. Such Standards Laboratories are closely monitored by National Standards Organizations.

In the United Kingdom, the appropriate National Standards Organization for validating Standards Laboratories is the National Physical Laboratory (in the United States of America, the equivalent body is the National Bureau of Standards). This has established a National Measurement Accreditation Service (NAMAS) that monitors both instrument calibration and mechanical testing laboratories. The formal structure for accrediting instrument calibration Standards Laboratories is known as the British Calibration Service (BCS), and that for accrediting testing facilities is known as the National Testing Laboratory Accreditation Scheme (NATLAS).

Although each country has its own structure for the maintenance of standards, each of these different frameworks tends to be equivalent in its effect. To achieve confidence in the goods and services that move across national boundaries, international agreements have established the equivalence of the different accreditation schemes in existence. As a result, NAMAS and the similar schemes operated by France, Germany, Italy, the USA, Australia and New Zealand enjoy mutual recognition.

The British Calibration Service lays down strict conditions that a Standards Laboratory has to meet before it is approved. These conditions control laboratory management, environment, equipment and documentation. The person appointed as head of the laboratory must be suitably qualified, and independence of operation of the laboratory must be guaranteed. The management structure must be such that any pressure to rush or skip calibration procedures for production reasons can be resisted. As far as the laboratory environment is concerned, proper temperature and humidity control must be provided, and high standards of cleanliness and housekeeping must be maintained. All equipment used for calibration purposes must be maintained to reference standards, and supported by calibration certificates that establish this traceability. Finally, full documentation must be maintained. This should describe all calibration procedures, maintain an index system for recalibration of equipment, and include a full inventory of apparatus and traceability schedules. Having met these conditions, a Standards Laboratory becomes an accredited laboratory for providing calibration services and issuing calibration certificates. This accreditation is reviewed at approximately 12 monthly intervals to ensure that the laboratory is continuing to satisfy the conditions for approval laid down.

Measurement and Instrumentation Principles 69

Primary reference standards, as listed in Table 2.1, describe the highest level of accuracy that is achievable in the measurement of any particular physical quantity. All items of equipment used in Standards Laboratories as secondary reference standards have to be calibrated themselves against primary reference standards at appropriate intervals of time. This procedure is acknowledged by the issue of a calibration certificate in the standard way. National Standards Organizations maintain suitable facilities for this calibration, which in the case of the United Kingdom are at the National Physical Laboratory. The equivalent National Standards Organization in the United States of America is the National Bureau of Standards. In certain cases, such primary reference standards can be located outside National Standards Organizations. For instance, the primary reference standard for dimension measurement is defined by the wavelength of the orange–red line of krypton light, and it can therefore be realized in any laboratory equipped with an interferometer. In certain cases (e.g. the measurement of viscosity), such primary reference standards are not available and reference standards for calibration are achieved by collaboration between several National Standards Organizations who perform measurements on identical samples under controlled conditions (ISO 5725, 1998).

What has emerged from the foregoing discussion is that calibration has a chain-like structure in which every instrument in the chain is calibrated against a more accurate instrument immediately above it in the chain, as shown in Figure 4.1. All of the elements in the calibration chain must be known so that the calibration of process instruments at the bottom of the chain is traceable to the fundamental measurement standards. This knowledge of the full chain of instruments involved in the calibration procedure is known as *traceability*, and is specified as a mandatory requirement in satisfying the BS EN ISO 9000 standard. Documentation must exist that shows that

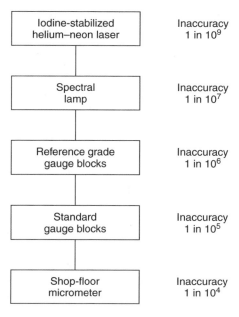

Fig. 4.2 Typical calibration chain for micrometers.

process instruments are calibrated by standard instruments that are linked by a chain of increasing accuracy back to national reference standards. There must be clear evidence to show that there is no break in this chain.

To illustrate a typical calibration chain, consider the calibration of micrometers (Figure 4.2). A typical shop floor micrometer has an uncertainty (inaccuracy) of less than 1 in 10^4. These would normally be calibrated in the instrumentation department or Standards Laboratory of a company against laboratory standard gauge blocks with a typical uncertainty of less than 1 in 10^5. A specialist calibration service company would provide facilities for calibrating these laboratory standard gauge blocks against reference-grade gauge blocks with a typical uncertainty of less than 1 in 10^6. More accurate calibration equipment still is provided by National Standards Organizations. The National Physical Laboratory maintains two sets of standards for this type of calibration, a working standard and a primary standard. Spectral lamps are used to provide a working reference standard with an uncertainty of less than 1 in 10^7. The primary standard is provided by an iodine-stabilized Helium–Neon laser that has a specified uncertainty of less than 1 in 10^9. All of the links in this calibration chain must be shown in any documentation that describes the use of micrometers in making quality-related measurements.

4.4 Calibration records

An essential element in the maintenance of measurement systems and the operation of calibration procedures is the provision of full documentation. This must give a full description of the measurement requirements throughout the workplace, the instruments used, and the calibration system and procedures operated. Individual calibration records for each instrument must be included within this. This documentation is a necessary part of the quality manual, although it may physically exist as a separate volume if this is more convenient. An overriding constraint on the style in which the documentation is presented is that it should be simple and easy to read. This is often greatly facilitated by a copious use of appendices.

The starting point in the documentation must be a statement of what measurement limits have been defined for each measurement system documented. Such limits are established by balancing the costs of improved accuracy against customer requirements, and also with regard to what overall quality level has been specified in the quality manual. The technical procedures required for this, which involve assessing the type and magnitude of relevant measurement errors, are described in Chapter 3. It is customary to express the final measurement limit calculated as ± 2 standard deviations, i.e. within 95% confidence limits (see Chapter 3 for an explanation of these terms).

The instruments specified for each measurement situation must be listed next. This list must be accompanied by full instructions about the proper use of the instruments concerned. These instructions will include details about any environmental control or other special precautions that must be taken to ensure that the instruments provide measurements of sufficient accuracy to meet the measurement limits defined. The proper training courses appropriate to plant personnel who will use the instruments must be specified.

Having disposed of the question about what instruments are used, the documentation must go on to cover the subject of calibration. Full calibration is not applied to every measuring instrument used in a workplace because BS EN ISO 9000 acknowledges that formal calibration procedures are not necessary for some equipment where it is uneconomic or technically unnecessary because the accuracy of the measurement involved has an insignificant effect on the overall quality target for a product. However, any equipment that is excluded from calibration procedures in this manner must be specified as such in the documentation. Identification of equipment that is in this category is a matter of informed judgement.

For instruments that are the subject of formal calibration, the documentation must specify what standard instruments are to be used for the purpose and define a formal procedure of calibration. This procedure must include instructions for the storage and handling of standard calibration instruments and specify the required environmental conditions under which calibration is to be performed. Where a calibration procedure for a particular instrument uses published standard practices, it is sufficient to include reference to that standard procedure in the documentation rather than to reproduce the whole procedure. Whatever calibration system is established, a formal review procedure must be defined in the documentation that ensures its continued effectiveness at regular intervals. The results of each review must also be documented in a formal way.

A standard format for the recording of calibration results should be defined in the documentation. A separate record must be kept for every instrument present in the workplace, irrespective of whether the instrument is normally in use or is just kept as a spare. A form similar to that shown in Figure 4.3 should be used that includes details of the instrument's description, the required calibration frequency, the date of each calibration and the calibration results on each occasion. Where appropriate, the documentation must also define the manner in which calibration results are to be recorded on the instruments themselves.

The documentation must specify procedures that are to be followed if an instrument is found to be outside the calibration limits. This may involve adjustment, redrawing its scale or withdrawing an instrument, depending upon the nature of the discrepancy and the type of instrument involved. Instruments withdrawn will either be repaired or scrapped. In the case of withdrawn instruments, a formal procedure for marking them as such must be defined to prevent them being accidentally put back into use.

Two other items must also be covered by the calibration document. The traceability of the calibration system back to national reference standards must be defined and supported by calibration certificates (see section 4.3). Training procedures must also be documented, specifying the particular training courses to be attended by various personnel and what, if any, refresher courses are required.

All aspects of these documented calibration procedures will be given consideration as part of the periodic audit of the quality control system that calibration procedures are instigated to support. Whilst the basic responsibility for choosing a suitable interval between calibration checks rests with the engineers responsible for the instruments concerned, the quality system auditor will require to see the results of tests that show that the calibration interval has been chosen correctly and that instruments are not going outside allowable measurement uncertainty limits between calibrations. Particularly

Type of instrument:	Company serial number:
Manufacturer's part number:	Manufacturer's serial number:
Measurement limit:	Date introduced:
Location:	
Instructions for use:	
Calibration frequency:	Signature of person responsible for calibration:

CALIBRATION RECORD		
Calibration date:	Calibration results	Calibrated by

Fig. 4.3 Typical format for instrument record sheets.

important in such audits will be the existence of procedures that are instigated in response to instruments found to be out of calibration. Evidence that such procedures are effective in avoiding degradation in the quality assurance function will also be required.

References and further reading

British Standards Society (1979) The operation of a company standards department, British Standards Society, London.

ISO 5725 (1998) Precision of test methods – determination of repeatability and reproducibility by inter-laboratory tests, International Organization for Standards, Geneva.

ISO 9000 (2000): Quality Management and Quality Assurance Standards, International Organization for Standards, Geneva (individual parts published as ISO 9001, ISO 9002, ISO 9003 and ISO 9004).

5

Measurement noise and signal processing

5.1 Sources of measurement noise

Chapter 3 has already provided a detailed analysis of error sources that arise during the measurement process of sensing the value of a physical variable and generating an output signal. However, further errors are often created in measurement systems when electrical signals from measurement sensors and transducers are corrupted by induced noise. This induced noise arises both within the measurement circuit itself and also during the transmission of measurement signals to remote points. The aim when designing measurement systems is always to reduce such induced noise voltage levels as far as possible. However, it is usually not possible to eliminate all such noise, and signal processing has to be applied to deal with any noise that remains.

Noise voltages can exist either in serial mode or common mode forms. Serial mode noise voltages act in series with the output voltage from a measurement sensor or transducer, which can cause very significant errors in the output measurement signal. The extent to which series mode noise corrupts measurement signals is measured by a quantity known as the *signal-to-noise ratio*. This is defined as:

$$\text{Signal-to-noise ratio} = 20 \log_{10}\left(\frac{V_s}{V_n}\right)$$

where V_s is the mean voltage level of the signal and V_n is the mean voltage level of the noise. In the case of a.c. noise voltages, the root-mean squared value is used as the mean.

Common mode noise voltages are less serious, because they cause the potential of both sides of a signal circuit to be raised by the same level, and thus the level of the output measurement signal is unchanged. However, common mode voltages do have to be considered carefully, since they can be converted into series mode voltages in certain circumstances.

Noise can be generated from sources both external and internal to the measurement system. Induced noise from external sources arises in measurement systems for a number of reasons that include their proximity to mains-powered equipment and cables (causing noise at the mains frequency), proximity to fluorescent lighting circuits (causing noise at twice the mains frequency), proximity to equipment operating at audio and radio frequencies (causing noise at corresponding frequency), switching of nearby d.c. and a.c. circuits, and corona discharge (both of the latter causing induced spikes and transients). Internal noise includes thermoelectric potentials, shot noise and potentials due to electrochemical action.

5.1.1 Inductive coupling

The primary mechanism by which external devices such as mains cables and equipment, fluorescent lighting and circuits operating at audio or radio frequencies generate noise is through inductive coupling. If signal-carrying cables are close to such external cables or equipment, a significant mutual inductance M can exist between them, as shown in Figure 5.1(a), and this can generate a series mode noise voltage of several millivolts given by $V_n = M\dot{I}$, where $\dot{I}$ is the rate of change of current in the mains circuit.

5.1.2 Capacitive (electrostatic) coupling

Capacitive coupling, also known as electrostatic coupling, can also occur between the signal wires in a measurement circuit and a nearby mains-carrying conductor. The magnitude of the capacitance between each signal wire and the mains conductor is represented by the quantities C_1 and C_2 in Figure 5.1(b). In addition to these capacitances, a capacitance can also exist between the signal wires and earth, represented by C_3 and C_4 in the figure. It can be shown (Cook, 1979) that the series mode noise voltage V_n is zero if the coupling capacitances are perfectly balanced, i.e. if $C_1 = C_2$ and $C_3 = C_4$. However, exact balance is unlikely in practice, since the signal wires are not perfectly straight, causing the distances and thus the capacitances to the mains cable and to earth to vary. Thus, some series mode noise voltage induced by capacitive coupling usually exists.

5.1.3 Noise due to multiple earths

As far as possible, measurement signal circuits are isolated from earth. However, leakage paths often exist between measurement circuit signal wires and earth at both

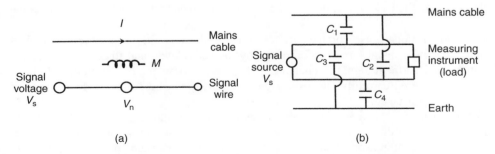

Fig. 5.1 Noise induced by coupling: (a) inductive coupling; (b) capacitive (electrostatic) coupling.

the source (sensor) end of the circuit and also the load (measuring instrument) end. This does not cause a problem as long as the earth potential at both ends is the same. However, it is common to find that other machinery and equipment carrying large currents is connected to the same earth plane. This can cause the potential to vary between different points on the earth plane. This situation, which is known as *multiple earths*, can cause a series mode noise voltage in the measurement circuit.

5.1.4 Noise in the form of voltage transients

When motors and other electrical equipment (both a.c. and d.c.) are switched on and off, large changes of power consumption suddenly occur in the electricity supply system. This can cause voltage transients ('spikes') in measurement circuits connected to the same power supply. Such noise voltages are of large magnitude but short time duration. *Corona discharge* can also cause voltage transients on the mains power supply. This occurs when the air in the vicinity of high voltage d.c. circuits becomes ionized and discharges to earth at random times.

5.1.5 Thermoelectric potentials

Whenever metals of two different types are connected together, a thermoelectric potential (sometimes called a *thermal e.m.f.*) is generated according to the temperature of the joint. This is known as the *thermoelectric effect* and is the physical principle on which temperature-measuring thermocouples operate (see Chapter 14). Such thermoelectric potentials are only a few millivolts in magnitude and so the effect is only significant when typical voltage output signals of a measurement system are of a similar low magnitude.

One such situation is where one e.m.f.-measuring instrument is used to monitor the output of several thermocouples measuring the temperatures at different points in a process control system. This requires a means of automatically switching the output of each thermocouple to the measuring instrument in turn. Nickel–iron reed-relays with copper connecting leads are commonly used to provide this switching function. This introduces a thermocouple effect of magnitude $40\,\mu V/°C$ between the reed-relay and the copper connecting leads. There is no problem if both ends of the reed relay are at the same temperature because then the thermoelectric potentials will be equal and opposite and so cancel out. However, there are several recorded instances where, because of lack of awareness of the problem, poor design has resulted in the two ends of a reed-relay being at different temperatures and causing a net thermoelectric potential. The serious error that this introduces is clear. For a temperature difference between the two ends of only $2°C$, the thermoelectric potential is $80\,\mu V$, which is very large compared with a typical thermocouple output level of $400\,\mu V$.

Another example of the difficulties that thermoelectric potentials can create becomes apparent in considering the following problem that was reported in a current-measuring system. This system had been designed such that the current flowing in a particular part of a circuit was calculated by applying it to an accurately calibrated wire-wound resistance of value $100\,\Omega$ and measuring the voltage drop across the resistance. In calibration of the system, a known current of 20 μA was applied to the resistance and a voltage of 2.20 mV was measured by an accurate high-impedance instrument. Simple application of Ohm's law reveals that such a voltage reading indicates a current value of 22 μA. What then was the explanation for this discrepancy? The answer once again is a thermoelectric potential. Because the designer was not aware of thermoelectric potentials, the circuit had been constructed such that one side of the standard resistance was close to a power transistor, creating a difference in temperature between the two ends of the resistor of 2°C. The thermoelectric potential associated with this was sufficient to account for the 10% measurement error found.

5.1.6 Shot noise

Shot noise occurs in transistors, integrated circuits and other semiconductor devices. It consists of random fluctuations in the rate of transfer of carriers across junctions within such devices.

5.1.7 Electrochemical potentials

These are potentials that arise within measurement systems due to electrochemical action. Poorly soldered joints are a common source.

5.2 Techniques for reducing measurement noise

Prevention is always better than cure, and much can be done to reduce the level of measurement noise by taking appropriate steps when designing the measurement system.

5.2.1 Location and design of signal wires

Both the mutual inductance and capacitance between signal wires and other cables are inversely proportional to the square of the distance between the wires and the cable. Thus, noise due to inductive and capacitive coupling can be minimized by ensuring that signal wires are positioned as far away as possible from such noise sources. A minimum separation of 0.3 m is essential, and a separation of at least 1 m is preferable. Noise due to inductive coupling is also substantially reduced if each pair of signal wires is twisted together along its length. This design is known as a *twisted pair*, and is illustrated in Figure 5.2. In the first loop, wire A is closest to the noise source and has a voltage V_1 induced in it, whilst wire B has an induced noise voltage V_2. For loop 2, wire B is closest to the noise source and has an induced voltage V_1 whilst wire A has an induced voltage V_2. Thus the total voltage induced in wire A is $V_1 + V_2$ and in wire B it is $V_2 + V_1$ over these two loops. This pattern continues for all the loops and hence the two wires have an identical voltage induced in them.

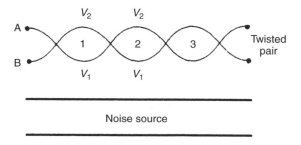

Fig. 5.2 Cancellation of induced noise.

5.2.2 Earthing

Noise due to multiple earths can be avoided by good earthing practices. In particular, this means keeping earths for signal wires and earths for high-current equipment entirely separate. Recommended practice is to install four completely isolated earth circuits as follows:

Power earth: provides a path for fault currents due to power faults.
Logic earth: provides a common line for all logic circuit potentials.
Analogue earth (ground): provides a common reference for all analogue signals.
Safety earth: connected to all metal parts of equipment to protect personnel should power lines come into contact with metal enclosures.

5.2.3 Shielding

Shielding consists of enclosing the signal wires in an earthed, metal shield that is itself isolated electrically from the signal wires. The shield should be earthed at only one point, preferably the signal source end. A shield consisting of braided metal eliminates 85% of noise due to capacitive coupling whilst a lapped metal foil shield eliminates noise almost entirely. The wires inside such a shield are normally formed as a twisted pair so that protection is also provided against induced noise due to nearby electromagnetic fields. Metal conduit is also sometimes used to provide shielding from capacitve-coupled noise, but the necessary supports for the conduit provide multiple earth points and lead to the problem of earth loops.

5.2.4 Other techniques

The *phase-locked loop* is often used as a signal-processing element to clean up poor quality signals. Although this is primarily a circuit for measuring the frequency of a signal, as described in Chapter 7; it is also useful for noise removal because its output waveform is a pure (i.e. perfectly clean) square wave at the same frequency as the input signal, irrespective of the amount of noise, modulation or distortion on the input signal.

Lock-in amplifiers (see section 5.5.10) are also commonly used to extract d.c. or slowly varying measurement signals from noise. The input measurement signal is modulated into a square-wave a.c. signal whose amplitude varies with the level of the input signal. This is normally achieved by either a relay or a field effect transistor. As a relay is subject to wear; the transistor is better. An alternative method is to use an analogue multiplier. Also, in the case of optical signals, the square wave can be produced by chopping the measurement signals using a set of windows in a rotating disc. This technique is frequently used with transducers like photodiodes that often generate large quantities of noise.

5.3 Introduction to signal processing

Signal processing is concerned with improving the quality of the reading or signal at the output of a measurement system, and one particular aim is to attenuate any noise in the measurement signal that has not been eliminated by careful design of the measurement system as discussed above. However, signal processing performs many other functions apart from dealing with noise, and the exact procedures that are applied depend on the nature of the raw output signal from a measurement transducer. Procedures of signal filtering, signal amplification, signal attenuation, signal linearization and bias removal are applied according to the form of correction required in the raw signal.

Traditionally, signal processing has been carried out by analogue techniques in the past, using various types of electronic circuit. However, the ready availability of digital computers in recent years has meant that signal processing has increasingly been carried out digitally, using software modules to condition the input measurement data.

Digital signal processing is inherently more accurate than analogue techniques, but this advantage is greatly reduced in the case of measurements coming from analogue sensors and transducers, because an analogue-to-digital conversion stage is necessary before the digital processing can be applied, thereby introducing conversion errors. Also, analogue processing remains the faster of the two alternatives in spite of recent advances in the speed of digital signal processing. Hence, both analogue and digital processing are considered in this chapter, with analogue processing being considered first because some preliminary analogue processing is often carried out even when the major part of the processing is carried out digitally.

5.4 Analogue signal filtering

Signal filtering consists of processing a signal to remove a certain band of frequencies within it. The band of frequencies removed can be either at the low-frequency end of the frequency spectrum, at the high-frequency end, at both ends, or in the middle of the spectrum. Filters to perform each of these operations are known respectively as low-pass filters, high-pass filters, band-pass filters and band-stop filters (also known as notch filters). All such filtering operations can be carried out by either analogue or digital methods.

The result of filtering can be readily understood if the analogy with a procedure such as sieving soil particles is considered. Suppose that a sample of soil A is passed

Measurement and Instrumentation Principles 79

through a system of two sieves of differing meshes such that the soil is divided into three parts, B, C and D, consisting of large, medium and small particles, as shown in Figure 5.3. Suppose that the system also has a mechanism for delivering one or more of the separated parts, B, C and D, as the system output. If the graded soil output consists of parts C and D, the system is behaving as a low-pass filter (rejecting large particles), whereas if it consists of parts B and C, the system is behaving as a high-pass filter (rejecting small particles). Other options are to deliver just part C (band-pass filter mode) or parts B and D together (band-stop filter mode). As any gardener knows, however, such perfect sieving is not achieved in practice and any form of graded soil output always contains a few particles of the wrong size.

Signal filtering consists of selectively passing or rejecting low-, medium- and high-frequency signals from the frequency spectrum of a general signal. The range of frequencies passed by a filter is known as the *pass-band*, the range not passed is known as the *stop-band*, and the boundary between the two ranges is known as the *cut-off frequency*. To illustrate this, consider a signal whose frequency spectrum is such that all frequency components in the frequency range from zero to infinity have equal magnitude. If this signal is applied to an ideal filter, then the outputs for a low-pass filter, high-pass filter, band-pass filter and band-stop filter respectively are shown

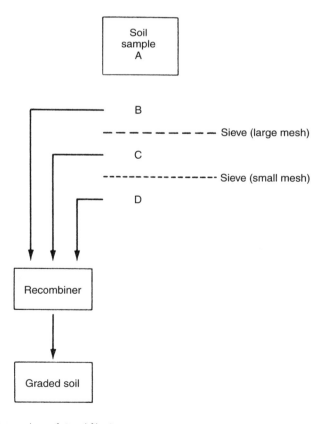

Fig. 5.3 Soil sieving analogy of signal filtering.

80 Measurement noise and signal processing

in Figure 5.4. Note that for the latter two types, the bands are defined by a pair of frequencies rather than by a single cut-off frequency.

Just as in the case of the soil sieving analogy presented above, the signal filtering mechanism is not perfect, with unwanted frequency components not being erased completely but only attenuated by varying degrees instead, i.e. the filtered signal always retains some components (of a relatively low magnitude) in the unwanted frequency range. There is also a small amount of attenuation of frequencies within the pass-band that increases as the cut-off frequency is approached. Figure 5.5 shows the typical output characteristics of a practical *constant-k** filter designed respectively for high-pass, low-pass, band-pass and band-stop filtering. Filter design is concerned with trying to obtain frequency rejection characteristics that are as close to the ideal as possible. However, improvement in characteristics is only achieved at the expense of greater complexity in the design. The filter chosen for any given situation is therefore a compromise between performance, complexity and cost.

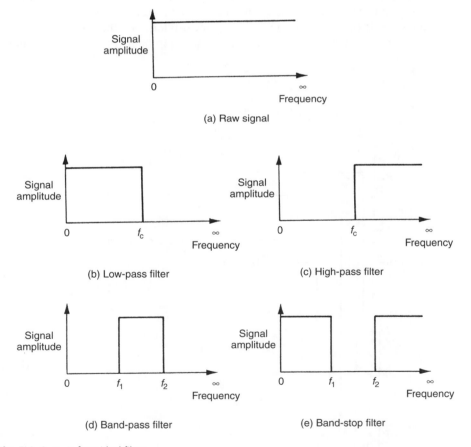

Fig. 5.4 Outputs from ideal filters.

---

* 'Constant-k' is a term used to describe a common class of passive filters, as discussed in the following section.

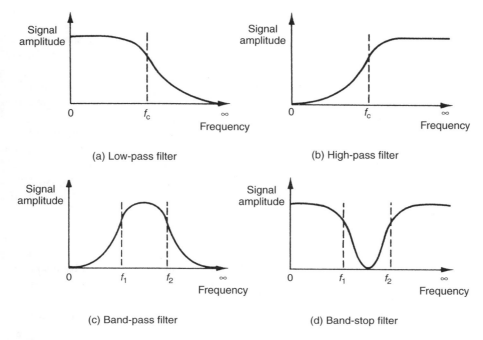

Fig. 5.5 Outputs from practical constant-k filters.

In the majority of measurement situations, the physical quantity being measured has a value that is either constant or only changing slowly with time. In these circumstances, the most common types of signal corruption are high-frequency noise components, and the type of signal processing element required is a low-pass filter. In a few cases, the measured signal itself has a high frequency, for instance when mechanical vibrations are being monitored, and the signal processing required is the application of a high-pass filter to attenuate low-frequency noise components. Band-stop filters can be used where a measurement signal is corrupted by noise at a particular frequency. Such noise is frequently due to mechanical vibrations or proximity of the measurement circuit to other electrical apparatus.

Analogue filters exist in two forms, passive and active, as discussed below. Nowadays, active filters are used more commonly than passive ones. Equivalent digital filters are discussed later in section 5.6.4.

5.4.1 Passive analogue filters

The very simplest passive filters are circuits that consist only of resistors and capacitors. Unfortunately, these only have a mild filtering effect. This is adequate for circuits like tone controls in radio receivers but unsuitable for the sort of signal processing requirements met in most measurement applications. In such cases, it is normal to use a network of impedances, such as those labelled Z_1 and Z_2 in Figure 5.6(a). Design formulae require the use of a mixture of capacitive and inductive impedances for Z_1 and Z_2. Ideally, these impedances should be either pure capacitances or pure inductances (i.e. components with zero resistance) so that there is no dissipation of energy in the filter. However, this ideal cannot always be achieved in practice since, although capacitors effectively have zero resistance, it is impossible to manufacture resistance-less inductors.

The detailed design of passive filters is quite complex and the reader is referred to specialist texts (e.g. Blinchikoff, 1976) for full details. The coverage below is therefore only a summary, and filter design formulae are quoted without full derivation.

Each element of the network shown in Figure 5.6(a) can be represented by either a T-section or π-section as shown in Figures 5.6(b) and 5.6(c) respectively. To obtain proper matching between filter sections, it is necessary for the input impedance of each section to be equal to the load impedance for that section. This value of impedance is known as the characteristic impedance (Z_0). For a T-section of filter, the characteristic impedance is calculated from:

$$Z_0 = \sqrt{Z_1 Z_2 [1 + (Z_1/4Z_2)]} \tag{5.1}$$

The frequency attenuation characteristics of the filter can be determined by inspecting this expression for Z_0. Frequency values for which Z_0 is real lie in the pass-band of the filter and frequencies for which Z_0 is imaginary lie in its stop-band.

Let $Z_1 = j\omega L$ and $Z_2 = 1/j\omega C$, where L is an inductance value, C is a capacitance value and ω is the angular frequency in radian/s, which is related to the frequency

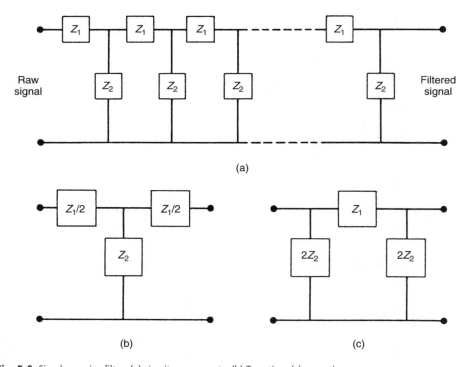

Fig. 5.6 Simple passive filter: (a) circuit components; (b) T-section; (c) π-section.

f according to $\omega = 2\pi f$. Substituting these values into the expression for Z_0 above gives:

$$Z_0 = \sqrt{\frac{L}{C}(1 - 0.25\omega^2 LC)}$$

For frequencies where $\omega < \sqrt{4/LC}$, Z_0 is real, and for higher frequencies, Z_0 is imaginary. These values of impedance therefore give a *low-pass* filter (see Figure 5.7(a)) with cut-off frequency f_c given by:

$$f_c = \omega_c/2\pi = 1/(2\pi\sqrt{LC})$$

A *high-pass* filter (see Figure 5.7(b)) can be synthesized with exactly the same cut-off frequency if the impedance values chosen are:

$$Z_1 = 1/j\omega C \quad \text{and} \quad Z_2 = j\omega L$$

It should be noted in both of these last two examples that the product $Z_1 Z_2$ could be represented by a constant k that is independent of frequency. Because of this, such filters are known by the name of *constant-k* filters. A constant-k *band-pass* filter (see Figure 5.7(c)) can be realized with the following choice of impedance values, where a is a constant and the other parameters are as before:

$$Z_1 = j\omega L = \frac{1}{j\omega C}; \quad Z_2 = \frac{(j\omega L_a)(a/j\omega C)}{j\omega L_a + (a/j\omega C)}$$

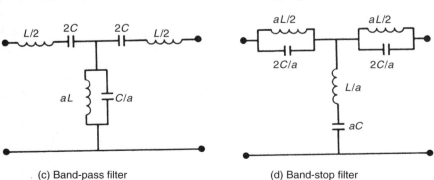

Fig. 5.7 Circuit components for passive filter T-sections.

The frequencies f_1 and f_2 defining the end of the pass-band are most easily expressed in terms of a frequency f_0 in the centre of the pass-band. The corresponding equations are:

$$f_0 = \frac{1}{2\pi\sqrt{LC}}; \quad f_1 = f_0[\sqrt{1+a} - \sqrt{a}]; \quad f_2 = f_0[\sqrt{1+a} + \sqrt{a}]$$

For a constant-k *band-stop* filter (see Figure 5.7(d)), the appropriate impedance values are:

$$Z_1 = \frac{(j\omega L_a)(a/j\omega C)}{j\omega L_a + (a/j\omega C)}; \quad Z_2 = \frac{1}{a}\left(j\omega L + \frac{1}{j\omega C}\right)$$

The frequencies defining the ends of the stop-band are again normally defined in terms of the frequency f_0 in the centre of the stop-band:

$$f_0 = \frac{1}{2\pi\sqrt{LC}}; \quad f_1 = f_0\left(1 - \frac{a}{4}\right); \quad f_2 = f_0\left(1 + \frac{a}{4}\right)$$

As has already been mentioned, a practical filter does not eliminate frequencies in the stop-band but merely attenuates them by a certain amount. The attenuation coefficient, α, at a frequency in the stop band, f, for a single T-section of a low-pass filter is given by:

$$\alpha = 2\cosh^{-1}(f/f_c) \tag{5.2}$$

The relatively poor attenuation characteristics are obvious if we evaluate this expression for a value of frequency close to the cut-off frequency given by $f = 2f_c$. Then $\alpha = 2\cosh^{-1}(2) = 2.64$. Further away from the cut-off frequency, for $f = 20f_c$, $\alpha = 2\cosh^{-1}(20) = 7.38$.

Improved attenuation characteristics can be obtained by putting several T-sections in cascade. If perfect matching is assumed, then two T-sections give twice the attenuation of one section, i.e. at frequencies of $2f_c$ and $20f_c$, α for two sections would have a value of 5.28 and 14.76 respectively.

The discussion so far has assumed that the inductances are resistance-less and that there is perfect matching between filter sections. However, it has already been noted that such ideal conditions cannot be achieved in practice, and this has several consequences. Inspection of the expression for the characteristic impedance (5.1) reveals frequency-dependent terms. Thus, the condition that the load impedance is equal to the input impedance for a section is only satisfied at one particular frequency. It is usual to match the impedances at zero frequency for a low-pass filter and at infinite frequency for a high-pass filter. This ensures that the frequency where the input and load impedances are matched is comfortably within the pass-band of the filter. Frequency dependency is one of the reasons for the degree of attenuation in the pass-band shown in the practical filter characteristics of Figure 5.5, the other reason being the presence of resistive components in the inductors of the filter. The effect of this in a practical filter is that the value of α at the cut-off frequency is 1.414 whereas the value predicted theoretically for an ideal filter (equation 5.2) is zero. Cascading filter sections together increases this attenuation in the pass-band as well as increasing attenuation of frequencies in the stop-band.

This problem of matching successive sections in a cascaded filter seriously degrades the performance of constant-k filters and this has resulted in the development of other types such as m-derived and n-derived composite filters. These produce less attenuation within the pass-band and greater attenuation outside it than constant-k filters, although this is only achieved at the expense of greater filter complexity and cost. The reader interested in further consideration of these is directed to consult one of the specialist texts recommended in the Further Reading section at the end of this chapter.

5.4.2 Active analogue filters

In the foregoing discussion on passive filters, the two main difficulties noted were those of obtaining resistance-less inductors and achieving proper matching between signal source and load through the filter sections. A further problem is that the inductors required by passive filters are bulky and relatively expensive. Active filters overcome all of these problems and so they are now used more commonly than passive filters.

The major component in an active filter is an electronic amplifier. The filter characteristics are defined by amplifier input and feedback components that consist of resistors and capacitors but not inductors. The fact that the necessary characteristics can be obtained using only resistors and capacitors, without requiring inductors, is a particular advantage of this class of filters. The circuits shown in Figure 5.8 produce the four types of filter characteristics discussed earlier. These are all known as second order filters, because the input–output relationship across each filter is described by a second order differential equation.

The characteristics of each filter in terms of attenuation behaviour in the pass- and stop-bands is determined by the choice of circuit components in Figure 5.8. A common set of design formulae is given below, although detailed derivation is not given. Further information on the derivation of these formulae can be found in specialist texts (e.g. Stephenson, 1985):

(a) *Low-pass filter*

$$\omega_0 = \sqrt{\frac{1}{R_1 R_2 C_1 C_2}}$$

$G = 1 + (R_4/R_3)$

where ω_0 is the cut-off frequency and G is the filter gain (at d.c.).

(b) *High-pass filter*

$$\omega_0 = \sqrt{\frac{1}{R_1 R_2 C_1 C_2}}$$

$G = 1 + (R_4/R_3)$

where ω_0 is the cut-off frequency and G is the filter gain (at infinite frequency).

(c) *Band-pass filter*

$$\omega_0 = \frac{\sqrt{2}}{R_1 C}$$

$$\omega_1 = \omega_0 - \frac{4-G}{2R_1 C}$$

$$\omega_2 = \omega_0 + \frac{4-G}{2R_1 C}$$

G = filter gain (at frequency ω_0) = $1 + R_3/R_2$ where ω_1 and ω_2 are frequencies at ends of the pass-band, and ω_0 is centre frequency of the pass-band.

(d) *Band-stop filter*

$$\omega_0 = \sqrt{\frac{1}{R_4 C^2}\left(\frac{1}{R_1} + \frac{1}{R_2}\right)}$$

$$\omega_1 = \omega_0 - \frac{1}{R_4 C}$$

$$\omega_2 = \omega_0 + \frac{1}{R_4 C}$$

G = filter gain (at d.c. and also high frequency) = $-R_6/R_3$ where ω_1 and ω_2 are frequencies at ends of the stop-band, and ω_0 is centre frequency of the stop-band.

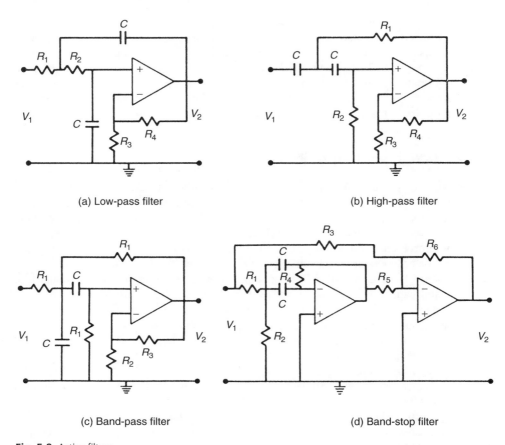

Fig. 5.8 Active filters.

Active filters with parameters derived from the above formulae are general purpose and suitable for most applications. However, many other design formulae exist for the parameters of filters with circuit structures as given in Figure 5.8, and these yield filters with special names and characteristics. *Butterworth filters*, for example, optimize the pass-band attenuation characteristics at the expense of stop-band performance. Another form, *Chebyshev filters*, have very good stop-band attenuation characteristics but poorer pass-band performance. Again, the reader is referred to the specialist texts (e.g. Stephenson, 1985) for more information.

5.5 Other analogue signal processing operations

To complete the discussion on analogue signal processing techniques, mention must also be made of certain other special-purpose devices and circuits used to manipulate signals. These are listed below.

5.5.1 Signal amplification

Signal amplification is carried out when the typical signal output level of a measurement transducer is considered to be too low. Amplification by analogue means is carried out by an operational amplifier. This is normally required to have a high input impedance so that its loading effect on the transducer output signal is minimized. In some circumstances, such as when amplifying the output signal from accelerometers and some optical detectors, the amplifier must also have a high-frequency response, to avoid distortion of the output reading.

The operational amplifier is an electronic device that has two input terminals and one output terminal, the two inputs being known as the inverting input and non-inverting input respectively. When connected as shown in Figure 5.9. The raw (unprocessed) signal V_i is connected to the inverting input through a resistor R_1 and the non-inverting input is connected to ground. A feedback path is provided from the output terminal through a resistor R_2 to the inverting input terminal. Assuming ideal operational amplifier characteristics, the processed signal V_0 at the output terminal is then related to the voltage V_i at the input terminal by the expression:

$$V_0 = \frac{R_2 V_i}{R_1} \tag{5.3}$$

The amount of signal amplification is therefore defined by the relative values of R_1 and R_2. This ratio between R_1 and R_2 in the amplifier configuration is often known as the *amplifier gain* or *closed-loop gain*. If, for instance, $R_1 = 1\,\text{M}\Omega$ and $R_2 = 10\,\text{M}\Omega$, an amplification factor of 10 is obtained (i.e. gain = 10). It is important to note that, in this standard way of connecting the operational amplifier (often known as the *inverting configuration*), the sign of the processed signal is inverted. This can be corrected for if necessary by feeding the signal through a further amplifier set up for unity gain ($R_1 = R_2$). This inverts the signal again and returns it to its original sign.

Instrumentation amplifier

For some applications requiring the amplification of very low level signals, a special type of amplifier known as an *instrumentation amplifier* is used. This consists of a circuit containing three standard operational amplifiers, as shown in Figure 5.10. The advantage of the instrumentation amplifier compared with a standard operational amplifier is that its differential input impedance is much higher. In consequence, its

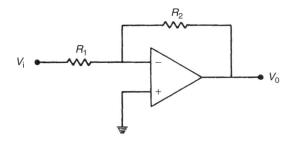

Fig. 5.9 Operational amplifier connected for signal amplification.

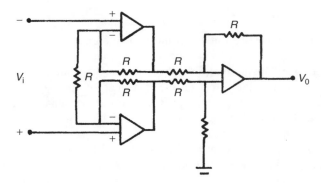

Fig. 5.10 Instrumentation amplifier.

common mode rejection capability* is much better. This means that, if a twisted wire pair is used to connect a transducer to the differential inputs of the amplifier, any induced noise will contaminate each wire equally and will be rejected by the common mode rejection capacity of the amplifier. The mechanism of common mode noise rejection is explained more fully in section 5.5.3.

5.5.2 Signal attenuation

One method of attenuating signals by analogue means is to use a potentiometer connected in a voltage-dividing circuit, as shown in Figure 5.11. For the potentiometer slider positioned a distance of x along the resistance element of total length L, the voltage level of the processed signal V_0 is related to the voltage level of the raw signal V_i by the expression:

$$V_0 = \frac{xV_i}{L}$$

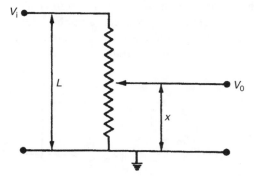

Fig. 5.11 Potentiometer in voltage-dividing circuit.

---

* Common mode rejection describes the ability of the amplifier to reject equal-magnitude signals that appear on both of its inputs.

Unfortunately, the potentiometer is unsuitable as a signal attenuator when it is followed by devices or circuits with a relatively low impedance, since these load the potentiometer circuit and distort the input–output (V_0/V_i) relationship above. In such cases, an operational amplifier is used as an attenuator instead. This is connected in the same way as the amplifier shown in Figure 5.9, except that R_1 is chosen to be greater than R_2. Equation 5.3 still holds and therefore, if R_1 is chosen to be $10\,\text{M}\Omega$ and R_2 as $1\,\text{M}\Omega$, an attenuation factor of ten is achieved (gain $= 0.1$). Use of an operational amplifier as an attenuating device is a more expensive solution than using a potentiometer, but, apart from being relatively unaffected by the circuit that is connected to its output, it has further advantages in terms of its small size and low power consumption.

5.5.3 Differential amplification

Figure 5.12 shows a common amplifier configuration that is used to amplify the small difference that may exist between two voltage signals V_A and V_B. These may represent, for example, the pressures either side of an obstruction device put in a pipe to measure the volume flow rate of fluid flowing through it (see Chapter 16). The output voltage V_0 is given by:

$$V_0 = \frac{R_3}{R_1}(V_B - V_A)$$

A differential amplifier is also very useful for removing common mode noise voltages. Suppose V_A and V_B in Figure 5.12 are signal wires such that $V_A = +V_s$ volts and $V_B = 0$ volts. Let us assume that the measurement circuit has been corrupted by a common mode noise voltage V_n such that the voltages on the $+V_s$ and $0\,\text{V}$ signal wires become $(V_s + V_n)$ and (V_n). The inputs to the amplifier V_1 and V_2 and the output V_0 can then be written as:

$$V_1 = \frac{R_3}{R_1}(V_s + V_n); \quad V_2 = \frac{R_4}{R_2 + R_4}V_n; \quad V_0 = V_2\left(1 + \frac{R_3}{R_1}\right) - V_1$$

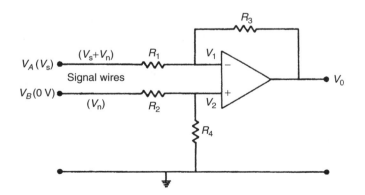

Fig. 5.12 Operational amplifier connected in differential amplification mode.

90 Measurement noise and signal processing

Hence:

$$V_0 = \left(\frac{R_4}{R_2 + R_4} V_n\right)\left(1 + \frac{R_3}{R_1}\right) - \frac{R_3}{R_1}(V_s + V_n)$$

$$= V_n\left(\frac{R_4}{R_2 + R_4} + \frac{R_3 R_4}{R_1(R_2 + R_4)}\right) - \frac{R_3}{R_1}(V_s + V_n)$$

$$= V_n\left(\frac{R_4(1 + R_3/R_1)}{R_2(1 + R_4/R_2)} - \frac{R_3}{R_1}\right) - \frac{R_3}{R_1} V_s \tag{5.4}$$

If the resistance values are chosen carefully such that $R_4/R_2 = R_3/R_1$, then equation (5.4) simplifies to:

$$V_0 = -\frac{R_3}{R_1} V_s$$

i.e. the noise voltage V_n has been removed.

5.5.4 Signal linearization

Several types of transducer used in measuring instruments have an output that is a non-linear function of the measured quantity input. In many cases, this non-linear signal can be converted to a linear one by special operational amplifier configurations that have an equal and opposite non-linear relationship between the amplifier input and output terminals. For example, light intensity transducers typically have an exponential relationship between the output signal and the input light intensity, i.e.:

$$V_0 = K e^{-\alpha Q} \tag{5.5}$$

where Q is the light intensity, V_0 is the voltage level of the output signal, and K and α are constants. If a diode is placed in the feedback path between the input and output terminals of the amplifier as shown in Figure 5.13, the relationship between the amplifier output voltage V_0 and input voltage V_1 is given by:

$$V_0 = C \log_e(V_1) \tag{5.6}$$

If the output of the light transducer with characteristic given by equation (5.5) is conditioned by an amplifier of characteristic given by equation (5.6), the voltage level

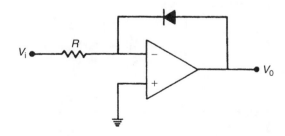

Fig. 5.13 Operational amplifier connected for signal linearization.

of the processed signal is given by:

$$V_0 = C \log_e(K) - \alpha C Q \qquad (5.7)$$

Expression (5.7) shows that the output signal now varies linearly with light intensity Q but with an offset of $C \log_e(K)$. This offset would normally be removed by further signal conditioning, as described below.

5.5.5 Bias (zero drift) removal

Sometimes, either because of the nature of the measurement transducer itself, or as a result of other signal conditioning operations, a bias (zero drift) exists in the output signal. This can be expressed mathematically for a physical quantity x and measurement signal y as:

$$y = Kx + C \qquad (5.8)$$

where C represents a bias in the output signal that needs to be removed by signal processing. The bias removal circuit shown in Figure 5.14 is a differential amplifier in which a potentiometer is used to produce a variable voltage V_p equal to the bias on the input voltage V_i. The differential amplification action thus removes the bias. Referring to the circuit, for $R_1 = R_2$ and $R_3 = R_4$, the output V_0 is given by:

$$V_0 = (R_3/R_1)(V_p - V_i) \qquad (5.9)$$

where V_i is the unprocessed signal y equal to $(Kx + C)$ and V_p is the output voltage from a potentiometer supplied by a known reference voltage V_{ref}, that is set such that $V_p = C$. Now, substituting these values for V_i and V_p into equation (5.9) and referring the quantities back into equation (5.8) gives:

$$y = K'x \qquad (5.10)$$

where the new constant K' is related to K according to $K' = -K(R_3/R_1)$. It is clear that a straight line relationship now exists between the measurement signal y and the measured quantity x. Thus, the unwanted bias has been removed.

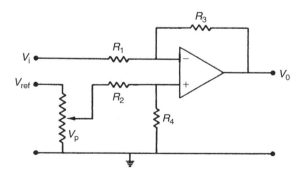

Fig. 5.14 Bias removal circuit.

5.5.6 Signal integration

Connected in the configuration shown in Figure 5.15, an operational amplifier is able to integrate the input signal V_i such that the output signal V_0 is given by:

$$V_0 = -\frac{1}{RC} \int V_i \, dt$$

This circuit is used whenever there is a requirement to integrate the output signal from a transducer.

5.5.7 Voltage follower (pre-amplifier)

The voltage follower, also known as a pre-amplifier, is a unity gain amplifier circuit with a short circuit in the feedback path, as shown in Figure 5.16, such that:

$$V_0 = V_i$$

It has a very high input impedance and its main application is to reduce the load on the measured system. It also has a very low output impedance that is very useful in some impedance-matching applications.

5.5.8 Voltage comparator

The output of a voltage comparator switches between positive and negative values according to whether the difference between the two input signals to it is positive or negative. An operational amplifier connected as shown in Figure 5.17 gives an output that switches between positive and negative saturation levels according to whether $(V_1 - V_2)$ is greater than or less than zero. Alternatively, the voltage of a single input

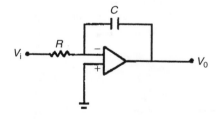

Fig. 5.15 Operational amplifier connected for signal integration.

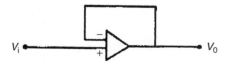

Fig. 5.16 Operational amplifier connected as voltage follower (pre-amplifier).

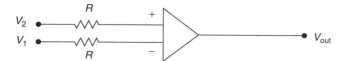

Fig. 5.17 Comparison between two voltage signals.

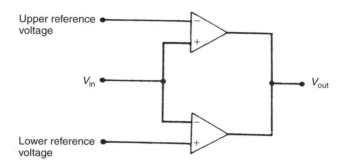

Fig. 5.18 Comparison of input signal against reference value.

signal can be compared against positive and negative reference levels with the circuit shown in Figure 5.18.

In practice, operational amplifiers have drawbacks as voltage comparators for several reasons. These include non-compatibility between output voltage levels and industry-standard logic circuits, propagation delays and slow recovery. In consequence, various other special-purpose integrated circuits have been developed for voltage comparison.

5.5.9 Phase-sensitive detector

One function of a phase-sensitive detector is to measure the phase difference between two signals that have the same frequency. For two input signals of amplitude V_1 and V_2 and frequency f, the output is given by $V_1 V_2 \cos \phi$, where ϕ is the phase difference between the signals. In many cases, the phase difference is adjusted to zero ($\cos \phi = 1$) so that the output is a maximum.

A phase-sensitive detector can also be used as a cross-correlator to enhance the quality of measurement signals that have a poor signal-to-noise ratio. This ability is also exploited in the use of phase-sensitive detectors to demodulate amplitude-modulated (AM) signals. For these roles, the detector requires firstly a clean reference voltage at the same frequency as the measurement signal and secondly, phase-control circuits to make the phases of the reference and measurement signals coincide. Commercial instruments known as *lock-in amplifiers* (see next section) are available that combine a phase-sensitive detector with the other components required to provide the demodulation function.

Phase-sensitive detectors are known by several alternative names, two examples of which are synchronous demodulator and synchronous detector. They can also exist physically in a number of alternative forms that include both transformer-based and fully electronic circuits (see Olsen, 1974, pp. 431–435).

5.5.10 Lock-in amplifier

A lock-in amplifier is used to demodulate small signals and extract them from noise when they are transmitted on an a.c. carrier. As shown in Figure 5.19, it consists of a phase-sensitive detector, an element to generate a square-wave reference signal at the same frequency as the a.c. carrier, a tuned narrow-band amplifier, a phase-control circuit, and a low-pass filter at the output. The role of the phase-sensitive detector is to selectively rectify only signals that are in phase with the a.c. carrier transmitting the measurement signal. This is achieved by using the phase-control circuit to lock the reference square-wave signal in phase with the carrier. Because any noise will be of random phase and generally out of phase with the measurement signal, it is rejected by the phase-sensitive detector because this only transmits in-phase signals. The phase-detector produces positive and negative outputs in response to noise that randomly lag and lead the measurement signal on the carrier, and therefore the noise component in the output goes to zero if the output is averaged over a period of time by a low-pass filter. Finally, the low-pass filter acts as a demodulator that regenerates the original measurement signal by extracting it from the carrier.

5.5.11 Signal addition

The most common mechanism for summing two or more input signals is the use of an operational amplifier connected in signal-inversion mode, as shown in Figure 5.20. For input signal voltages V_1, V_2 and V_3, the output voltage V_0 is given by:

$$V_0 = -(V_1 + V_2 + V_3)$$

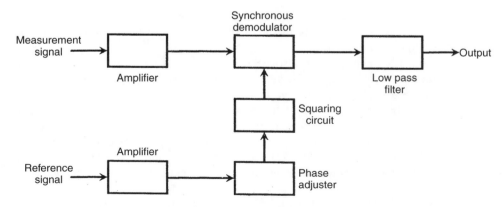

Fig. 5.19 Lock-in amplifier.

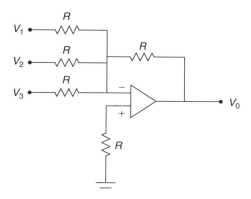

Fig. 5.20 Operational amplifier connected for signal addition.

5.5.12 Signal multiplication

Great care must be taken when choosing a signal multiplier because, whilst many circuits exist for multiplying two analogue signals together, most of them are two-quadrant types that only work for signals of a single polarity, i.e. both positive or both negative. Such schemes are unsuitable for general analogue signal processing, where the signals to be multiplied may be of changing polarity.

For analogue signal processing, a four-quadrant multiplier is required. Two forms of such a multiplier are easily available, the Hall-effect multiplier and the translinear multiplier.

5.6 Digital signal processing

Digital techniques achieve much greater levels of accuracy in signal processing than equivalent analogue methods. However, the time taken to process a signal digitally is longer than that required to carry out the same operation by analogue techniques, and the equipment required is more expensive. Therefore, some care is needed in making the correct choice between digital and analogue methods.

Whilst digital signal processing elements in a measurement system can exist as separate units, it is more usual to find them as an integral part of an intelligent instrument (see Chapter 9). However, the construction and mode of operation of such processing elements are the same irrespective of whether they are part of an intelligent instrument of not. The hardware aspect of a digital signal-processing element consists of a digital computer and analogue interface boards. The actual form that signal processing takes depends on the software program executed by the processor. However, before consideration is given to this, some theoretical aspects of signal sampling need to be discussed.

5.6.1 Signal sampling

Digital computers require signals to be in digital form whereas most instrumentation transducers have an output signal in analogue form. Analogue-to-digital conversion

96 Measurement noise and signal processing

is therefore required at the interface between analogue transducers and the digital computer, and digital-to-analogue conversion is often required at a later stage to convert the processed signals back into analogue form. The process of analogue-to-digital conversion consists of sampling the analogue signal at regular intervals of time. Each sample of the analogue voltage is then converted into an equivalent digital value. This conversion takes a certain finite time, during which the analogue signal can be changing in value. The next sample of the analogue signal cannot be taken until the conversion of the last sample to digital form is completed. The representation within a digital computer of a continuous analogue signal is therefore a sequence of samples whose pattern only approximately follows the shape of the original signal. This pattern of samples taken at successive, equal intervals of time is known as a discrete signal. The process of conversion between a continuous analogue signal and a discrete digital one is illustrated for a sine wave in Figure 5.21.

The raw analogue signal in Figure 5.21 has a frequency of approximately 0.75 cycles per second. With the rate of sampling shown, which is approximately 11 samples per second, reconstruction of the samples matches the original analogue signal very well. If the rate of sampling was decreased, the fit between the reconstructed samples and the original signal would be less good. If the rate of sampling was very much less than the frequency of the raw analogue signal, such as 1 sample per second, only the samples marked 'X' in Figure 5.21 would be obtained. Fitting a line through these 'X's incorrectly estimates a signal whose frequency is approximately 0.25 cycles per second. This phenomenon, whereby the process of sampling transmutes a high-frequency signal into a lower frequency one, is known as *aliasing*. To avoid aliasing, it is necessary theoretically for the sampling rate to be at least twice the highest frequency in the analogue signal sampled. In practice, sampling rates of between 5 and 10 times the highest frequency signal are normally chosen so that the discrete sampled signal is a close approximation to the original analogue signal in amplitude as well as frequency.

Problems can arise in sampling when the raw analogue signal is corrupted by high-frequency noise of unknown characteristics. It would be normal practice to choose the sampling interval as, say, a ten-times multiple of the frequency of the measurement

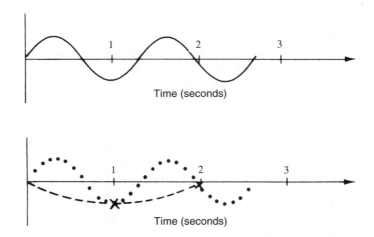

Fig. 5.21 Conversion of continuous analogue signal to discrete sampled signal.

component in the raw signal. If such a sampling interval is chosen, aliasing can in certain circumstances transmute high-frequency noise components into the same frequency range as the measurement component in the signal, thus giving erroneous results. This is one of the circumstances mentioned earlier, where prior analogue signal conditioning in the form of a low-pass filter must be carried out before processing the signal digitally.

One further factor that affects the quality of a signal when it is converted from analogue to digital form is *quantization*. Quantization describes the procedure whereby the continuous analogue signal is converted into a number of discrete levels. At any particular value of the analogue signal, the digital representation is either the discrete level immediately above this value or the discrete level immediately below this value. If the difference between two successive discrete levels is represented by the parameter Q, then the maximum error in each digital sample of the raw analogue signal is $\pm Q/2$. This error is known as the quantization error and is clearly proportional to the resolution of the analogue-to-digital converter, i.e. to the number of bits used to represent the samples in digital form.

5.6.2 Sample and hold circuit

A sample and hold circuit is normally an essential element at the interface between an analogue sensor or transducer and an analogue-to-digital converter. It holds the input signal at a constant level whilst the analogue-to-digital conversion process is taking place. This prevents the conversion errors that would probably result if variations in the measured signal were allowed to pass through to the converter. The operational amplifier circuit shown in Figure 5.22 provides this sample and hold function. The input signal is applied to the circuit for a very short time duration with switch S_1 closed and S_2 open, after which S_1 is opened and the signal level is then held until, when the next sample is required, the circuit is reset by closing S_2.

5.6.3 Analogue-to-digital converters

Important factors in the design of an analogue-to-digital converter are the speed of conversion and the number of digital bits used to represent the analogue signal level. The minimum number of bits used in analogue-to-digital converters is eight. The use

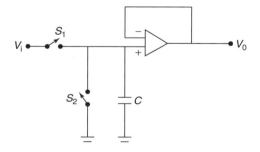

Fig. 5.22 Operational amplifier connected as 'sample and hold' circuit.

of eight bits means that the analogue signal can be represented to a resolution of 1 part in 256 if the input signal is carefully scaled to make full use of the converter range. However, it is more common to use either 10 bit or 12 bit analogue-to-digital converters, which give resolutions respectively of 1 part in 1024 and 1 part in 4096. Several types of analogue-to-digital converter exist. These differ in the technique used to effect signal conversion, in operational speed, and in cost.

The simplest type of analogue-to-digital converter is the *counter analogue-to-digital converter*, as shown in Figure 5.23. This, like most types of analogue-to-digital converter, does not convert continuously, but in a stop-start mode triggered by special signals on the computer's control bus. At the start of each conversion cycle, the counter is set to zero. The digital counter value is converted to an analogue signal by a digital-to-analogue converter (a discussion of digital-to-analogue converters follows in the next section), and a comparator then compares this analogue counter value with the unknown analogue signal. The output of the comparator forms one of the inputs to an AND logic gate. The other input to the AND gate is a sequence of clock pulses. The comparator acts as a switch that can turn on and off the passage of pulses from the clock through the AND gate. The output of the AND gate is connected to the input of the digital counter. Following reset of the counter at the start of the conversion cycle, clock pulses are applied continuously to the counter through the AND gate, and the analogue signal at the output of the digital-to-analogue converter gradually increases in magnitude. At some point in time, this analogue signal becomes equal in magnitude to the unknown signal at the input to the comparator. The output of the comparator changes state in consequence, closing the AND gate and stopping further increments of the counter. At this point, the value held in the counter is a digital representation of the level of the unknown analogue signal.

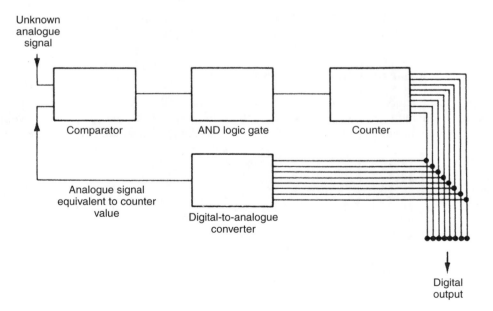

Fig. 5.23 Counter analogue–digital converter circuit.

5.6.4 Digital-to-analogue (D/A) conversion

Digital-to-analogue conversion is much simpler to achieve than analogue-to-digital conversion and the cost of building the necessary hardware circuit is considerably less. It is required wherever a digitally processed signal has to be presented to an analogue control actuator or an analogue signal display device. A common form of digital-to-analogue converter is illustrated in Figure 5.24. This is shown with 8 bits for simplicity of explanation, although in practice 10 and 12 bit D/A converters are used more frequently. This form of D/A converter consists of a resistor-ladder network on the input to an operational amplifier. The analogue output voltage from the amplifier is given by:

$$V_A = V_7 + \frac{V_6}{2} + \frac{V_5}{4} + \frac{V_4}{8} + \frac{V_3}{16} + \frac{V_2}{32} + \frac{V_1}{64} + \frac{V_0}{128}$$

$V_0 \cdots V_7$ are set at either the reference voltage level V_{ref} or at zero volts according to whether an associated switch is open or closed. Each switch is controlled by the logic level of one of the bits 0–7 of the 8 bit binary signal being converted. A particular switch is open if the relevant binary bit has a value of 0 and closed if the value is 1. Consider for example a digital signal with binary value of 11010100. The values of $V_7 \cdots V_0$ are therefore:

$$V_7 = V_6 = V_4 = V_2 = V_{ref}; \quad V_5 = V_3 = V_1 = V_0 = 0$$

The analogue output from the converter is then given by:

$$V_A = V_{ref} + \frac{V_{ref}}{2} + \frac{V_{ref}}{8} + \frac{V_{ref}}{32}$$

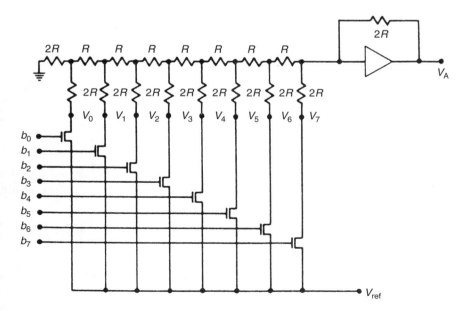

Fig. 5.24 Common form of digital–analogue converter.

5.6.5 Digital filtering

Digital signal processing can perform all of the filtering functions mentioned earlier in respect of analogue filters, i.e. low pass, high pass, band pass and band stop. However, the detailed design of digital filters requires a level of theoretical knowledge, including the use of z-transform theory, which is outside the scope of this book. The reader interested in digital filter design is therefore referred elsewhere (Lynn, 1989; Huelsman, 1993).

5.6.6 Autocorrelation

Autocorrelation is a special digital signal processing technique that has the ability to extract a measurement signal when it is completely swamped by noise, i.e. when the noise amplitude is larger than the signal amplitude. Unfortunately, phase information in the measurement signal is lost during the autocorrelation process, but the amplitude and frequency can be extracted accurately. For a measurement signal $s(t)$, the autocorrelation coefficient ϕ_s is the average value of the product of $s(t)$ and $s(t - \tau)$, where $s(t - \tau)$ is the value of the measurement signal delayed by a time τ. ϕ_s can be derived by the scheme shown in Figure 5.25, and mathematically it is given by:

$$\phi_s = \overline{s(t)s(t - \tau)}$$

The autocorrelation function $\phi_s(\tau)$ describes the relationship between ϕ_s and τ as τ varies:

$$\phi_s(\tau) = \lim_{T \to \infty} \left[\frac{1}{2T} \int_{-T}^{T} s(t)s(t - \tau) \, dt \right]$$

If the measurement signal is corrupted by a noise signal $n(t)$ (such that the total signal $y(t)$ at the output of the measurement system is given by $y(t) = s(t) + n(t)$), the noise can be represented by an autocorrelation function of the form $\phi_n(\tau)$ where:

$$\phi_n(\tau) = \lim_{T \to \infty} \left[\frac{1}{2T} \int_{-T}^{T} n(t)n(t - \tau) \, dt \right]$$

If $n(t)$ only consists of random noise, $\phi_n(\tau)$ has a large value close to $\tau = 0$, but, away from $\tau = 0$, $\phi_n(\tau)$ decreases to a very small value. The autocorrelation function for the combined signal plus noise is given by $\phi_s(\tau) + \phi_n(\tau)$. For $\tau \gg 0$, $\phi_n(\tau) \to 0$ and thus $\phi_s(\tau) + \phi_n(\tau) \to \phi_s(\tau)$. Thus, at large time delays, the amplitude and period of the signal can be found from the amplitude and period of the autocorrelation function

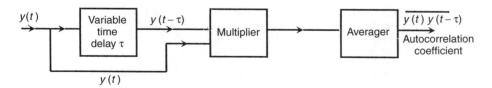

Fig. 5.25 Scheme to derive autocorrelation coefficient.

of the signal at the output of the measurement system. Further details can be found in Healey, (1975).

5.6.7 Other digital signal processing operations

Once a satisfactory digital representation in discrete form of an analogue signal has been obtained, many signal processing operations become trivial. For signal amplification and attenuation, all samples have to be multiplied or divided by a fixed constant. Bias removal involves simply adding or subtracting a fixed constant from each sample of the signal. Signal linearization requires *a priori* knowledge of the type of non-linearity involved, in the form of a mathematical equation that expresses the relationship between the output measurements from an instrument and the value of the physical quantity being measured. This can be obtained either theoretically through knowledge of the physical laws governing the system or empirically using input–output data obtained from the measurement system under controlled conditions. Once this relationship has been obtained, it is used to calculate the value of the measured physical quantity corresponding to each discrete sample of the measurement signal. Whilst the amount of computation involved in this is greater than for the trivial cases of signal amplification etc. already mentioned, the computational burden is still relatively small in most measurement situations.

References and further reading

Blinchikoff, H.J. (1976) *Filtering in the Time and Frequency Domains*, Wiley, New York.
Cook, B.J. (1979) *Journal of Measurement and Control*, **12**(8), 1979, pp. 326–335.
Healey, M. (1975) *Principles of Automatic Control*, Hodder and Stoughton.
Huelsman, L.P. (1993) *Active and Passive Analog Filter Design*, McGraw-Hill, New York.
Lynn, P.A. (1989) *The Analysis and Processing of Signals*, Macmillan, London.
Olsen, G.H. (1974) *A Handbook for Engineers and Scientists*, Butterworth, London.
Stephenson, F.W. (1985) *RC Active Filter Design Handbook*, Wiley, New York.

6

Electrical indicating and test instruments

The magnitude of voltage signals can be measured by various electrical indicating and test instruments, such as meters (both analogue and digital), the cathode ray oscilloscope and the digital storage oscilloscope. As well as signal-level voltages, many of these instruments can also measure higher-magnitude voltages, and this is indicated where appropriate.

6.1 Digital meters

All types of digital meter are basically modified forms of the *digital voltmeter* (DVM), irrespective of the quantity that they are designed to measure. Digital meters designed to measure quantities other than voltage are in fact digital voltmeters that contain appropriate electrical circuits to convert current or resistance measurement signals into voltage signals. *Digital multimeters* are also essentially digital voltmeters that contain several conversion circuits, thus allowing the measurement of voltage, current and resistance within one instrument.

Digital meters have been developed to satisfy a need for higher measurement accuracies and a faster speed of response to voltage changes than can be achieved with analogue instruments. They are technically superior to analogue meters in almost every respect. However, they have a greater cost due to the higher manufacturing costs compared with analogue meters. The binary nature of the output reading from a digital instrument can be readily applied to a display that is in the form of discrete numerals. Where human operators are required to measure and record signal voltage levels, this form of output makes an important contribution to measurement reliability and accuracy, since the problem of analogue meter parallax error is eliminated and the possibility of gross error through misreading the meter output is greatly reduced. The availability in many instruments of a direct output in digital form is also very useful in the rapidly expanding range of computer control applications. Quoted inaccuracy figures are between $\pm 0.005\%$ (measuring d.c. voltages) and $\pm 2\%$. Additional advantages of digital meters are their very high input impedance (10 MΩ compared with 1–20 kΩ for analogue meters), the ability to measure signals of frequency up to 1 MHz and the common inclusion of features such as automatic ranging, which prevents overload and reverse polarity connection etc.

The major part of a digital voltmeter is the circuitry that converts the analogue voltage being measured into a digital quantity. As the instrument only measures d.c. quantities in its basic mode, another necessary component within it is one that performs a.c.–d.c. conversion and thereby gives it the capacity to measure a.c. signals. After conversion, the voltage value is displayed by means of indicating tubes or a set of solid-state light-emitting diodes. Four-, five- or even six-figure output displays are commonly used, and although the instrument itself may not be inherently more accurate than some analogue types, this form of display enables measurements to be recorded with much greater accuracy than that obtainable by reading an analogue meter scale.

Digital voltmeters differ mainly in the technique used to effect the analogue-to-digital conversion between the measured analogue voltage and the output digital reading. As a general rule, the more expensive and complicated conversion methods achieve a faster conversion speed. Some common types of DVM are discussed below.

6.1.1 Voltage-to-time conversion digital voltmeter

This is the simplest form of DVM and is a ramp type of instrument. When an unknown voltage signal is applied to the input terminals of the instrument, a negative-slope ramp waveform is generated internally and compared with the input signal. When the two are equal, a pulse is generated that opens a gate, and at a later point in time a second pulse closes the gate when the negative ramp voltage reaches zero. The length of time between the gate opening and closing is monitored by an electronic counter, which produces a digital display according to the level of the input voltage signal. Its main drawbacks are non-linearities in the shape of the ramp waveform used and lack of noise rejection, and these problems lead to a typical inaccuracy of $\pm 0.05\%$. It is relatively cheap, however.

6.1.2 Potentiometric digital voltmeter

This uses a servo principle, in which the error between the unknown input voltage level and a reference voltage is applied to a servo-driven potentiometer that adjusts the reference voltage until it balances the unknown voltage. The output reading is produced by a mechanical drum-type digital display driven by the potentiometer. This is also a relatively cheap form of DVM that gives excellent performance for its price.

6.1.3 Dual-slope integration digital voltmeter

This is another relatively simple form of DVM that has better noise-rejection capabilities than many other types and gives correspondingly better measurement accuracy (inaccuracy as low as $\pm 0.005\%$). Unfortunately, it is quite expensive. The unknown voltage is applied to an integrator for a fixed time T_1, following which a reference voltage of opposite sign is applied to the integrator, which discharges down to a zero output in an interval T_2 measured by a counter. The output–time relationship for the integrator is shown in Figure 6.1, from which the unknown voltage V_i can be calculated

104 Electrical indicating and test instruments

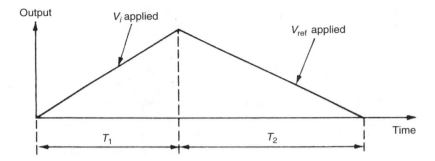

Fig. 6.1 Output–time relationship for integrator in a dual-slope digital voltmeter (DVM).

geometrically from the triangle as:

$$V_i = V_{\text{ref}}(T_1/T_2) \qquad (6.1)$$

6.1.4 Voltage-to-frequency conversion digital voltmeter

In this instrument, the unknown voltage signal is fed via a range switch and an amplifier into a converter circuit whose output is in the form of a train of voltage pulses at a frequency proportional to the magnitude of the input signal. The main advantage of this type of DVM is its ability to reject a.c. noise.

6.1.5 Digital multimeter

This is an extension of the DVM. It can measure both a.c. and d.c. voltages over a number of ranges through inclusion within it of a set of switchable amplifiers and attenuators. It is widely used in circuit test applications as an alternative to the analogue multimeter, and includes protection circuits that prevent damage if high voltages are applied to the wrong range.

6.2 Analogue meters

Analogue meters are relatively simple and inexpensive and are often used instead of digital instruments, especially when cost is of particular concern. Whilst digital instruments have the advantage of greater accuracy and much higher input impedance, analogue instruments suffer less from noise and isolation problems. In addition, because analogue instruments are usually passive instruments that do not need a power supply, this is often very useful in measurement applications where a suitable mains power supply is not readily available. Many examples of analogue meter also remain in use for historical reasons.

Analogue meters are electromechanical devices that drive a pointer against a scale. They are prone to measurement errors from a number of sources that include inaccurate scale marking during manufacture, bearing friction, bent pointers and ambient temperature variations. Further human errors are introduced through parallax error (not reading the scale from directly above) and mistakes in interpolating between scale markings. Quoted inaccuracy figures are between ±0.1% and ±3%. Various types of analogue meter are used as discussed below.

6.2.1 Moving-coil meters

A moving-coil meter is a very commonly used form of analogue voltmeter because of its sensitivity, accuracy and linear scale, although it only responds to d.c. signals. As shown schematically in Figure 6.2, it consists of a rectangular coil wound round a soft iron core that is suspended in the field of a permanent magnet. The signal being measured is applied to the coil and this produces a radial magnetic field. Interaction between this induced field and the field produced by the permanent magnet causes a torque, which results in rotation of the coil. The amount of rotation of the coil is measured by attaching a pointer to it that moves past a graduated scale. The theoretical torque produced is given by:

$$T = BIhwN \qquad (6.2)$$

where B is the flux density of the radial field, I is the current flowing in the coil, h is the height of the coil, w is the width of the coil and N is the number of turns in the coil. If the iron core is cylindrical and the air gap between the coil and pole faces of the permanent magnet is uniform, then the flux density B is constant, and equation (6.2) can be rewritten as:

$$T = KI \qquad (6.3)$$

i.e. the torque is proportional to the coil current and the instrument scale is linear.

As the basic instrument operates at low current levels of one milliamp or so, it is only suitable for measuring voltages up to around 2 volts. If there is a requirement to measure higher voltages, the measuring range of the instrument can be increased by placing a resistance in series with the coil, such that only a known proportion of

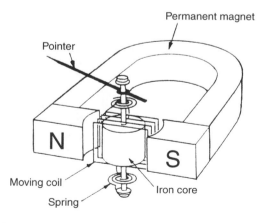

Fig. 6.2 Moving-coil meter.

the applied voltage is measured by the meter. In this situation the added resistance is known as a *shunting resistor*.

Whilst Figure 6.2 shows the traditional moving-coil instrument with a long U-shaped permanent magnet, many newer instruments employ much shorter magnets made from recently developed magnetic materials such as Alnico and Alcomax. These materials produce a substantially greater flux density, which, besides allowing the magnet to be smaller, has additional advantages in allowing reductions to be made in the size of the coil and in increasing the usable range of deflection of the coil to about 120°. Some versions of the instrument also have either a specially shaped core or specially shaped magnet pole faces to cater for special situations where a non-linear scale such as a logarithmic one is required.

6.2.2 Moving-iron meter

As well as measuring d.c. signals, the moving-iron meter can also measure a.c. signals at frequencies up to 125 Hz. It is the cheapest form of meter available and, consequently, this type of meter is also commonly used for measuring voltage signals. The signal to be measured is applied to a stationary coil, and the associated field produced is often amplified by the presence of an iron structure associated with the fixed coil. The moving element in the instrument consists of an iron vane that is suspended within the field of the fixed coil. When the fixed coil is excited, the iron vane turns in a direction that increases the flux through it.

The majority of moving-iron instruments are either of the attraction type or of the repulsion type. A few instruments belong to a third combination type. The attraction type, where the iron vane is drawn into the field of the coil as the current is increased, is shown schematically in Figure 6.3(a). The alternative repulsion type is sketched in Figure 6.3(b). For an excitation current I, the torque produced that causes the vane to

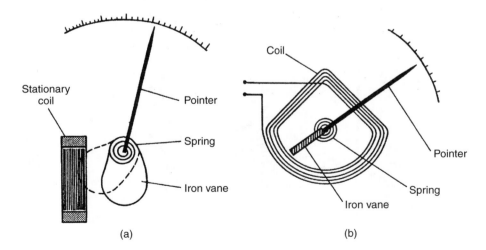

Fig. 6.3 Moving-iron meter: (a) attraction type; (b) repulsion type.

turn is given by:

$$T = \frac{I^2}{2} \frac{dM}{d\theta}$$

where M is the mutual inductance and θ is the angular deflection. Rotation is opposed by a spring that produces a backwards torque given by:

$$T_s = K\theta$$

At equilibrium, $T = T_s$, and θ is therefore given by:

$$\theta = \frac{I^2}{2K} \frac{dM}{d\theta} \tag{6.4}$$

The instrument thus has a square-law response where the deflection is proportional to the square of the signal being measured, i.e. the output reading is a root-mean-squared (r.m.s.) quantity.

The instrument can typically measure voltages in the range of 0 to 30 volts. However, it can be modified to measure higher voltages by placing a resistance in series with it, as in the case of moving coil meters. A series resistance is particularly beneficial in a.c. signal measurements because it compensates for the effect of coil inductance by reducing the total resistance/inductance ratio, and hence measurement accuracy is improved. A switchable series resistance is often provided within the casing of the instrument to facilitate range extension. However, when the voltage measured exceeds about 300 volts, it becomes impractical to use a series resistance within the case of the instrument because of heat-dissipation problems, and an external resistance is used instead.

6.2.3 Electrodynamic meters

Electrodynamic meters (or dynamometers) can measure both d.c. signals and a.c. signals up to a frequency of 2 kHz. As illustrated in Figure 6.4, the instrument has a moving circular coil that is mounted in the magnetic field produced by two separately wound, series-connected, circular stator coils. The torque is dependent upon the mutual inductance between the coils and is given by:

$$T = I_1 I_2 \frac{dM}{d\theta} \tag{6.5}$$

where I_1 and I_2 are the currents flowing in the fixed and moving coils, M is the mutual inductance and θ represents the angular displacement between the coils.

When used as an ammeter, the measured current is applied to both coils. The torque is thus proportional to *current*^{2}. If the measured current is a.c., the meter is unable to follow the alternating torque values and it displays instead the mean value of *current*^{2}. By suitable drawing of the scale, the position of the pointer shows the squared root of this value, i.e. the r.m.s. current.

Electrodynamic meters are typically expensive but have the advantage of being more accurate than moving-coil and moving-iron instruments. Voltage, current and power can

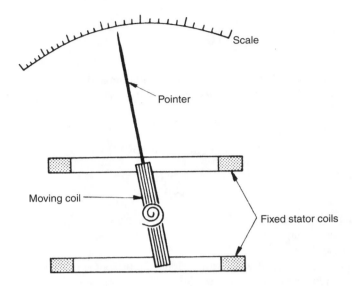

Fig. 6.4 Electrodynamic meter.

all be measured if the fixed and moving coils are connected appropriately. When used for voltage measurement, the instrument can typically measure voltages in the range of 0 to 30 volts. However, it can be modified to measure higher voltages by placing a resistance in series with it, as in the case of moving-coil and moving-iron meters. Also, as in the moving-iron meter, a series resistance is particularly beneficial in a.c. signal measurements because it compensates for the effect of coil inductance by reducing the total resistance/inductance ratio, and hence measurement accuracy is improved. This series resistance can either be inside or outside the instrument case, as discussed above for the case of moving-iron meters.

6.2.4 Clamp-on meters

These are used for measuring circuit currents and voltages in a non-invasive manner that avoids having to break the circuit being measured. The meter clamps onto a current-carrying conductor, and the output reading is obtained by transformer action. The principle of operation is illustrated in Figure 6.5, where it can be seen that the clamp-on jaws of the instrument act as a transformer core and the current-carrying conductor acts as a primary winding. Current induced in the secondary winding is rectified and applied to a moving-coil meter. Although it is a very convenient instrument to use, the clamp-on meter has low sensitivity and the minimum current measurable is usually about 1 amp.

6.2.5 Analogue multimeter

The analogue multimeter is a multi-function instrument that can measure current and resistance as well as d.c. and a.c. voltage signals. Basically, the instrument consists of a moving-coil meter with a switchable bridge rectifier to allow it to measure a.c. signals,

Measurement and Instrumentation Principles 109

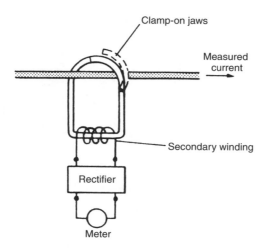

Fig. 6.5 Clamp-on meter.

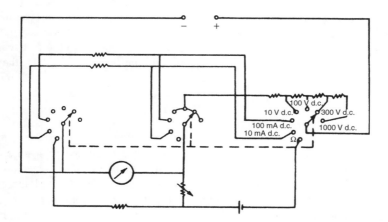

Fig. 6.6 Analogue multimeter.

as shown in Figure 6.6. A set of rotary switches allows the selection of various series and shunt resistors, which make the instrument capable of measuring both voltage and current over a number of ranges. An internal power source is also provided to allow it to measure resistances as well. Whilst this instrument is very useful for giving an indication of voltage levels, the compromises in its design that enable it to measure so many different quantities necessarily mean that its accuracy is not as good as instruments that are purpose designed to measure just one quantity over a single measuring range.

6.2.6 Measuring high-frequency signals

One major limitation in using analogue meters for a.c. voltage measurement is that the maximum frequency measurable directly is low, 2 kHz for the dynamometer voltmeter

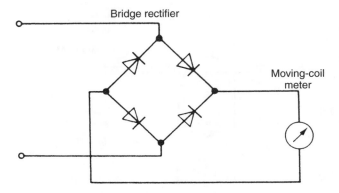

Fig. 6.7 Measurement of high-frequency voltage signals.

and only 100 Hz in the case of the moving-iron instrument. A partial solution to this limitation is to rectify the voltage signal and then apply it to a moving-coil meter, as shown in Figure 6.7. This extends the upper measurable frequency limit to 20 kHz. However, the inclusion of the bridge rectifier makes the measurement system particularly sensitive to environmental temperature changes, and non-linearities significantly affect measurement accuracy for voltages that are small relative to the full-scale value.

An alternative solution to the upper frequency limitation is provided by the *thermocouple meter* (see below).

6.2.7 Thermocouple meter

The principle of operation of the thermocouple meter is shown in Figure 6.8. The measured a.c. voltage signal is applied to a small element. This heats up and the resulting temperature rise is measured by a thermocouple. The d.c. voltage generated in the thermocouple is applied to a moving-coil meter. The output meter reading is an r.m.s. quantity that varies in a non-linear fashion with the magnitude of the measured voltage. Very high-frequency voltage signals up to 50 MHz can be measured by this method.

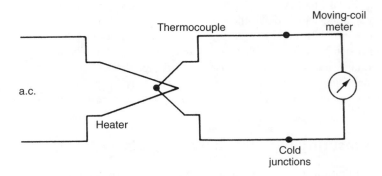

Fig. 6.8 Thermocouple meter.

6.2.8 Electronic analogue voltmeters

Electronic voltmeters differ from all other forms of analogue voltmeters in being active rather than passive instruments. They have important advantages compared with other analogue instruments. Firstly, they have a high input impedance that avoids the circuit-loading problems associated with many applications of electromechanical instruments. Secondly, they have an amplification capability that enables them to measure small signal levels accurately.

The standard electronic voltmeter for d.c. measurements consists of a simple direct-coupled amplifier and a moving-coil meter, as shown in Figure 6.9(a). For measurement of very low-level voltages of a few microvolts, a more sophisticated circuit, known as a chopper amplifier, is used, as shown in Figure 6.9(b). In this, the d.c. input is chopped at a low frequency of around 250 Hz, passed through a blocking capacitor, amplified, passed through another blocking capacitor to remove drift, demodulated, filtered and applied to a moving-coil meter.

Three versions of electronic voltmeter exist for measuring a.c. signals. The *average-responding type* is essentially a direct-coupled d.c. electronic voltmeter with an additional rectifying stage at the input. The output is a measure of the average value of the measured voltage waveform. The second form, known as a *peak-responding type*, has a half-wave rectifier at the input followed by a capacitor. The final part of the circuit consists of an amplifier and moving-coil meter. The capacitor is charged to the peak value of the input signal, and therefore the amplified signal applied to the moving-coil meter gives a reading of the peak voltage in the input waveform. Finally, a third type is available, known as an *r.m.s.-responding type*, which gives an output reading in terms of the r.m.s. value of the input waveform. This type is essentially a thermocouple meter in which an amplification stage has been inserted at the input.

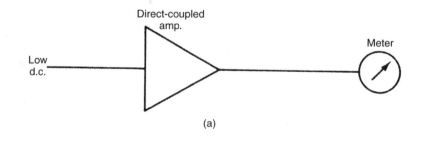

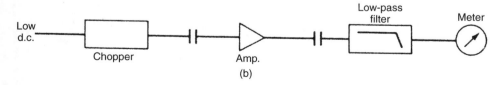

Fig. 6.9 D.c. electronic voltmeter: (a) simple form; (b) including chopper amplifier.

6.2.9 Calculation of meter outputs for non-standard waveforms

The examples below provide an exercise in calculating the output reading from various types of analogue voltmeter. These examples also serve as a useful reminder of the mode of operation of each type of meter and the form that the output takes.

Example 6.1
Calculate the reading that would be observed on a moving-coil ammeter when it is measuring the current in the circuit shown in Figure 6.10.

Solution
A moving-coil meter measures mean current.

$$I_{mean} = \frac{1}{2\pi} \left(\int_0^\pi \frac{5\omega t}{\pi} d\omega t + \int_\pi^{2\pi} 5 \sin(\omega t) \, d\omega t \right)$$

$$= \frac{1}{2\pi} \left(\left[\frac{5(\omega t)^2}{2\pi} \right]_0^\pi + 5[-\cos(\omega t)]_\pi^{2\pi} \right)$$

$$= \frac{1}{2\pi} \left(\frac{5\pi^2}{2\pi} - 0 - 5 - 5 \right) = \frac{1}{2\pi} \left(\frac{5\pi}{2} - 10 \right) = \frac{5}{2\pi} \left(\frac{\pi}{2} - 2 \right)$$

$$= -0.342 \text{ amps}$$

Example 6.2
Calculate the reading that would be observed on a moving-iron ammeter when it is measuring the current in the circuit shown in Figure 6.10.

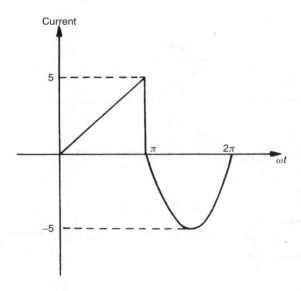

Fig. 6.10 Circuit for example 6.1.

Solution
A moving-iron meter measures r.m.s. current.

$$I_{\text{r.m.s.}}^2 = \frac{1}{2\pi}\left(\int_0^\pi \frac{25\,(\omega t)^2}{\pi^2}\,d\omega t + \int_\pi^{2\pi} 25\sin^2(\omega t)\,d\omega t\right)$$

$$= \frac{1}{2\pi}\left(\int_0^\pi \frac{25\,(\omega t)^2}{\pi^2}\,d\omega t + \int_\pi^{2\pi} \frac{25\,(1-\cos 2\omega t)}{2}\,d\omega t\right)$$

$$= \frac{25}{2\pi}\left(\left[\frac{(\omega t)^3}{3\pi^2}\right]_0^\pi + \left[\frac{\omega t}{2} - \frac{\sin 2\omega t}{4}\right]_\pi^{2\pi}\right) = \frac{25}{2\pi}\left(\frac{\pi}{3} + \frac{2\pi}{2} - \frac{\pi}{2}\right)$$

$$= \frac{25}{2\pi}\left(\frac{\pi}{3} + \frac{\pi}{2}\right) = \frac{25}{2}\left(\frac{1}{3} + \frac{1}{2}\right) = 10.416$$

Thus, $I_{\text{r.m.s.}} = \sqrt{(I_{\text{r.m.s.}}^2)} = 3.23$ amps

Example 6.3
A dynamometer ammeter is connected in series with a 500 Ω resistor, a rectifying device and a 240 V r.m.s. alternating sinusoidal power supply. The rectifier behaves as a resistance of 200 Ω to current flowing in one direction and as a resistance of 2 kΩ to current in the opposite direction. Calculate the reading on the meter.

Solution

$$V_{\text{peak}} = \sqrt{V_{\text{r.m.s.}}(2)} = 339.4\text{ V}$$

For $0 < \omega t < \pi$, $R = 700\,\Omega$ and for $\pi < \omega t < 2\pi$, $R = 2500\,\Omega$. Thus:

$$I_{\text{r.m.s.}}^2 = \frac{1}{2\pi}\left(\int_0^\pi \frac{(339.4\sin\omega t)^2}{700^2}\,d\omega t + \int_\pi^{2\pi} \frac{(339.4\sin\omega t)^2}{2500^2}\,d\omega t\right)$$

$$= \frac{339.4^2}{2\pi\,10^4}\left(\int_0^\pi \frac{\sin^2\omega t}{49}\,d\omega t + \int_\pi^{2\pi} \frac{\sin^2\omega t}{625}\,d\omega t\right)$$

$$= \frac{339.4^2}{4\pi\,10^4}\left(\int_0^\pi \frac{(1-\cos 2\omega t)}{49}\,d\omega t + \int_\pi^{2\pi} \frac{(1-\cos 2\omega t)}{625}\,d\omega t\right)$$

$$= \frac{339.4^2}{4\pi\,10^4}\left(\left[\frac{\omega t}{49} - \frac{\sin 2\omega t}{98}\right]_0^\pi + \left[\frac{\omega t}{625} - \frac{\sin 2\omega t}{1250}\right]_\pi^{2\pi}\right)$$

$$= \frac{339.4^2}{4\pi\,10^4}\left(\frac{\pi}{49} + \frac{\pi}{625}\right) = 0.0634$$

Hence, $I_{\text{r.m.s.}} = \sqrt{0.0634} = 0.25$ amp.

6.3 Cathode ray oscilloscope

The cathode ray oscilloscope is probably the most versatile and useful instrument available for signal measurement. In its basic form, it is an analogue instrument and is often called an *analogue oscilloscope* to distinguish it from digital storage oscilloscopes which have emerged more recently (these are discussed in section 6.4). The analogue oscilloscope is widely used for voltage measurement, especially as an item of test equipment for circuit fault-finding, and it is able to measure a very wide range of both a.c. and d.c. voltage signals. Besides measuring voltage levels, it can also measure other quantities such as the frequency and phase of a signal. It can also indicate the nature and magnitude of noise that may be corrupting the measurement signal. The more expensive models can measure signals at frequencies up to 500 MHz and even the cheapest models can measure signals up to 20 MHz. One particularly strong merit of the oscilloscope is its high input impedance, typically 1 MΩ, which means that the instrument has a negligible loading effect in most measurement situations. As a test instrument, it is often required to measure voltages whose frequency and magnitude are totally unknown. The set of rotary switches that alter its timebase so easily, and the circuitry that protects it from damage when high voltages are applied to it on the wrong range, make it ideally suited for such applications. However, it is not a particularly accurate instrument and is best used where only an approximate measurement is required. In the best instruments, inaccuracy can be limited to $\pm 1\%$ of the reading but inaccuracy can approach $\pm 10\%$ in the cheapest instruments. Further disadvantages of oscilloscopes include their fragility (being built around a cathode ray tube) and their moderately high cost.

The most important aspects in the specification of an oscilloscope are its bandwidth, its rise time and its accuracy. The bandwidth is defined as the range of frequencies over which the oscilloscope amplifier gain is within 3 dB* of its peak value, as illustrated in Figure 6.11. The -3 dB point is where the gain is 0.707 times its maximum value.

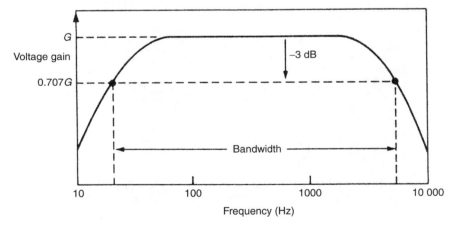

Fig. 6.11 Bandwidth.

---

* The decibel, commonly written dB, is used to express the ratio between two quantities. For two voltage levels V_1 and V_2, the difference between the two levels is expressed in decibels as $20 \log_{10} (V_1/V_2)$. It follows from this that $20 \log_{10} (0.7071) = -3$ dB.

In most oscilloscopes, the amplifier is direct coupled, which means that it amplifies d.c. voltages by the same factor as low-frequency a.c. ones. For such instruments, the minimum frequency measurable is zero and the bandwidth can be interpreted as the maximum frequency where the sensitivity (deflection/volt) is within 3 dB of the peak value. In all measurement situations, the oscilloscope chosen for use must be such that the maximum frequency to be measured is well within the bandwidth. The −3 dB specification means that an oscilloscope with a specified inaccuracy of ±2% and bandwidth of 100 MHz will have an inaccuracy of ±5% when measuring 30 MHz signals, and this inaccuracy will increase still further at higher frequencies. Thus, when applied to signal-amplitude measurement, the oscilloscope is only usable at frequencies up to about 0.3 times its specified bandwidth.

The rise time is the transit time between the 10% and 90% levels of the response when a step input is applied to the oscilloscope. Oscilloscopes are normally designed such that:

$$\text{Bandwidth} \times \text{Rise time} = 0.35$$

Thus, for a bandwidth of 100 MHz, rise time $= 0.35/100\,000\,000 = 3.5$ ns.

An oscilloscope is a relatively complicated instrument that is constructed from a number of subsystems, and it is necessary to consider each of these in turn in order to understand how the complete instrument functions.

6.3.1 Cathode ray tube

The cathode ray tube, shown in Figure 6.12, is the fundamental part of an oscilloscope. The cathode consists of a barium and strontium oxide coated, thin, heated filament from which a stream of electrons is emitted. The stream of electrons is focused onto a well-defined spot on a fluorescent screen by an electrostatic focusing system that consists of a series of metal discs and cylinders charged at various potentials. Adjustment of this focusing mechanism is provided by controls on the front panel of an oscilloscope. An *intensity* control varies the cathode heater current and therefore the rate of emission of electrons, and thus adjusts the intensity of the display on the screen. These and other typical controls are shown in the illustration of the front panel of a simple oscilloscope given in Figure 6.13.

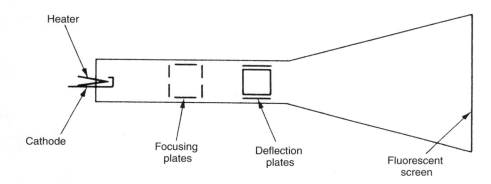

Fig. 6.12 Cathode ray tube.

116 Electrical indicating and test instruments

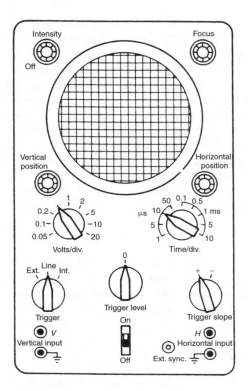

Fig. 6.13 Controls of a simple oscilloscope.

Application of potentials to two sets of deflector plates mounted at right angles to one another within the tube provide for deflection of the stream of electrons, such that the spot where the electrons are focused on the screen is moved. The two sets of deflector plates are normally known as the horizontal and vertical deflection plates, according to the respective motion caused to the spot on the screen. The magnitude of any signal applied to the deflector plates can be calculated by measuring the deflection of the spot against a crossed-wire graticule etched on the screen.

In the oscilloscope's most common mode of usage measuring time-varying signals, the unknown signal is applied, via an amplifier, to the y-axis (vertical) deflector plates and a timebase to the x-axis (horizontal) deflector plates. In this mode of operation, the display on the oscilloscope screen is in the form of a graph with the magnitude of the unknown signal on the vertical axis and time on the horizontal axis.

6.3.2 Channel

One channel describes the basic subsystem of an electron source, focusing system and deflector plates. This subsystem is often duplicated one or more times within the cathode ray tube to provide a capability of displaying two or more signals at the same time on the screen. The common oscilloscope configuration with two channels can therefore display two separate signals simultaneously.

6.3.3 Single-ended input

This type of input only has one input terminal plus a ground terminal per oscilloscope channel and, consequently, only allows signal voltages to be measured relative to ground. It is normally only used in simple oscilloscopes.

6.3.4 Differential input

This type of input is provided on more expensive oscilloscopes. Two input terminals plus a ground terminal are provided for each channel, which allows the potentials at two non-grounded points in a circuit to be compared. This type of input can also be used in single-ended mode to measure a signal relative to ground by using just one of the input terminals plus ground.

6.3.5 Timebase circuit

The purpose of a timebase is to apply a voltage to the horizontal deflector plates such that the horizontal position of the spot is proportional to time. This voltage, in the form of a ramp known as a sweep waveform, must be applied repetitively, such that the motion of the spot across the screen appears as a straight line when a d.c. level is applied to the input channel. Furthermore, this timebase voltage must be synchronized with the input signal in the general case of a time-varying signal, such that a steady picture is obtained on the oscilloscope screen. The length of time taken for the spot to traverse the screen is controlled by a *time/div* switch, which sets the length of time taken by the spot to travel between two marked divisions on the screen, thereby allowing signals at a wide range of frequencies to be measured.

Each cycle of the sweep waveform is initiated by a pulse from a pulse generator. The input to the pulse generator is a sinusoidal signal known as a triggering signal, with a pulse being generated every time the triggering signal crosses a preselected slope and voltage level condition. This condition is defined by the *trigger level* and *trigger slope* switches. The former selects the voltage level on the trigger signal, commonly zero, at which a pulse is generated, whilst the latter selects whether pulsing occurs on a positive- or negative-going part of the triggering waveform.

Synchronization of the sweep waveform with the measured signal is most easily achieved by deriving the trigger signal from the measured signal, a procedure that is known as *internal triggering*. Alternatively, *external triggering* can be applied if the frequencies of the triggering signal and measured signals are related by an integer constant such that the display is stationary. External triggering is necessary when the amplitude of the measured signal is too small to drive the pulse generator, and it is also used in applications where there is a requirement to measure the phase difference between two sinusoidal signals of the same frequency. It is very convenient to use the 50 Hz line voltage for external triggering when measuring signals at mains frequency, and this is often given the name *line triggering*.

6.3.6 Vertical sensitivity control

This consists of a series of attenuators and pre-amplifiers at the input to the oscilloscope. These condition the measured signal to the optimum magnitude for input to the main amplifier and vertical deflection plates, thus enabling the instrument to measure a very wide range of different signal magnitudes. Selection of the appropriate input amplifier/attenuator is made by setting a *volts/div* control associated with each oscilloscope channel. This defines the magnitude of the input signal that will cause a deflection of one division on the screen.

6.3.7 Display position control

This allows the position at which a signal is displayed on the screen to be controlled in two ways. The horizontal position is adjusted by a *horizontal position* knob on the oscilloscope front panel and similarly a *vertical position* knob controls the vertical position. These controls adjust the position of the display by biasing the measured signal with d.c. voltage levels.

6.4 Digital storage oscilloscopes

Digital storage oscilloscopes consist of a conventional analogue cathode ray oscilloscope with the added facility that the measured analogue signal can be converted to digital format and stored in computer memory within the instrument. This stored data can then be reconverted to analogue form at the frequency necessary to refresh the analogue display on the screen. This produces a non-fading display of the signal on the screen.

The signal displayed by a digital oscilloscope consists of a sequence of individual dots rather than a continuous line as displayed by an analogue oscilloscope. However, as the density of dots increases, the display becomes closer and closer to a continuous line, and the best instruments have displays that look very much like continuous traces. The density of the dots is entirely dependent upon the sampling rate at which the analogue signal is digitized and the rate at which the memory contents are read to reconstruct the original signal. Inevitably, the speed of sampling etc. is a function of cost, and the most expensive instruments give the best performance in terms of dot density and the accuracy with which the analogue signal is recorded and represented.

Besides their ability to display the magnitude of voltage signals and other parameters such as signal phase and frequency, some digital oscilloscopes can also compute signal parameters such as peak values, mean values and r.m.s. values. They are also ideally suited to capturing transient signals when set to single-sweep mode. This avoids the problem of the very careful synchronization that is necessary to capture such signals on an analogue oscilloscope. In addition, digital oscilloscopes often have facilities to output analogue signals to devices like chart recorders and output digital signals in a form that is compatible with standard interfaces like IEEE488 and RS232. Some now even have floppy disk drives to extend their storage ability. Fuller details on digital oscilloscopes can be found elsewhere Hickman, (1997).

References and further reading

Hickman, I. (1997) *Digital Storage Oscilloscopes*, Newnes.

7

Variable conversion elements

Outputs from measurement sensors that take the form of voltage signals can be measured using the voltage indicating and test instruments discussed in the last chapter. However, in many cases, the sensor output does not take the form of an electrical voltage. Examples of these other forms of sensor output include translational displacements and changes in various electrical parameters such as resistance, inductance, capacitance and current. In some cases, the output may alternatively take the form of variations in the phase or frequency of an a.c. signal.

For sensor outputs that are initially in some non-voltage form, conversion to a measurement signal that is in a more convenient form can be achieved by various types of variable conversion element in the measurement system. Bridge circuits are a particularly important type of variable conversion element, and these will be covered in some detail. Following this, the various alternative techniques for transducing the outputs of a measurement sensor will be covered.

7.1 Bridge circuits

Bridge circuits are used very commonly as a variable conversion element in measurement systems and produce an output in the form of a voltage level that changes as the measured physical quantity changes. They provide an accurate method of measuring resistance, inductance and capacitance values, and enable the detection of very small changes in these quantities about a nominal value. They are of immense importance in measurement system technology because so many transducers measuring physical quantities have an output that is expressed as a change in resistance, inductance or capacitance. The displacement-measuring strain gauge, which has a varying resistance output, is but one example of this class of transducers. Normally, excitation of the bridge is by a d.c. voltage for resistance measurement and by an a.c. voltage for inductance or capacitance measurement. Both null and deflection types of bridge exist, and, in a like manner to instruments in general, null types are mainly employed for calibration purposes and deflection types are used within closed-loop automatic control schemes.

7.1.1 Null-type, d.c. bridge (Wheatstone bridge)

A null-type bridge with d.c. excitation, commonly known as a Wheatstone bridge, has the form shown in Figure 7.1. The four arms of the bridge consist of the unknown resistance R_u, two equal value resistors R_2 and R_3 and a variable resistor R_v (usually a decade resistance box). A d.c. voltage V_i is applied across the points AC and the resistance R_v is varied until the voltage measured across points BD is zero. This null point is usually measured with a high sensitivity galvanometer.

To analyse the Whetstone bridge, define the current flowing in each arm to be $I_1 \ldots I_4$ as shown in Figure 7.1. Normally, if a high impedance voltage-measuring instrument is used, the current I_m drawn by the measuring instrument will be very small and can be approximated to zero. If this assumption is made, then, for $I_m = 0$:

$$I_1 = I_3 \quad \text{and} \quad I_2 = I_4$$

Looking at path ADC, we have a voltage V_i applied across a resistance $R_u + R_3$ and by Ohm's law:

$$I_1 = \frac{V_i}{R_u + R_3}$$

Similarly for path ABC:

$$I_2 = \frac{V_i}{R_v + R_2}$$

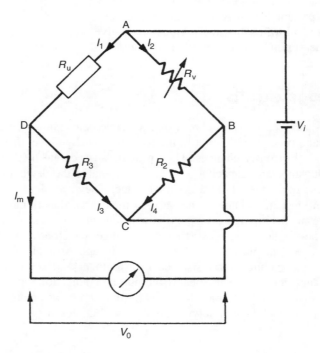

Fig. 7.1 Analysis of Wheatstone bridge.

Now we can calculate the voltage drop across AD and AB:

$$V_{AD} = I_1 R_v = \frac{V_i R_u}{R_u + R_3}; \quad V_{AB} = I_2 R_v = \frac{V_i R_v}{R_v + R_2}$$

By the principle of superposition,

$$V_0 = V_{BD} = V_{BA} + V_{AD} = -V_{AB} + V_{AD}$$

Thus:

$$V_0 = -\frac{V_i R_v}{R_v + R_2} + \frac{V_i R_u}{R_u + R_3} \tag{7.1}$$

At the null point $V_0 = 0$, so:

$$\frac{R_u}{R_u + R_3} = \frac{R_v}{R_v + R_2}$$

Inverting both sides:

$$\frac{R_u + R_3}{R_u} = \frac{R_v + R_2}{R_v} \quad \text{i.e.} \quad \frac{R_3}{R_u} = \frac{R_2}{R_v} \quad \text{or} \quad R_u = \frac{R_3 R_v}{R_2} \tag{7.2}$$

Thus, if $R_2 = R_3$, then $R_u = R_v$. As R_v is an accurately known value because it is derived from a variable decade resistance box, this means that R_u is also accurately known.

7.1.2 Deflection-type d.c. bridge

A deflection-type bridge with d.c. excitation is shown in Figure 7.2. This differs from the Wheatstone bridge mainly in that the variable resistance R_v is replaced by a fixed resistance R_1 of the same value as the nominal value of the unknown resistance R_u. As the resistance R_u changes, so the output voltage V_0 varies, and this relationship between V_0 and R_u must be calculated.

This relationship is simplified if we again assume that a high impedance voltage measuring instrument is used and the current drawn by it, I_m, can be approximated to zero. (The case when this assumption does not hold is covered later in this section.) The analysis is then exactly the same as for the preceding example of the Wheatstone bridge, except that R_v is replaced by R_1. Thus, from equation (7.1), we have:

$$V_0 = V_i \left(\frac{R_u}{R_u + R_3} - \frac{R_1}{R_1 + R_2} \right) \tag{7.3}$$

When R_u is at its nominal value, i.e. for $R_u = R_1$, it is clear that $V_0 = 0$ (since $R_2 = R_3$). For other values of R_u, V_0 has negative and positive values that vary in a non-linear way with R_u.

Example 7.1
A certain type of pressure transducer, designed to measure pressures in the range 0–10 bar, consists of a diaphragm with a strain gauge cemented to it to detect diaphragm

122 Variable conversion elements

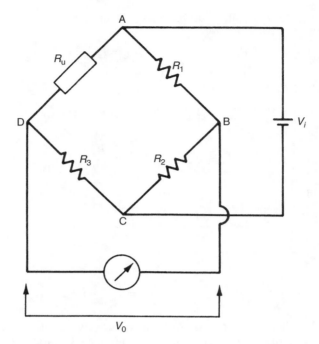

Fig. 7.2 Deflection-type d.c. bridge.

deflections. The strain gauge has a nominal resistance of 120 Ω and forms one arm of a Wheatstone bridge circuit, with the other three arms each having a resistance of 120 Ω. The bridge output is measured by an instrument whose input impedance can be assumed infinite. If, in order to limit heating effects, the maximum permissible gauge current is 30 mA, calculate the maximum permissible bridge excitation voltage. If the sensitivity of the strain gauge is 338 mΩ/bar and the maximum bridge excitation voltage is used, calculate the bridge output voltage when measuring a pressure of 10 bar.

Solution
This is the type of bridge circuit shown in Figure 7.2 in which the components have the following values:

$$R_1 = R_2 = R_3 = 120\, \Omega$$

Defining I_1 to be the current flowing in path ADC of the bridge, we can write:

$$V_i = I_1(R_u + R_3)$$

At balance, $R_u = 120$ and the maximum value allowable for I_1 is 0.03 A.
Hence:

$$V_i = 0.03(120 + 120) = 7.2 \text{ V}$$

Thus, the maximum bridge excitation voltage allowable is 7.2 volts.

For a pressure of 10 bar applied, the resistance change is 3.38 Ω, i.e. R_u is then equal to 123.38 Ω.

Applying equation (7.3), we can write:

$$V_0 = V_i \left(\frac{R_u}{R_u + R_3} - \frac{R_1}{R_1 + R_2} \right) = 7.2 \left(\frac{123.38}{243.38} - \frac{120}{240} \right) = 50\,\text{mV}$$

Thus, if the maximum permissible bridge excitation voltage is used, the output voltage is 50 mV when a pressure of 10 bar is measured.

The non-linear relationship between output reading and measured quantity exhibited by equation (7.3) is inconvenient and does not conform with the normal requirement for a linear input–output relationship. The method of coping with this non-linearity varies according to the form of primary transducer involved in the measurement system.

One special case is where the change in the unknown resistance R_u is typically small compared with the nominal value of R_u. If we calculate the new voltage V'_0 when the resistance R_u in equation (7.3) changes by an amount δR_u, we have:

$$V'_0 = V_i \left(\frac{R_u + \delta R_u}{R_u + \delta R_u + R_3} - \frac{R_1}{R_1 + R_2} \right) \qquad (7.4)$$

The change of voltage output is therefore given by:

$$\delta V_0 = V'_0 - V_0 = \frac{V_i \delta R_u}{R_u + \delta R_u + R_3}$$

If $\delta R_u \ll R_u$, then the following linear relationship is obtained:

$$\frac{\delta V_0}{\delta R_u} = \frac{V_i}{R_u + R_3} \qquad (7.5)$$

This expression describes the measurement sensitivity of the bridge. Such an approximation to make the relationship linear is valid for transducers such as strain gauges where the typical changes of resistance with strain are very small compared with the nominal gauge resistance.

However, many instruments that are inherently linear themselves at least over a limited measurement range, such as resistance thermometers, exhibit large changes in output as the input quantity changes, and the approximation of equation (7.5) cannot be applied. In such cases, specific action must be taken to improve linearity in the relationship between the bridge output voltage and the measured quantity. One common solution to this problem is to make the values of the resistances R_2 and R_3 at least ten times those of R_1 and R_u (nominal). The effect of this is best observed by looking at a numerical example.

Consider a platinum resistance thermometer with a range of 0°–50°C, whose resistance at 0°C is 500 Ω and whose resistance varies with temperature at the rate of 4 Ω/°C. Over this range of measurement, the output characteristic of the thermometer itself is nearly perfectly linear. (N.B. The subject of resistance thermometers is discussed further in Chapter 14.)

Taking first the case where $R_1 = R_2 = R_3 = 500\,\Omega$ and $V_i = 10$ V, and applying equation (7.3):

124 Variable conversion elements

At 0°C; $V_0 = 0$

At 25°C; $R_u = 600\,\Omega$ and $V_0 = 10\left(\dfrac{600}{1100} - \dfrac{500}{1000}\right) = 0.455$ V

At 50°C; $R_u = 700\,\Omega$ and $V_0 = 10\left(\dfrac{700}{1200} - \dfrac{500}{1000}\right) = 0.833$ V

This relationship between V_0 and R_u is plotted as curve (a) in Figure 7.3 and the non-linearity is apparent. Inspection of the manner in which the output voltage V_0 above changes for equal steps of temperature change also clearly demonstrates the non-linearity.

For the temperature change from 0 to 25°C, the change in V_0 is
$(0.455 - 0) = 0.455$ V

For the temperature change from 25 to 50°C, the change in V_0 is
$(0.833 - 0.455) = 0.378$ V

If the relationship was linear, the change in V_0 for the 25–50°C temperature step would also be 0.455 V, giving a value for V_0 of 0.910 V at 50°C.

Now take the case where $R_1 = 500\,\Omega$ but $R_2 = R_3 = 5000\,\Omega$ and let $V_i = 26.1$ V:

At 0°C; $V_0 = 0$

At 25°C; $R_u = 600\,\Omega$ and $V_0 = 26.1\left(\dfrac{600}{5600} - \dfrac{500}{5500}\right) = 0.424$ V

At 50°C; $R_u = 700\,\Omega$ and $V_0 = 26.1\left(\dfrac{700}{5700} - \dfrac{500}{5500}\right) = 0.833$ V

This relationship is shown as curve (b) in Figure 7.3 and a considerable improvement in linearity is achieved. This is more apparent if the differences in values for V_0 over the two temperature steps are inspected.

From 0 to 25°C, the change in V_0 is 0.424 V

From 25 to 50°C, the change in V_0 is 0.409 V

The changes in V_0 over the two temperature steps are much closer to being equal than before, demonstrating the improvement in linearity. However, in increasing the values of R_2 and R_3, it was also necessary to increase the excitation voltage from 10 V to 26.1 V to obtain the same output levels. In practical applications, V_i would normally be set at the maximum level consistent with the limitation of the effect of circuit heating in order to maximize the measurement sensitivity ($V_0/\delta R_u$ relationship). It would therefore not be possible to increase V_i further if R_2 and R_3 were increased, and the general effect of such an increase in R_2 and R_3 is thus a decrease in the sensitivity of the measurement system.

The importance of this inherent non-linearity in the bridge output relationship is greatly diminished if the primary transducer and bridge circuit are incorporated as elements within an intelligent instrument. In that case, digital computation is applied

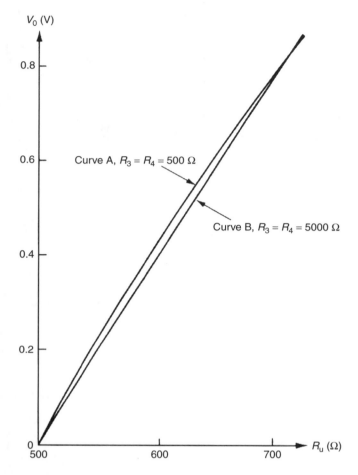

Fig. 7.3 Linearization of bridge circuit characteristic.

to produce an output in terms of the measured quantity that automatically compensates for the non-linearity in the bridge circuit.

Case where current drawn by measuring instrument is not negligible

For various reasons, it is not always possible to meet the condition that the impedance of the instrument measuring the bridge output voltage is sufficiently large for the current drawn by it to be negligible. Wherever the measurement current is not negligible, an alternative relationship between the bridge input and output must be derived that takes the current drawn by the measuring instrument into account.

Thévenin's theorem is again a useful tool for this purpose. Replacing the voltage source V_i in Figure 7.4(a) by a zero internal resistance produces the circuit shown in Figure 7.4(b), or the equivalent representation shown in Figure 7.4(c). It is apparent from Figure 7.4(c) that the equivalent circuit resistance consists of a pair of parallel resistors R_u and R_3 in series with the parallel resistor pair R_1 and R_2. Thus, R_{DB} is

126 Variable conversion elements

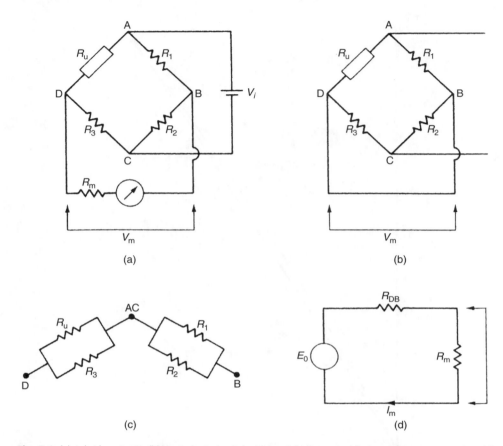

Fig. 7.4 (a) A bridge circuit; (b) equivalent circuit by Thévenin's theorem; (c) alternative representation; (d) equivalent circuit for alternative representation.

given by:

$$R_{DB} = -\frac{R_1 R_2}{R_1 + R_2} + \frac{R_u R_3}{R_u + R_3} \tag{7.6}$$

The equivalent circuit derived via Thévenin's theorem with the resistance R_m of the measuring instrument connected across the output is shown in Figure 7.4(d). The open-circuit voltage across DB, E_0, is the output voltage calculated earlier (equation 7.3) for the case of $R_m = 0$:

$$E_0 = V_i \left(\frac{R_u}{R_u + R_3} - \frac{R_1}{R_1 + R_2} \right) \tag{7.7}$$

If the current flowing is I_m when the measuring instrument of resistance R_m is connected across DB, then, by Ohm's law, I_m is given by:

$$I_m = \frac{E_0}{R_{DB} + R_m} \tag{7.8}$$

If V_m is the voltage measured across R_m, then, again by Ohm's law:

$$V_m = I_m R_m = \frac{E_0 R_m}{R_{DB} + R_m} \quad (7.9)$$

Substituting for E_0 and R_{DB} in equation (7.9), using the relationships developed in equations (7.6) and (7.7), we obtain:

$$V_m = \frac{V_i \left[R_u/(R_u + R_3) - R_1/(R_1 + R_2) \right] R_m}{R_1 R_2/(R_1 + R_2) + R_u R_3/(R_u + R_3) + R_m}$$

Simplifying:

$$V_m = \frac{V_i R_m (R_u R_2 - R_1 R_3)}{R_1 R_2 (R_u + R_3) + R_u R_3 (R_1 + R_2) + R_m (R_1 + R_2)(R_u + R_3)} \quad (7.10)$$

Example 7.2
A bridge circuit, as shown in Figure 7.5, is used to measure the value of the unknown resistance R_u of a strain gauge of nominal value 500 Ω. The output voltage measured across points DB in the bridge is measured by a voltmeter. Calculate the measurement sensitivity in volts/ohm change in R_u if

(a) the resistance R_m of the measuring instrument is neglected, and
(b) account is taken of the value of R_m.

Solution
For $R_u = 500\,\Omega$, $V_m = 0$.
To determine sensitivity, calculate V_m for $R_u = 501\,\Omega$.
(a) Applying equation (7.3): $V_m = V_i \left(\dfrac{R_u}{R_u + R_3} - \dfrac{R_1}{R_1 + R_2} \right)$

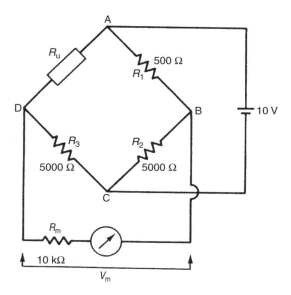

Fig. 7.5 Bridge circuit.

128 Variable conversion elements

Substituting in values: $V_m = 10 \left(\frac{501}{1001} - \frac{500}{1000} \right) = 5.00 \, \text{mV}$

Thus, if the resistance of the measuring circuit is neglected, the measurement sensitivity is 5.00 mV per ohm change in R_u.

(b) Applying equation (7.10) and substituting in values:

$$V_m = \frac{10 \times 10^4 \times 500(501 - 500)}{500^2(1001) + 500 \times 501(1000) + 10^4 \times 1000 \times 1001} = 4.76 \, \text{mV}$$

Thus, if proper account is taken of the 10 kΩ value of the resistance of R_m, the true measurement sensitivity is shown to be 4.76 mV per ohm change in R_u.

7.1.3 Error analysis

In the application of bridge circuits, the contribution of component-value tolerances to total measurement system accuracy limits must be clearly understood. The analysis below applies to a null-type (Wheatstone) bridge, but similar principles can be applied for a deflection-type bridge. The maximum measurement error is determined by first finding the value of R_u in equation (7.2) with each parameter in the equation set at that limit of its tolerance which produces the maximum value of R_u. Similarly, the minimum possible value of R_u is calculated, and the required error band is then the span between these maximum and minimum values.

Example 7.3
In the Wheatstone bridge circuit of Figure 7.1, R_v is a decade resistance box with a specified inaccuracy $\pm 0.2\%$ and $R_2 = R_3 = 500 \, \Omega \pm 0.1\%$. If the value of R_v at the null position is 520.4 Ω, determine the error band for R_u expressed as a percentage of its nominal value.

Solution
Applying equation (7.2) with $R_v = 520.4 \, \Omega + 0.2\% = 521.44 \, \Omega$, $R_3 = 5000 \, \Omega + 0.1\% = 5005 \, \Omega$, $R_2 = 5000 \, \Omega - 0.1\% = 4995 \, \Omega$ we get:

$$R_v = \frac{521.44 \times 5005}{4995} = 522.48 \, \Omega \, (= +0.4\%)$$

Applying equation (7.2) with $R_v = 520.4 \, \Omega - 0.2\% = 519.36 \, \Omega$, $R_3 = 5000 \, \Omega - 0.1\% = 4995 \, \Omega$, $R_2 = 5000 \, \Omega + 0.1\% = 5005 \, \Omega$, we get:

$$R_v = \frac{519.36 \times 4995}{5005} = 518.32 \, \Omega \, (= -0.4\%)$$

Thus, the error band for R_u is $\pm 0.4\%$.

The cumulative effect of errors in individual bridge circuit components is clearly seen. Although the maximum error in any one component is $\pm 0.2\%$, the possible error in the measured value of R_u is $\pm 0.4\%$. Such a magnitude of error is often not acceptable, and special measures are taken to overcome the introduction of error by component-value tolerances. One such practical measure is the introduction of apex balancing. This is one of many methods of bridge balancing that all produce a similar result.

Apex balancing

One form of apex balancing consists of placing an additional variable resistor R_5 at the junction C between the resistances R_2 and R_3, and applying the excitation voltage V_i to the wiper of this variable resistance, as shown in Figure 7.6.

For calibration purposes, R_u and R_v are replaced by two equal resistances whose values are accurately known, and R_5 is varied until the output voltage V_0 is zero. At this point, if the portions of resistance on either side of the wiper on R_5 are R_6 and R_7 (such that $R_5 = R_6 + R_7$), we can write:

$$R_3 + R_6 = R_2 + R_7$$

We have thus eliminated any source of error due to the tolerance in the value of R_2 and R_3, and the error in the measured value of R_u depends only on the accuracy of one component, the decade resistance box R_v.

Example 7.4
A potentiometer R_5 is put into the apex of the bridge shown in Figure 7.6 to balance the circuit. The bridge components have the following values:

$$R_u = 500\ \Omega,\ R_v = 500\ \Omega,\ R_2 = 515\ \Omega,\ R_3 = 480\ \Omega,\ R_5 = 100\ \Omega.$$

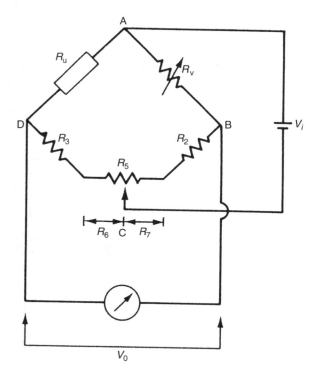

Fig. 7.6 Apex balancing.

Determine the required value of the resistances R_6 and R_7 of the parts of the potentiometer track either side of the slider in order to balance the bridge and compensate for the unequal values of R_2 and R_3.

Solution
For balance, $R_2 + R_7 = R_3 + R_6$; hence, $515 + R_7 = 480 + R_6$
Also, because R_6 and R_7 are the two parts of the potentiometer track R_5 whose resistance is 100 Ω:

$R_6 + R_7 = 100$;   thus $515 + R_7 = 480 + (100 - R_7)$;   i.e. $2R_7 = 580 - 515 = 65$
Thus, $R_7 = 32.5$; hence, $R_6 = 100 - 32.5 = 67.5\,\Omega$.

7.1.4 A.c. bridges

Bridges with a.c. excitation are used to measure unknown impedances. As for d.c. bridges, both null and deflection types exist, with null types being generally reserved for calibration duties.

Null-type impedance bridge

A typical null-type impedance bridge is shown in Figure 7.7. The null point can be conveniently detected by monitoring the output with a pair of headphones connected via an operational amplifier across the points BD. This is a much cheaper method of null detection than the application of an expensive galvanometer that is required for a d.c. Wheatstone bridge.

Referring to Figure 7.7, at the null point,

$$I_1 R_1 = I_2 R_2; \quad I_1 Z_u = I_2 Z_v$$

Thus:

$$Z_u = \frac{Z_v R_1}{R_2} \tag{7.11}$$

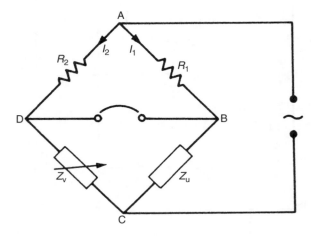

Fig. 7.7 Null-type impedance bridge.

If Z_u is capacitive, i.e. $Z_u = 1/j\omega C_u$, then Z_v must consist of a variable capacitance box, which is readily available. If Z_u is inductive, then $Z_u = R_u + j\omega L_u$.

Notice that the expression for Z_u as an inductive impedance has a resistive term in it because it is impossible to realize a pure inductor. An inductor coil always has a resistive component, though this is made as small as possible by designing the coil to have a high Q factor (Q factor is the ratio inductance/resistance). Therefore, Z_v must consist of a variable-resistance box and a variable-inductance box. However, the latter are not readily available because it is difficult and hence expensive to manufacture a set of fixed value inductors to make up a variable-inductance box. For this reason, an alternative kind of null-type bridge circuit, known as the *Maxwell bridge*, is commonly used to measure unknown inductances.

Maxwell bridge

A Maxwell bridge is shown in Figure 7.8. The requirement for a variable inductance box is avoided by introducing instead a second variable resistance. The circuit requires one standard fixed-value capacitor, two variable-resistance boxes and one standard fixed-value resistor, all of which are components that are readily available and inexpensive. Referring to Figure 7.8, we have at the null-output point:

$$I_1 Z_{AD} = I_2 Z_{AB}; \quad I_1 Z_{DC} = I_2 Z_{BC}$$

Thus:

$$\frac{Z_{BC}}{Z_{AB}} = \frac{Z_{DC}}{Z_{AD}} \quad \text{or} \quad Z_{BC} = \frac{Z_{DC} Z_{AB}}{Z_{AD}} \qquad (7.12)$$

The quantities in equation (7.12) have the following values:

$$\frac{1}{Z_{AD}} = \frac{1}{R_1} + j\omega C \quad \text{or} \quad Z_{AD} = \frac{R_1}{1 + j\omega C R_1}$$

$$Z_{AB} = R_3; \quad Z_{BC} = R_u + j\omega L_u; \quad Z_{DC} = R_2$$

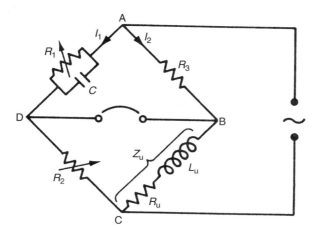

Fig. 7.8 Maxwell bridge.

Variable conversion elements

Substituting the values into equation (7.12):

$$R_u + j\omega L_u = \frac{R_2 R_3 (1 + j\omega C R_1)}{R_1}$$

Taking real and imaginary parts:

$$R_u = \frac{R_2 R_3}{R_1}; \quad L_u = R_2 R_3 C \tag{7.13}$$

This expression (7.13) can be used to calculate the quality factor (Q value) of the coil:

$$Q = \frac{\omega L_u}{R_u} = \frac{\omega R_2 R_3 C R_1}{R_2 R_3} = \omega C R_1$$

If a constant frequency ω is used:

$$Q \approx R_1$$

Thus, the Maxwell bridge can be used to measure the Q value of a coil directly using this relationship.

Example 7.5
In the Maxwell bridge shown in Figure 7.8, let the fixed-value bridge components have the following values: $R_3 = 5\,\Omega$; $C = 1\,\text{mF}$. Calculate the value of the unknown impedance (L_u, R_u) if $R_1 = 159\,\Omega$ and $R_2 = 10\,\Omega$ at balance.

Solution
Substituting values into the relations developed in equation (7.13) above:

$$R_u = \frac{R_2 R_3}{R_1} = \frac{10 \times 5}{159} = 0.3145\,\Omega; \quad L_u = R_2 R_3 C = \frac{10 \times 5}{1000} = 50\,\text{mH}$$

Example 7.6
Calculate the Q factor for the unknown impedance in example 7.5 above at a supply frequency of 50 Hz.

Solution

$$Q = \frac{\omega L_u}{R_u} = \frac{2\pi 50 (0.05)}{0.3145} = 49.9$$

Deflection-type a.c. bridge

A common deflection type of a.c. bridge circuit is shown in Figure 7.9.

For capacitance measurement:

$$Z_u = 1/j\omega C_u; \quad Z_1 = 1/j\omega C_1$$

For inductance measurement (making the simplification that the resistive component of the inductor is small and approximates to zero):

$$Z_u = j\omega L_u; \quad Z_1 = j\omega L_1$$

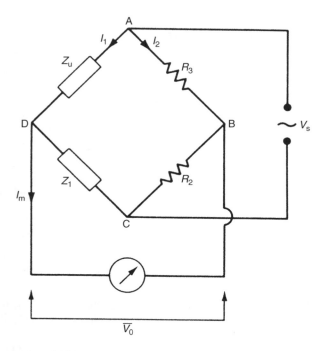

Fig. 7.9 Deflection-type a.c. bridge.

Analysis of the circuit to find the relationship between V_0 and Z_u is greatly simplified if one assumes that I_m is negligible. This is valid provided that the instrument measuring V_0 has a high impedance. For $I_m = 0$, currents in the two branches of the bridge, as defined in Figure 7.9, are given by:

$$I_1 = \frac{V_s}{Z_1 + Z_u}; \quad I_2 = \frac{V_s}{R_2 + R_3}$$

Also

$$V_{AD} = I_1 Z_u \quad \text{and} \quad V_{AB} = I_2 R_3$$

Hence:

$$V_0 = V_{BD} = V_{AD} - V_{AB} = V_s \left(\frac{Z_u}{Z_1 + Z_u} - \frac{R_3}{R_2 + R_3} \right)$$

Thus, for capacitances:

$$V_0 = V_s \left(\frac{1/C_u}{1/C_1 + 1/C_u} - \frac{R_3}{R_2 + R_3} \right) = V_s \left(\frac{C_1}{C_1 + C_u} - \frac{R_3}{R_2 + R_3} \right) \quad (7.14)$$

and for inductances:

$$V_0 = V_s \left(\frac{L_u}{L_1 + L_u} - \frac{R_3}{R_2 + R_3} \right) \quad (7.15)$$

Variable conversion elements

This latter relationship (7.15) is in practice only approximate since inductive impedances are never pure inductances as assumed but always contain a finite resistance (i.e. $Z_u = j\omega L_u + R$). However, the approximation is valid in many circumstances.

Example 7.7
A deflection bridge as shown in Figure 7.9 is used to measure an unknown capacitance, C_u. The components in the bridge have the following values:

$$V_s = 20\,V_{r.m.s.}, C_1 = 100\,\mu F, R_2 = 60\,\Omega, R_3 = 40\,\Omega$$

If $C_u = 100\,\mu F$, calculate the output voltage V_0.

Solution
From equation (7.14):

$$V_0 = V_s \left(\frac{C_1}{C_1 + C_u} - \frac{R_3}{R_2 + R_3} \right) = 20(0.5 - 0.4) = 2\,V_{r.m.s.}$$

Example 7.8
An unknown inductance L_u is measured using a deflection type of bridge as shown in Figure 7.9. The components in the bridge have the following values:

$$V_s = 10\,V_{r.m.s.}, L_1 = 20\,mH, R_2 = 100\,\Omega, R_3 = 100\,\Omega$$

If the output voltage V_0 is $1\,V_{r.m.s.}$, calculate the value of L_u.

Solution
From equation (7.15):

$$\frac{L_u}{L_1 + L_u} = \frac{V_0}{V_s} + \frac{R_3}{R_2 + R_3} = 0.1 + 0.5 = 0.6$$

Thus

$$L_u = 0.6(L_1 + L_u); \quad 0.4L_u = 0.6L_1; \quad L_u = \frac{0.6L_1}{0.4} = 30\,mH$$

7.2 Resistance measurement

Devices that convert the measured quantity into a change in resistance include the resistance thermometer, the thermistor, the wire-coil pressure gauge and the strain gauge. The standard devices and methods available for measuring change in resistance, which is measured in units of *ohms* (Ω), include the d.c. bridge circuit, the voltmeter–ammeter method, the resistance-substitution method, the digital voltmeter and the ohmmeter. Apart from the ohmmeter, these instruments are normally only used to measure medium values of resistance in the range of $1\,\Omega$ to $1\,M\Omega$. Special instruments are available for obtaining high-accuracy resistance measurements outside this range (see Baldwin (1973)).

7.2.1 D.c. bridge circuit

D.c. bridge circuits, as discussed earlier, provide the most commonly used method of measuring medium value resistance values. The best measurement accuracy is provided by the null-output-type Wheatstone bridge, and inaccuracy figures of less than ±0.02% are achievable with commercially available instruments. Deflection-type bridge circuits are simpler to use in practice than the null-output type, but their measurement accuracy is inferior and the non-linear output relationship is an additional difficulty. Bridge circuits are particularly useful in converting resistance changes into voltage signals that can be input directly into automatic control systems.

7.2.2 Voltmeter–ammeter method

The voltmeter–ammeter method consists of applying a measured d.c. voltage across the unknown resistance and measuring the current flowing. Two alternatives exist for connecting the two meters, as shown in Figure 7.10. In Figure 7.10(a), the ammeter measures the current flowing in both the voltmeter and the resistance. The error due to this is minimized when the measured resistance is small relative to the voltmeter resistance. In the alternative form of connection, Figure 7.10(b), the voltmeter measures the voltage drop across the unknown resistance and the ammeter. Here, the measurement error is minimized when the unknown resistance is large with respect to the ammeter resistance. Thus, method (a) is best for measurement of small resistances and method (b) for large ones.

Having thus measured the voltage and current, the value of the resistance is then calculated very simply by Ohm's law. This is a suitable method wherever the measurement inaccuracy of up to ±1% that it gives is acceptable.

7.2.3 Resistance-substitution method

In the voltmeter–ammeter method above, either the voltmeter is measuring the voltage across the ammeter as well as across the resistance, or the ammeter is measuring the current flow through the voltmeter as well as through the resistance. The measurement error caused by this is avoided in the resistance-substitution technique. In this method, the unknown resistance in a circuit is temporarily replaced by a variable resistance. The variable resistance is adjusted until the measured circuit voltage and current are the same as existed with the unknown resistance in place. The variable resistance at this point is equal in value to the unknown resistance.

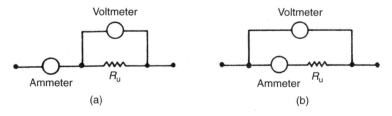

Fig. 7.10 Voltmeter–ammeter method of measuring resistance.

7.2.4 Use of the digital voltmeter to measure resistance

The digital voltmeter can also be used for measuring resistance if an accurate current source is included within it that passes current through the resistance. This can give a measurement inaccuracy as small as ±0.1%.

7.2.5 The ohmmeter

The ohmmeter is a simple instrument in which a battery applies a known voltage across a combination of the unknown resistance and a known resistance in series, as shown in Figure 7.11. Measurement of the voltage, V_m, across the known resistance, R, allows the unknown resistance, R_u, to be calculated from:

$$R_u = \frac{R(V_b - V_m)}{V_m}$$

where V_b is the battery voltage.

Ohmmeters are used to measure resistances over a wide range from a few milliohms up to 50 MΩ. The measurement inaccuracy is ±2% or greater, and ohmmeters are therefore more suitable for use as test equipment rather than in applications where high accuracy is required. Most of the available versions contain a switchable set of standard resistances, so that measurements of reasonable accuracy over a number of ranges can be made.

Most *digital and analogue multimeters* contain circuitry of the same form as in an ohmmeter, and hence can be similarly used to obtain approximate measurements of resistance.

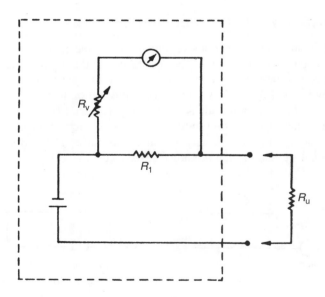

Fig. 7.11 Ohmmeter.

7.2.6 Codes for resistor values

When standard resistors are being used as part of bridge circuits, and also in other applications, it is often useful to know their approximate value. To satisfy this need, coded marks are made on resistors during manufacture. The two main styles of marking are a four-band colour system and an alphanumeric code.

In the *four-band coding system*, the resistance value and the maximum possible tolerance about that value are defined by a set of four coloured bands. These are displaced towards one end of the resistor, as shown in Figure 7.12, with band one defined as the band that is closest to the end of the resistor.

Alphanumeric coding indicates the resistance value using two, three or four numbers plus one letter. The letter acts both as a decimal point and also as a multiplier for the value specified by the numbers in the code. The letters R, K, M, G, T define multipliers of $\times 1$, $\times 10^3$, $\times 10^6$, $\times 10^9$, $\times 10^{12}$ respectively. For example: 6M8 means 6.8×10^6, i.e. $6.8\,M\Omega$. 50R04 means $59.04\,\Omega$. A separate letter indicating the tolerance is given after the value coding. The meaning of tolerance codes is as follows:

$B = \pm 0.1\%$; $C = \pm 0.25\%$; $D = \pm 0.5\%$; $F = \pm 1\%$; $G = \pm 2\%$;
$J = \pm 5\%$; $K = \pm 10\%$; $M = \pm 20\%$; $N = \pm 30\%$

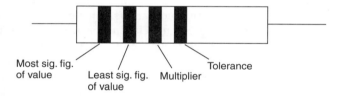

Most sig. fig. of value / Least sig. fig. of value / Multiplier / Tolerance

Codes for bands 1–3				Code for band 4	
Black	0	Green	5	Brown	∓1%
Brown	1	Blue	6	Red	∓2%
Red	2	Purple	7	Gold	∓5%
Orange	3	Grey	8	Silver	∓10%
Yellow	4	White	9		

Example

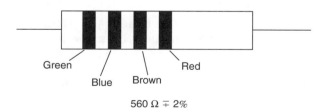

Green Blue Brown Red

$560\,\Omega \mp 2\%$

Fig. 7.12 Four-band resistance-value marking system with example.

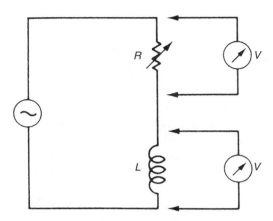

Fig. 7.13 Approximate method of measuring inductance.

7.3 Inductance measurement

The main device that has an output in the form of a change in inductance is the inductive displacement sensor. Inductance is measured in *henry* (H). It can only be measured accurately by an a.c. bridge circuit, and various commercial inductance bridges are available. However, when such a commercial inductance bridge is not immediately available, the following method can be applied to give an approximate measurement of inductance.

This approximate method consists of connecting the unknown inductance in series with a variable resistance, in a circuit excited with a sinusoidal voltage, as shown in Figure 7.13. The variable resistance is adjusted until the voltage measured across the resistance is equal to that measured across the inductance. The two impedances are then equal, and the value of the inductance L can be calculated from:

$$L = \frac{\sqrt{(R^2 - r^2)}}{2\pi f}$$

where R is the value of the variable resistance, r is the value of the inductor resistance and f is the excitation frequency.

7.4 Capacitance measurement

Devices that have an output in the form of a change in capacitance include the capacitive level gauge, the capacitive displacement sensor, the capacitive moisture meter and the capacitive hygrometer. Capacitance is measured in units of *Farads* (F). Like inductance, capacitance can only be measured accurately by an a.c. bridge circuit, and various types of capacitance bridge are available commercially. In circumstances where a proper capacitance bridge is not immediately available, and if an approximate measurement of capacitance is acceptable, one of the following two methods can be considered.

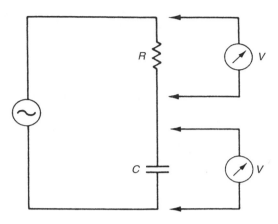

Fig. 7.14 Approximate method of measuring capacitance.

The first of these, shown in Figure 7.14, consists of connecting the unknown capacitor in series with a known resistance in a circuit excited at a known frequency. An a.c. voltmeter is used to measure the voltage drop across both the resistor and the capacitor. The capacitance value is then given by:

$$C = \frac{V_r}{2\pi f R V_c}$$

where V_r and V_c are the voltages measured across the resistance and capacitance respectively, f is the excitation frequency and R is the known resistance.

An alternative approximate method of measurement is to measure the time constant of the capacitor connected in an RC circuit.

7.4.1 Alphanumeric codes for capacitor values

As for resistors, it is often useful to know the approximate value of capacitors used in an a.c. bridge circuit and also in other applications. This need is satisfied by putting an alphanumeric code on capacitors during manufacture. The code consists of one letter and several numbers. The letter acts both as a decimal point and also as a multiplier for the value specified by the numbers in the code. The letters p, n, µ, m and F define multipliers of $\times 10^{-12}$, $\times 10^{-9}$, $\times 10^{-6}$, $\times 10^{-3}$, $\times 1$ respectively.

For example, p10 means 0.1 pF, 333p means 333 pF and 15n means 15 nF.

A separate letter indicating the tolerance is given after the value coding. The meaning of tolerance codes is as follows:

$B = \pm 0.1\%$ $C = \pm 0.25\%$ $D = \pm 0.5\%$ $F = \pm 1\%$ $G = \pm 2\%$
$J = \pm 5\%$ $K = \pm 10\%$ $M = \pm 20\%$ $N = \pm 30\%$

Hence, 333pK means a 333 pF capacitor with a tolerance of $\pm 10\%$.

7.5 Current measurement

Current measurement is needed for devices like the thermocouple-gauge pressure sensor and the ionization gauge that have an output in the form of a varying electrical current. It is often also needed in signal transmission systems that convert the measured signal into a varying current. Any of the digital and analogue voltmeters discussed in the last chapter can measure current if the meter is placed in series with the current-carrying circuit, and the same frequency limits apply for the measured signal as they do for voltage measurement. The upper frequency limit for a.c. current measurement can be raised by rectifying the current prior to measurement or by using a thermocouple meter. To minimize the loading effect on the measured system, any current-measuring instrument must have a small resistance. This is opposite to the case of voltage measurement where the instrument is required to have a high resistance for minimal circuit loading.

Besides the requirement to measure signal-level currents, many measurement applications also require higher-magnitude electrical currents to be measured. Hence, the following discussion covers the measurement of currents at both signal level and higher magnitudes.

For d.c. current measurement, moving-coil meters can measure in the milliamp range up to 1 ampere, dynamometer ammeters can measure up to several amps and moving-iron meters can measure up to several hundred amps directly. Similar measurement ranges apply when moving-iron and dynamometer-type instruments are used to measure a.c. currents.

To measure larger currents with electromechanical meters, it is necessary to insert a shunt resistance into the circuit and measure the voltage drop across it. Apart from the obvious disturbance of the measured system, one particular difficulty that results from this technique is the large power dissipation in the shunt. In the case of a.c. current measurement, care must also be taken to match the resistance and reactance of the shunt to that of the measuring instrument so that frequency and waveform distortion in the measured signal are avoided.

Current transformers provide an alternative method of measuring high-magnitude currents that avoids the difficulty of designing a suitable shunt. Different versions of these exist for transforming both d.c. and a.c. currents. A d.c. current transformer is shown in Figure 7.15. The central d.c. conductor in the instrument is threaded through two magnetic cores that carry two high impedance windings connected in series opposition. It can be shown (Baldwin, 1973) that the current flowing in the windings when excited with an a.c. voltage is proportional to the d.c. current in the central conductor. This output current is commonly rectified and then measured by a moving-coil instrument.

An a.c. current transformer typically has a primary winding consisting of only a few copper turns wound on a rectangular or ring-shaped core. The secondary winding on the other hand would normally have several hundred turns according to the current step-down ratio required. The output of the secondary winding is measured by any suitable current-measuring instrument. The design of current transformers is substantially different from that of voltage transformers. The rigidity of its mechanical construction has to be sufficient to withstand the large forces arising from short-circuit currents, and special attention has to be paid to the insulation between its

Measurement and Instrumentation Principles 141

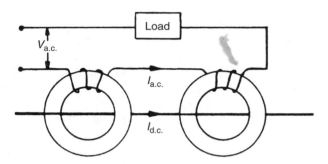

Fig. 7.15 Current transformer.

windings for similar reasons. A low-loss core material is used and flux densities are kept as small as possible to reduce losses. In the case of very high currents, the primary winding often consists of a single copper bar that behaves as a single-turn winding. The clamp-on meter, described in the last chapter, is a good example of this.

Apart from electromechanical meters, all the other instruments for measuring voltage discussed in Chapter 6 can be applied to current measurement by using them to measure the voltage drop across a known resistance placed in series with the current-carrying circuit. The digital voltmeter and electronic meters are widely applied for measuring currents accurately by this method, and the cathode ray oscilloscope is frequently used to obtain approximate measurements in circuit-test applications. Finally, mention must also be made of the use of digital and analogue multimeters for current measurement, particularly in circuit-test applications. These instruments include a set of switchable dropping resistors and so can measure currents over a wide range. Protective circuitry within such instruments prevents damage when high currents are applied on the wrong input range.

7.6 Frequency measurement

Frequency measurement is required as part of those devices that convert the measured physical quantity into a frequency change, such as the variable-reluctance velocity transducer, stroboscopes, the vibrating-wire force sensor, the resonant-wire pressure sensor, the turbine flowmeter, the Doppler-shift ultrasonic flowmeter, the transit-time ultrasonic flowmeter, the vibrating level sensor, the quartz moisture meter and the quartz thermometer. In addition, the output relationship in some forms of a.c. bridge circuit used for measuring inductance and capacitance requires accurate measurement of the bridge excitation frequency.

Frequency is measured in units of *hertz* (Hz). The digital counter-timer is the most common instrument for measuring frequency. The oscilloscope is also commonly used for obtaining approximate measurements of frequency, especially in circuit test and fault-diagnosis applications. Within the audio frequency range, the Wien bridge is a further instrument that is sometimes used.

7.6.1 Digital counter-timers

A digital counter-timer is the most accurate and flexible instrument available for measuring frequency. Inaccuracy can be reduced down to 1 part in 10^8, and all frequencies between d.c. and several gigahertz can be measured. The essential component within a counter-timer instrument is an oscillator that provides a very accurately known and stable reference frequency, which is typically either 100 kHz or 1 MHz. This is often maintained in a temperature-regulated environment within the instrument to guarantee its accuracy. The oscillator output is transformed by a pulse-shaper circuit into a train of pulses and applied to an electronic gate, as shown in Figure 7.16. Successive pulses at the reference frequency alternately open and close the gate. The input signal of unknown frequency is similarly transformed into a train of pulses and applied to the gate. The number of these pulses that get through the gate during the time that it is open during each gate cycle is proportional to the frequency of the unknown signal.

The accuracy of measurement obviously depends upon how far the unknown frequency is above the reference frequency. As it stands therefore, the instrument can only accurately measure frequencies that are substantially above 1 MHz. To enable the instrument to measure much lower frequencies, a series of decade frequency dividers are provided within it. These increase the time between the reference frequency pulses by factors of ten, and a typical instrument can have gate pulses separated in time by between 1 µs and 1 second.

Improvement in the accuracy of low-frequency measurement can be obtained by modifying the gating arrangements such that the signal of unknown frequency is made to control the opening and closing of the gate. The number of pulses at the reference frequency that pass through the gate during the open period is then a measure of the frequency of the unknown signal.

7.6.2 Phase-locked loop

A phase-locked loop is a circuit consisting of a phase-sensitive detector, a voltage controlled oscillator (VCO), and amplifiers, connected in a closed-loop system as shown in Figure 7.17. In a VCO, the oscillation frequency is proportional to the applied voltage. Operation of a phase-locked loop is as follows. The phase-sensitive

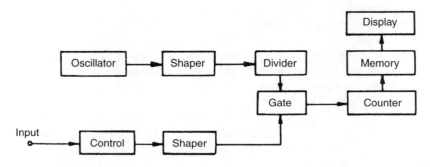

Fig. 7.16 Digital counter-timer system.

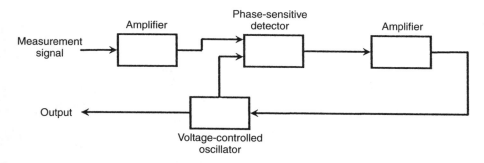

Fig. 7.17 Phase-locked loop.

detector compares the phase of the amplified input signal with the phase of the VCO output. Any phase difference generates an error signal, which is amplified and fed back to the VCO. This adjusts the frequency of the VCO until the error signal goes to zero, and thus the VCO becomes locked to the frequency of the input signal. The d.c. output from the VCO is then proportional to the input signal frequency.

7.6.3 Cathode ray oscilloscope

The cathode ray oscilloscope can be used in two ways to measure frequency. Firstly, the internal timebase can be adjusted until the distance between two successive cycles of the measured signal can be read against the calibrated graticule on the screen. Measurement accuracy by this method is limited, but can be optimized by measuring between points in the cycle where the slope of the waveform is steep, generally where it is crossing through from the negative to the positive part of the cycle. Calculation of the unknown frequency from this measured time interval is relatively simple. For example, suppose that the distance between two cycles is 2.5 divisions when the internal timebase is set at 10 ms/div. The cycle time is therefore 25 ms and hence the frequency is 1000/25, i.e. 40 Hz. Measurement accuracy is dependent upon how accurately the distance between two cycles is read, and it is very difficult to reduce the error level below ±5% of the reading.

The alternative way of using an oscilloscope to measure frequency is to generate *Lisajous patterns*. These are produced by applying a known reference-frequency sine wave to the y input (vertical deflection plates) of the oscilloscope and the unknown-frequency sinusoidal signal to the x input (horizontal deflection plates). A pattern is produced on the screen according to the frequency ratio between the two signals, and if the numerator and denominator in the ratio of the two signals both represent an integral number of cycles, the pattern is stationary. Examples of these patterns are shown in Figure 7.18, which also shows that phase difference between the waveforms has an effect on the shape. Frequency measurement proceeds by adjusting the reference frequency until a steady pattern is obtained on the screen and then calculating the unknown frequency according to the frequency ratio that the pattern obtained represents.

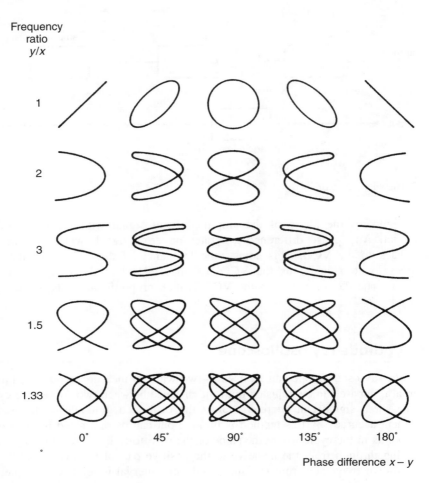

Fig. 7.18 Lisajous patterns.

7.6.4 The Wien bridge

The Wien bridge, shown in Figure 7.19, is a special form of a.c. bridge circuit that can be used to measure frequencies in the audio range. An alternative use of the instrument is as a source of audio frequency signals of accurately known frequency. A simple set of headphones is often used to detect the null-output balance condition. Other suitable instruments for this purpose are the oscilloscope and the electronic voltmeter. At balance, the unknown frequency is calculated according to:

$$f = \frac{1}{2\pi R_3 C_3}$$

The instrument is very accurate at audio frequencies, but at higher frequencies errors due to losses in the capacitors and stray capacitance effects become significant.

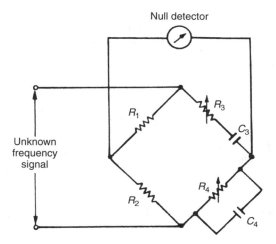

Fig. 7.19 Wien bridge.

7.7 Phase measurement

Instruments that convert the measured variable into a phase change in a sinusoidal electrical signal include the transit-time ultrasonic flowmeter, the radar level sensor, the LVDT and the resolver. The most accurate instrument for measuring the phase difference between two signals is the electronic counter-timer. However, two other methods also exist that are less accurate but are nevertheless very useful in some circumstances. One method involves plotting the signals on an X–Y plotter and the other uses a dual beam oscilloscope.

7.7.1 Electronic counter-timer

In principle, the phase difference between two sinusoidal signals can be determined by measuring the time that elapses between the two signals crossing the time axis. However, in practice, this is inaccurate because the zero crossings are susceptible to noise contamination. The normal solution to this problem is to amplify/attenuate the two signals so that they have the same amplitude and then measure the time that elapses between the two signals crossing some non-zero threshold value.

The basis of this method of phase measurement is a digital counter-timer with a quartz-controlled oscillator providing a frequency standard that is typically 10 MHz. The crossing points of the two signals through the reference threshold voltage level are applied to a gate that starts and then stops pulses from the oscillator into an electronic counter, as shown in Figure 7.20. The elapsed time, and hence phase difference, between the two input signals is then measured in terms of the counter display.

7.7.2 X–Y plotter

This is a useful technique for approximate phase measurement but is limited to low frequencies because of the very limited bandwidth of an X–Y plotter. If two input

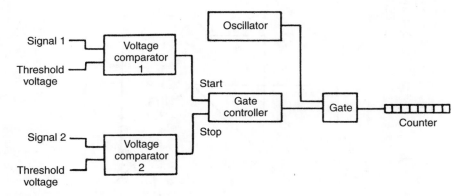

Fig. 7.20 Phase measurement with digital counter-timer.

signals of equal magnitude are applied to the X and Y inputs of a plotter, the plot obtained is an ellipse, as shown in Figure 7.21. If the X and Y inputs are given by:

$$V_X = V \sin(\omega t); \quad V_Y = V \sin(\omega t + \phi)$$

At $t = 0$, $V_X = 0$ and $V_Y = V \sin \phi$. Thus, from Figure 7.21, for $V_X = 0$, $V_Y = \pm h$:

$$\sin \phi = \pm h/V \tag{7.16}$$

Solution of equation (7.4) gives four possible values for ϕ but the ambiguity about which quadrant ϕ is in can usually be solved by observing the two signals plotted against time on a dual-beam oscilloscope.

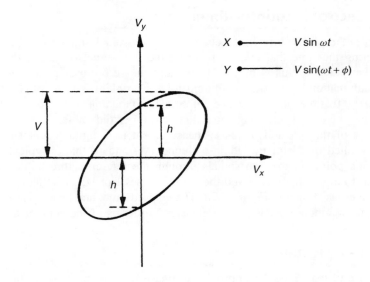

Fig. 7.21 Phase measurement using X–Y plotter.

7.7.3 Oscilloscope

Approximate measurement of the phase difference between signals can be made using a dual-beam oscilloscope. The two signals are applied to the two oscilloscope inputs and a suitable timebase chosen such that the time between the crossing points of the two signals can be measured. The phase difference of both low- and high-frequency signals can be measured by this method, the upper frequency limit measurable being dictated by the bandwidth of the oscilloscope (which is normally very high).

7.7.4 Phase-sensitive detector

The phase-sensitive detector described earlier in section 5.5.9 can be used to measure the phase difference between two signals that have an identical frequency. This can be exploited in measurement devices like the varying-phase output resolver (see Chapter 20).

7.8 Self-test questions

7.1 If the elements in the d.c. bridge circuit shown in Figure 7.2 have the following values: $R_u = 110\,\Omega$, $R_1 = 100\,\Omega$, $R_2 = 1000\,\Omega$, $R_3 = 1000\,\Omega$, $V_i = 10\,\text{V}$, calculate the output voltage V_0 if the impedance of the voltage-measuring instrument is assumed to be infinite.

7.2 Suppose that the resistive components in the d.c. bridge shown in Figure 7.2 have the following nominal values: $R_u = 3\,\text{k}\Omega$; $R_1 = 6\,\text{k}\Omega$; $R_2 = 8\,\text{k}\Omega$; $R_3 = 4\,\text{k}\Omega$. The actual value of each resistance is related to the nominal value according to $R_{\text{actual}} = R_{\text{nominal}} + \partial R$ where ∂R has the following values: $\partial R_u = 30\,\Omega$; $\partial R_1 = -20\,\Omega$; $\partial R_2 = 40\,\Omega$; $\partial R_3 = -50\,\Omega$. Calculate the open-circuit bridge output voltage if the bridge supply voltage V_i is 50 V.

7.3 (a) Suppose that the unknown resistance R_u in Figure 7.2 is a resistance thermometer whose resistance at 100°C is 500 Ω and whose resistance varies with temperature at the rate of 0.5 Ω/°C for small temperature changes around 100°C. Calculate the sensitivity of the total measurement system for small changes in temperature around 100°C, given the following resistance and voltage values measured at 15°C by instruments calibrated at 15°C : $R_1 = 500\,\Omega$; $R_2 = R_3 = 5000\,\Omega$; $V_i = 10\,\text{V}$.

(b) If the resistance thermometer is measuring a fluid whose true temperature is 104°C, calculate the error in the indicated temperature if the ambient temperature around the bridge circuit is 20°C instead of the calibration temperature of 15°C, given the following additional information:

Voltage-measuring instrument zero drift coefficient $= +1.3\,\text{mV}/°\text{C}$

Voltage-measuring instrument sensitivity drift coefficient $= 0$

Resistances R_1, R_2 and R_3 have a positive temperature coefficient of $+0.2\%$ of nominal value/°C

Voltage source V_i is unaffected by temperature changes.

148 Variable conversion elements

7.4 Four strain gauges of resistance $120\,\Omega$ each are arranged into a d.c. bridge configuration such that each of the four arms in the bridge has one strain gauge in it. The maximum permissible current in each strain gauge is 100 mA. What is the maximum bridge supply voltage allowable, and what power is dissipated in each strain gauge with that supply voltage?

7.5 (a) Suppose that the variables shown in Figure 7.2 have the following values: $R_1 = 100\,\Omega$, $R_2 = 100\,\Omega$, $R_3 = 100\,\Omega$; $V_i = 12\,\text{V}$. R_u is a resistance thermometer with a resistance of $100\,\Omega$ at $100°C$ and a temperature coefficient of $+0.3\,\Omega/°C$ over the temperature range from $50°C$ to $150°C$ (i.e. the resistance increases as the temperature goes up). Draw a graph of bridge output voltage V_0 for ten-degree steps in temperature between $100°C$ and $150°C$ (calculating V_0 according to equation 7.3).

(b) Draw a graph of V_0 for similar temperature values if $R_2 = R_3 = 1000\,\Omega$ and all other components have the same values as given in part (a) above. Notice that the line through the data points is straighter than that drawn in part (a) but the output voltage is much less at each temperature point.

7.6 The unknown resistance R_u in a d.c. bridge circuit, connected as shown in Figure 7.4(a), is a resistance thermometer. The thermometer has a resistance of $350\,\Omega$ at $50°C$ and its temperature coefficient is $+1\,\Omega/°C$ (the resistance increases as the temperature rises). The components of the system have the following values: $R_1 = 350\,\Omega$, $R_2 = R_3 = 2\,\text{k}\Omega$, $R_m = 20\,\text{k}\Omega$, $V_i = 5\,\text{V}$. What is the output voltage reading when the temperature is $100°C$? (Hint–use equation 7.10.)

7.7 In the d.c. bridge circuit shown in Figure 7.22, the resistive components have the following values: $R_1 = R_2 = 120\,\Omega$; $R_3 = 117\,\Omega$; $R_4 = 123\,\Omega$; $R_A = R_P = 1000\,\Omega$.

(a) What are the resistance values of the parts of the potentiometer track either side of the slider when the potentiometer is adjusted to balance the bridge?

(b) What then is the effective resistance of each of the two left-hand arms of the bridge when the bridge is balanced?

7.8 A Maxwell bridge, designed to measure the unknown impedance (R_u, L_u) of a coil, is shown in Figure 7.8.

(a) Derive an expression for R_u and L_u under balance conditions.

(b) If the fixed bridge component values are $R_3 = 100\,\Omega$ and $C = 20\,\mu\text{F}$, calculate the value of the unknown impedance if $R_1 = 3183\,\Omega$ and $R_2 = 50\,\Omega$ at balance.

(c) Calculate the Q factor for the coil if the supply frequency is 50 Hz.

7.9 The deflection bridge shown in Figure 7.9 is used to measure an unknown inductance L_u. The components in the bridge have the following values: $V_s = 30\,\text{V}_{\text{r.m.s.}}$, $L_1 = 80\,\text{mH}$, $R_2 = 70\,\Omega$, $R_3 = 30\,\Omega$. If $L_u = 50\,\text{mH}$, calculate the output voltage V_0.

7.10 An unknown capacitance C_u is measured using a deflection bridge as shown in Figure 7.9. The components of the bridge have the following values: $V_s = 10\,\text{V}_{\text{r.m.s.}}$, $C_1 = 50\,\mu\text{F}$, $R_2 = 80\,\Omega$, $R_3 = 20\,\Omega$. If the output voltage is $3\,\text{V}_{\text{r.m.s.}}$, calculate the value of C_u.

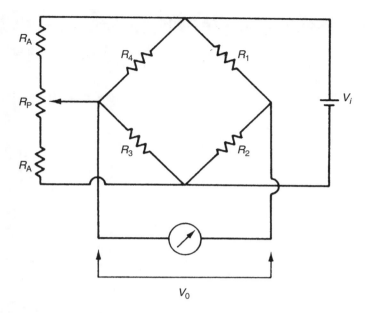

Fig. 7.22 D.c. bridge with apex balancing.

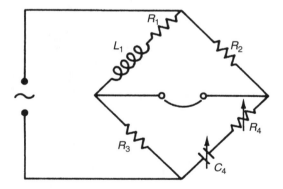

Fig. 7.23 Hays bridge.

7.11 A Hays bridge is often used for measuring the inductance of high-Q coils and has the configuration shown in Figure 7.23.
   (a) Obtain the bridge balance conditions.
   (b) Show that if the Q value of an unknown inductor is high, the expression for the inductance value when the bridge is balanced is independent of frequency.
   (c) If the Q value is high, calculate the value of the inductor if the bridge component values at balance are as follows: $R_2 = R_3 = 1000\,\Omega$; $C = 0.02\,\mu F$.

Variable conversion elements

References and further reading

Baldwin, C.T. (1973) *Fundamentals of Electrical Measurements*, Prentice-Hall, London.
Buckingham, H. and Price, E.M. (1966) *Principles of Electrical Measurements*, English Universities Press, London.
Smith, R.J. and Dorf, R.C. (1992) *Circuits, Devices and Systems*, Wiley, New York.

8

Signal transmission

There is a necessity in many measurement systems to transmit measurement signals over quite large distances from the point of measurement to the place where the signals are recorded and/or used in a process control system. This creates several problems for which a solution must be found. Of the many difficulties associated with long distance signal transmission, contamination of the measurement signal by noise is the most serious. Many sources of noise exist in industrial environments, such as radiated electromagnetic fields from electrical machinery and power cables, induced fields through wiring loops, and spikes (large transient voltages) on the a.c. power supply. Signals can be transmitted electrically, pneumatically, optically, or by radio-telemetry, in either analogue or digital format. Optical data transmission can be further divided into fibre-optic transmission and optical wireless transmission, according to whether a fibre-optic cable or just a plain air path is used as the transmission medium. These various options are explored in the following sections.

8.1 Electrical transmission

The simplest method of electrical transmission is to transmit the measurement signal as a varying analogue voltage. However, this can cause the measurement signal to become corrupted by noise. If noise causes a problem, the signal can either be transmitted in the form of a varying current, or else it can be superimposed on an a.c. carrier system.

8.1.1 Transmission as varying voltages

As most signals already exist in an electrical form as varying analogue voltages, the simplest mode of transmission is to maintain the signals in the same form. However, electrical transmission suffers problems of signal attenuation, and also exposes signals to corruption through induced noise. Therefore, special measures have to be taken to overcome these problems.

Because the output signal levels from many types of measurement transducer are very low, *signal amplification* prior to transmission is essential if a reasonable signal-to-noise ratio is to be obtained after transmission. Amplification at the input to the transmission system is also required to compensate for the attenuation of the signal that results from the resistance of the signal wires. The means of amplifying signals have already been discussed in section 5.1.

It is also usually necessary to provide *shielding* for the signal wires. Shielding consists of surrounding the signal wires in a cable with a metal shield that is connected to earth. This provides a high degree of noise protection, especially against capacitive-induced noise due to the proximity of signal wires to high-current power conductors. A fuller discussion on noise sources and the procedures followed to prevent the corruption of measurement voltage signals can be found in Chapter 5.

8.1.2 Current loop transmission

The signal-attenuation effect of conductor resistances can be minimized if varying voltage signals are transmitted as varying current signals. This technique, which also provides high immunity to induced noise, is known as current loop transmission and uses currents in the range between 4 mA and 20 mA* to represent the voltage level of the analogue signal. It requires a voltage-to-current converter of the form shown in Figure 8.1, which is commonly known as a *4–20 mA current loop interface*. Two voltage-controlled current sources are used, one providing a constant 4 mA output that is used as the power supply current and the other providing a variable 0–16 mA output that is scaled and proportional to the input voltage level. The net output current therefore varies between 4 mA and 20 mA, corresponding to analogue signal levels between zero and the maximum value. The use of a positive, non-zero current level to represent a zero value of the transmitted signal enables transmission faults to be readily identified. If the transmitted current is zero, this automatically indicates the presence of a transmission fault, since the minimum value of current that represents a proper signal is 4 mA.

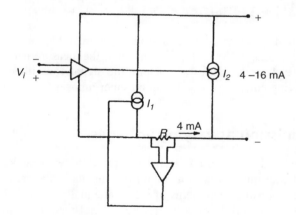

Fig. 8.1 Voltage-to-current converter (current loop interface).

---

* The 4–20 mA standard was agreed in 1972, prior to which a variety of different current ranges were used for signal transmission.

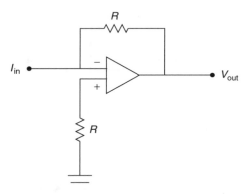

Fig. 8.2 Current-to-voltage converter.

Current-to-voltage conversion is usually required at the termination of the transmission line to change the transmitted currents back to voltages. An operational amplifier, connected as shown in Figure 8.2, is suitable for this purpose. The output voltage V is simply related to the input current I by $V = IR$.

8.1.3 Transmission using an a.c. carrier

Another solution to the problem of noise corruption in low level d.c. voltage signals is to transfer the signal onto an a.c. carrier system before transmission and extract it from the carrier at the end of the transmission line. Both amplitude modulation (AM) and frequency modulation (FM) can be used for this.

AM consists of translating the varying voltage signal into variations in the amplitude of a carrier sine wave at a frequency of several kHz. An a.c. bridge circuit is commonly used for this, as part of the system for transducing the outputs of sensors that have a varying resistance (R), capacitance (C) or inductance (L) form of output. Referring back to equations (7.14), and (7.15) in Chapter 7, for a sinusoidal bridge excitation voltage of $V_s = V \sin(\omega t)$, the output can be represented by $V_0 = FV \sin(\omega t)$. V_0 is a sinusoidal voltage at the same frequency as the bridge excitation frequency and its amplitude FV represents the magnitude of the sensor input (R, C or L) to the bridge. For example, in the case of equation (6.15):

$$FV = \left(\frac{L_u}{L_1 + L_u} - \frac{R_3}{R_2 + R_3} \right) V$$

After shifting the d.c. signal onto a high-frequency a.c. carrier, a high-pass filter can be applied to the AM signal. This successfully rejects noise in the form of low-frequency drift voltages and mains interference. At the end of the transmission line, demodulation is carried out to extract the measurement signal from the carrier.

FM achieves even better noise rejection than AM and involves translating variations in an analogue voltage signal into frequency variations in a high-frequency carrier signal. A suitable voltage-to-frequency conversion circuit is shown in Figure 8.3, in which the analogue voltage signal input is integrated and applied to the input of a

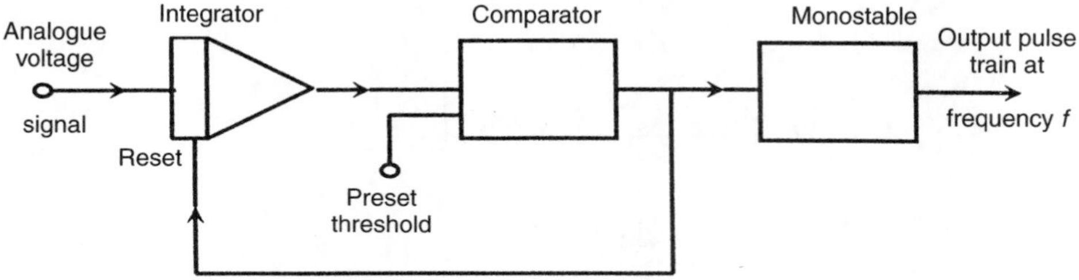

154 Signal transmission

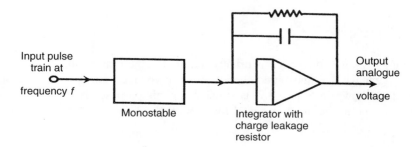

Fig. 8.3 Voltage-to-frequency converter.

Fig. 8.4 Frequency-to-voltage converter.

comparator that is preset to a certain threshold voltage level. When this threshold level is reached, the comparator generates an output pulse that resets the integrator and is also applied to a monostable. This causes the frequency f of the output pulse train to be proportional to the amplitude of the input analogue voltage.

At the end of the transmission line, the FM signal is usually converted back to an analogue voltage by a frequency-to-voltage converter. A suitable conversion circuit is shown in Figure 8.4, in which the input pulse train is applied to an integrator that charges up for a specified time. The charge on the integrator decays through a leakage resistor, and a balance voltage is established between the input charge on the integrator and the decaying charge at the output. This output balance voltage is proportional to the input pulse train at frequency f.

8.2 Pneumatic transmission

In recent years, pneumatic transmission tends to have been replaced by other alternatives in most new implementations of instrumentation systems, although many examples can still be found in operation in the process industries. Pneumatic transmission consists of transmitting analogue signals as a varying pneumatic pressure level that is usually in the range of 3–15 p.s.i. (Imperial units are still commonly used in process industries, though the equivalent range in SI units is 207–1034 mbar, which is often rounded to 200–1000 mbar in metric systems). A few systems also use alternative ranges of 3–27 p.s.i. or 6–48 p.s.i. Frequently, the initial signal is in the form of a varying voltage level that is converted into a corresponding pneumatic pressure.

However, in some examples of pneumatic transmission, the signal is in varying current form to start with, and a current to pressure converter is used to convert the 4–20 mA current signals into pneumatic signals prior to transmission. Pneumatic transmission has the advantage of being intrinsically safe, and provides similar levels of noise immunity to current loop transmission. However, one disadvantage of using air as the transmission medium is that transmission speed is much less than electrical or optical transmission. A further potential source of error would arise if there were a pressure gradient along the transmission tube. This would introduce a measurement error because air pressure changes with temperature.

Pneumatic transmission is found particularly in pneumatic control systems where sensors or actuators or both are pneumatic. Typical pneumatic sensors are the pressure thermometer (see Chapter 14) and the motion-sensing nozzle-flapper (see Chapter 19), and a typical actuator is a pneumatic cylinder that converts pressure into linear motion. A pneumatic amplifier is often used to amplify the pneumatic signal to a suitable level for transmission.

8.3 Fibre-optic transmission

Light has a number of advantages over electricity as a medium for transmitting information. For example, it is intrinsically safe, and noise corruption of signals by neighbouring electromagnetic fields is almost eliminated. The most common form of optical transmission consists of transmitting light along a fibre-optic cable, although wireless transmission also exists as described in section 8.4.

Apart from noise reduction, optical signal attenuation along a fibre-optic link is much less than electric signal attenuation along an equivalent length of metal conductor. However, there is an associated cost penalty because of the higher cost of a fibre-optic system compared with the cost of metal conductors. In short fibre-optic links, cost is dominated by the terminating transducers that are needed to transform electrical signals into optical ones and vice versa. However, as the length of the link increases, the cost of the fibre-optic cable itself becomes more significant.

Fibre-optic cables are used for signal transmission in three distinct ways. Firstly, relatively short fibre-optic cables are used as part of various instruments to transmit light from conventional sensors to a more convenient location for processing, often in situations where space is very short at the point of measurement. Secondly, longer fibre-optic cables are used to connect remote instruments to controllers in instrumentation networks. Thirdly, even longer links are used for data transmission systems in telephone and computer networks. These three application classes have different requirements and tend to use different types of fibre-optic cable.

Signals are normally transmitted along a fibre-optic cable in digital format, although analogue transmission is sometimes used. If there is a requirement to transmit more than one signal, it is more economical to multiplex the signals onto a single cable rather than transmit the signals separately on multiple cables. *Multiplexing* involves switching the analogue signals in turn, in a synchronized sequential manner, into an analogue-to-digital converter that outputs onto the transmission line. At the other end of the transmission line, a digital-to-analogue converter transforms the digital signal back into analogue form and it is then switched in turn onto separate analogue signal lines.

8.3.1 Principles of fibre optics

The central part of a fibre optic system is a light transmitting cable containing at least one, but more often a bundle, of glass or plastic fibres. This is terminated at each end by a transducer, as shown in Figure 8.5. At the input end, the transducer converts the signal from the electrical form in which most signals originate into light. At the output end, the transducer converts the transmitted light back into an electrical form suitable for use by data recording, manipulation and display systems. These two transducers are often known as the transmitter and receiver respectively.

Fibre-optic cable consists of an inner cylindrical core surrounded by an outer cylindrical cladding, as shown in Figure 8.6. The refractive index of the inner material is greater than that of the outer material, and the relationship between the two refractive indices affects the transmission characteristics of light along the cable. The amount of attenuation of light as it is travels along the cable varies with the wavelength of the light transmitted. This characteristic is very non-linear and a graph of attenuation against wavelength shows a number of peaks and troughs. The position of these peaks and troughs varies according to the material used for the fibres. It should be noted that fibre manufacturers rarely mention these non-linear attenuation characteristics and quote the value of attenuation that occurs at the most favourable wavelength.

Two forms of cable exist, known as monomode and multimode. Monomode cables have a small diameter core, typically 6 μm, whereas multimode cables have a much larger core, typically between 50 μm and 200 μm in diameter. Both glass and plastic in different combinations are used in various forms of cable. One option is to use different types of glass fibre for both the core and the cladding. A second, and cheaper, option is to have a glass fibre core and a plastic cladding. This has the additional advantage of being less brittle than the all-glass version. Finally, all-plastic cables also exist, where two types of plastic fibre with different refractive indices are used. This is the cheapest form of all but it has the disadvantage of having high attenuation characteristics, making it unsuitable for transmission of light over medium to large distances.

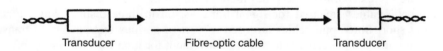

Fig. 8.5 Fibre-optic cables and transducers.

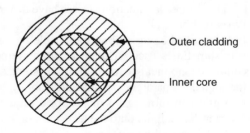

Fig. 8.6 Cross-section through fibre-optic cable.

Protection is normally given to the cable by enclosing it in the same types of insulating and armouring materials that are used for copper cables. This protects the cable against various hostile operating environments and also against mechanical damage. When suitably protected, fibre-optic cables can even withstand being engulfed in flames.

A *fibre-optic transmitter* usually consists of a light-emitting diode (LED). This converts an electrical signal into light and transmits it into the cable. The LED is particular suitable for this task as it has an approximately linear relationship between the input current and the light output. The type of LED chosen must closely match the attenuation characteristics of the light path through the cable and the spectral response of the receiving transducer. An important characteristic of the transmitter is the proportion of its power that is coupled into the fibre-optic cable: this is more important than its absolute output power. This proportion is maximized by making purpose-designed LED transmitters that have a spherical lens incorporated into the chip during manufacture. This produces an approximately parallel beam of light into the cable with a typical diameter of 400 µm.

The proportion of light entering the fibre-optic cable is also governed by the quality of the end face of the cable and the way it is bonded to the transmitter. A good end face can be produced by either polishing or cleaving. Polishing involves grinding the fibre end down with progressively finer polishing compounds until a surface of the required quality is obtained. Attachment to the transmitter is then normally achieved by gluing. This is a time-consuming process but uses cheap materials. Cleaving makes use of special kits that nick the fibre, break it very cleanly by applying mechanical force and then attach it to the transmitter by crimping. This is a much faster method but cleaving kits are quite expensive. Both methods produce good results.

The proportion of light transmitted into the cable is also dependent on the proper alignment of the transmitter with the centre of the cable. The effect of misalignment depends on the relative diameters of the cable. Figure 8.7 shows the effect on the proportion of power transmitted into the cable for the cases of (a) cable diameter > beam diameter, (b) cable diameter = beam diameter and (c) cable diameter < beam diameter. This shows that some degree of misalignment can be tolerated except where the beam and cable diameters are equal. The cost of producing exact alignment of the transmitter and cable is very high, as it requires the LED to be exactly aligned in its housing, the fibre to be exactly aligned in its connector and the housing to be exactly aligned with the connector. Therefore, great cost savings can be achieved wherever some misalignment can be tolerated in the specification for the cable.

The *fibre-optic receiver* is the device that converts the optical signal back into electrical form. It is usually either a PIN diode or phototransistor. Phototransistors have good sensitivity but only have a low bandwidth. On the other hand, PIN diodes have a much higher bandwidth but a lower sensitivity. If both high bandwidth and high sensitivity are required, then special avalanche photodiodes are used, but at a severe cost penalty. The same considerations about losses at the interface between the cable and receiver apply as for the transmitter, and both polishing and cleaving are used to prepare the fibre ends.

The output voltages from the receiver are very small and amplification is always necessary. The system is very prone to noise corruption at this point. However, the

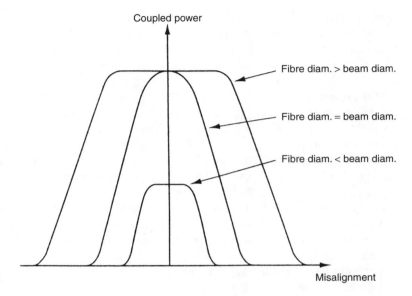

Fig. 8.7 Effect of transmitter alignment on light power transmitted.

development of receivers that incorporate an amplifier are finding great success in reducing the scale of this noise problem.

8.3.2 Transmission characteristics

Monomode cables have very simple transmission characteristics because the core has a very small diameter and light can only travel in a straight line down it. On the other hand, multimode cables have quite complicated transmission characteristics because of the relatively large diameter of the core.

Whilst the transmitter is designed to maximize the amount of light that enters the cable in a direction that is parallel to its length, some light will inevitably enter multimode cables at other angles. Light that enters a multimode cable at any angle other than normal to the end face will be refracted in the core. It will then travel in a straight line until it meets the boundary between the core and cladding materials. At this boundary, some of the light will be reflected back into the core and some will be refracted in the cladding.

For materials of refractive indices n_1 and n_2, as shown in Figure 8.8, light entering from the external medium with refractive index n_0 at an angle α_0 will be refracted at an angle α_1 in the core and, when it meets the core-cladding boundary, part will be reflected at an angle β_1 back into the core and part will be refracted at an angle β_2 in the cladding. α_1 and α_0 are related by Snell's law according to:

$$n_0 \sin \alpha_0 = n_1 \sin \alpha_1 \tag{8.1}$$

Similarly, β_1 and β_2 are related by:

$$n_1 \sin \beta_1 = n_2 \sin \beta_2 \tag{8.2}$$

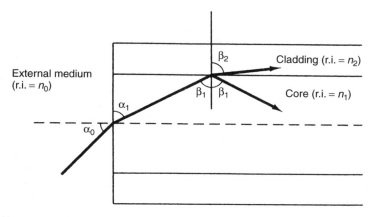

Fig. 8.8 Transmission of light through cable.

Light that enters the cladding is lost and contributes to the attenuation of the transmitted signal in the cable. However, observation of equation (8.1) shows how this loss can be prevented. If $\beta_2 = 90°$, then the refracted ray will travel along the boundary between the core and cladding and if $\beta_2 > 90°$, all of the beam will be reflected back into the core. The case where $\beta_2 = 90°$, corresponding to incident light at an angle α_c, is therefore the critical angle for total internal reflection to occur at the core/cladding boundary. The condition for this is that $\sin \beta_2 = 1$.

Setting $\sin \beta_2 = 1$ in equation (8.1):

$$\frac{n_1 \sin \beta_1}{n_2} = 1.$$

Thus:

$$\sin \beta_1 = \frac{n_2}{n_1}$$

Inspection of Figure 8.8 shows that $\cos \alpha_1 = \sin \beta_1$.

Hence:

$$\sin \alpha_1 = \sqrt{1 - \cos^2 \alpha_1} = \sqrt{1 - \sin^2 \beta_1} = \sqrt{1 - (n_2/n_1)^2}$$

From equation (8.1):

$$\sin \alpha_c = \sin \alpha_0 = \frac{n_1}{n_0} \sin \alpha_1.$$

Thus:

$$\sin \alpha_c = \frac{n_1}{n_0} \sqrt{1 - \left(\frac{n_2}{n_1}\right)^2}$$

Therefore, provided that the angle of incidence of the light into the cable is greater than the critical angle given by $\theta = \sin^{-1} \alpha_c$, all of the light will be internally reflected at the core/cladding boundary. Further reflections will occur as the light passes down the fibres and it will thus travel in a zigzag fashion to the end of the cable.

Whilst attenuation has been minimized, there is a remaining problem that the transmission time of the parts of the beam which travel in this zigzag manner will be greater than light which enters the fibre at 90° to the face and so travels in a straight line to the other end. In practice, the incident light rays to the cable will be spread over the range given by $\sin^{-1} \alpha_c < \theta < 90°$ and so the transmission times of these separate parts of the beam will be distributed over a corresponding range. These differential delay characteristics of the light beam are known as modal dispersion. The practical effect is that a step change in light intensity at the input end of the cable will be received over a finite period of time at the output.

It is possible to largely overcome this latter problem in multimode cables by using cables made solely from glass fibres in which the refractive index changes gradually over the cross-section of the core rather than abruptly at the core/cladding interface as in the step index cable discussed so far. This special type of cable is known as graded index cable and it progressively bends light incident at less than 90° to its end face rather than reflecting it off the core/cladding boundary. Although the parts of the beam away from the centre of the cable travel further, they also travel faster than the beam passing straight down the centre of the cable because the refractive index is lower away from the centre. Hence, all parts of the beam are subject to approximately the same propagation delay. In consequence, a step change in light intensity at the input produces an approximately step change of light intensity at the output. The alternative solution is to use a monomode cable. This propagates light in a single mode only, which means that time dispersion of the signal is almost eliminated.

8.3.3 Multiplexing schemes

Various types of branching network and multiplexing schemes have been proposed, some of which have been implemented as described in Grattan (1989). Wavelength division multiplexing is particularly well suited to fibre-optic applications, and the technique is now becoming well established. A single fibre is capable of propagating a large number of different wavelengths without cross-interference, and multiplexing thus allows a large number of distributed sensors to be addressed. A single optical light source is often sufficient for this, particularly if the modulated parameter is not light intensity.

8.4 Optical wireless telemetry

Wireless telemetry allows signal transmission to take place without laying down a physical link in the form of electrical or fibre-optic cable. This can be achieved using either radio or light waves to carry the transmitted signal across a plain air path between a transmitter and a receiver.

Optical wireless transmission was first developed in the early 1980s. It consists of a light source (usually infrared) transmitting encoded data information across an open, unprotected air path to a light detector. Three distinct modes of optical telemetry are possible, known as point-to-point, directed and diffuse:

- *Point-to-point telemetry* uses a narrowly focused, fine beam of light, which is commonly used for transmission between adjacent buildings. A data transmission speed of 5 Mbit/s is possible at the maximum transmission distance of 1000 m. However, if the transmission distance is limited to 200 m, a transmission speed of 20 Mbit/s is possible. Point-to-point telemetry is commonly used to connect electrical or fibre-optic ethernet networks in adjacent buildings.
- *Directed telemetry* transmits a slightly divergent beam of light that is directed towards reflective surfaces, such as the walls and ceilings in a room. This produces a wide area of coverage and means that the transmitted signal can be received at a number of points. However, the maximum transmission rate possible is only 1 Mbit/s at the maximum transmission distance of 70 m. If the transmission distance is limited to 20 m, a transmission speed of 10 Mbit/s is possible.
- *Diffuse telemetry* is similar to directed telemetry but the beam is even more divergent. This increases the area of coverage but reduces transmission speed and range. At the maximum range of 20 m, the maximum speed of transmission is 500 kbit/s, though this increases to 2 Mbit/s at a reduced range of 10 m.

In practice, implementations of optical wireless telemetry are relatively uncommon. Where optical transmission is favoured because of its immunity to electromagnetic noise, fibre-optic transmission is usually preferred since optical wireless transmission is susceptible to random interruption when data is transmitted across an open, unprotected air path. This preference for fibre-optic transmission exists despite its much greater cost than optical wireless transmission. Similarly, when the difficulty of laying a physical cable link determines that wireless transmission is used, it is normal to use radio rather than optical transmission. This preference arises because radio transmission is much less prone to interference than optical transmission, since radio waves can pass through most materials. However, there are a few instances where radio transmission is subject to interference from neighbouring radio frequency systems operating at a similar wavelength and, in such circumstances, optical transmission is sometimes a better option.

8.5 Radio telemetry (radio wireless transmission)

Radio telemetry is normally used over transmission distances up to 400 miles, though this can be extended by special techniques to provide communication through space over millions of miles. However, radio telemetry is also commonly used over quite short distances to transmit signals where physical electrical or fibre-optic links are difficult to install or maintain. This occurs particularly when the source of the signals is mobile. The great advantage that radio telemetry has over optical wireless transmission through an air medium is that radio waves are attenuated much less by obstacles between the energy transmitter and receiver. Hence, as noted above, radio telemetry usually performs better than optical wireless telemetry and is therefore used much more commonly.

In radio telemetry, data are usually transmitted in a frequency modulated (FM) format according to the scheme shown in Figure 8.9. This scheme actually involves two separate stages of frequency modulation, and the system is consequently known

Signal transmission

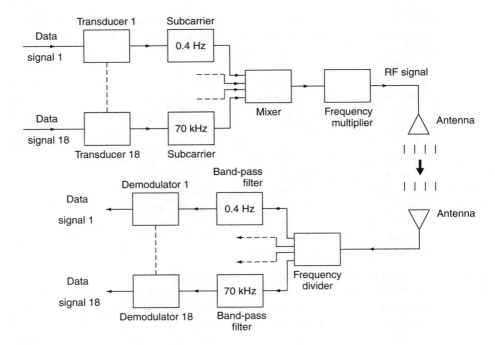

Fig. 8.9 Radio transmission using FM/FM system.

as an FM/FM system. Eighteen data channels are provided over the frequency range from 0.4 kHz to 70 kHz, as given in Table 8.1. Each channel is known as a subcarrier frequency and can be used to transmit data for a different physical variable. Thus, the system can transmit information on 18 different variables simultaneously.

A voltage-to-frequency converter is used in the first FM stage to convert each analogue voltage signal into a varying frequency around the centre frequency of the subcarrier assigned for that channel. The 18 channels are then mixed into a single signal spanning the frequency range 0.4 kHz to 70 kHz. For transmission, the length of the antenna has to be one-quarter or one-half of the wavelength. At 10 kHz, which is a typical subcarrier frequency in an 18-channel system, the wavelength is 30 km. Hence, an antenna for transmission at this frequency is totally impractical. In consequence, a second FM stage is used to translate the 0.4 kHz to 70 kHz signal into the radio frequency range as modulations on a typical carrier frequency of 217.5 MHz.* At this

Table 8.1 Standard subcarrier frequencies for FM channels

Band	1	2	3	4	5	6	7	8	9
Centre frequency (kHz)	0.4	0.56	0.73	0.96	1.3	1.7	2.3	3.0	3.9
Band	10	11	12	13	14	15	16	17	18
Centre frequency (kHz)	5.4	7.35	10.5	14.5	22.0	30.0	40.0	52.5	70.0

(Maximum frequency deviation allowed is ±7.5%)

---

* Particular frequencies are allocated for industrial telemetry. These are subject to national agreements and vary in different countries.

frequency, the wavelength is 1.38 m, and so a transmission antenna of length 0.69 m or 0.345 m would be suitable. The signal is received by an antenna of identical length some distance away. A frequency divider is then used to convert the signal back to one across the 0.4 kHz to 70 kHz subcarrier frequency spectrum, following which a series of band-pass filters are applied to extract the 18 separate frequency bands containing the measurement data. Finally, a demodulator is applied to each channel to return each signal into varying voltage form.

The inaccuracy of radio telemetry is typically $\pm 1\%$. Thus, measurement uncertainty in transmitting a temperature measurement signal with a range of 0–100°C over one channel would be $\pm 1\%$, i.e. $\pm 1°C$. However, if there are unused transmission channels available, the signal could be divided into two ranges (0–50°C and 50–100°C) and transmitted over two channels, reducing the measurement uncertainty to $\pm 0.5°C$. By using ten channels for one variable, a maximum measurement uncertainty of $\pm 0.1°C$ could be achieved.

In theory, radio telemetry is very reliable because, although the radio frequency waveband is relatively crowded, specific frequencies within it are allocated to specific usages under national agreements that are normally backed by legislation. Interference is avoided by licensing each frequency to only one user in a particular area, and limiting the transmission range through limits on the power level of transmitted signals, such that there is no interference to other licensed users of the same frequency in other areas. Unfortunately, interference can still occur in practice, due both to adverse atmospheric conditions extending the transmission range beyond that expected into adjoining areas, and also due to unauthorized transmissions by other parties at the wavelengths licensed to registered users. There is a legal solution to this latter problem, although some time may elapse before the offending transmission is successfully stopped.

8.6 Digital transmission protocols

Digital transmission has very significant advantages compared with analogue transmission because the possibility of signal corruption during transmission is greatly reduced. Many different protocols exist for digital signal transmission, and these are considered in detail in Chapter 10. However, the protocol that is normally used for the transmission of data from a measurement sensor or circuit is asynchronous serial transmission, with other forms of transmission being reserved for use in instrumentation and computer networks. Asynchronous transmission involves converting an analogue voltage signal into a binary equivalent, using an analogue-to-digital converter as discussed in section 6.4.3. This is then transmitted as a sequence of voltage pulses of equal width that represent binary '1' and '0' digits. Commonly, a voltage level of $+6V$ is used to represent binary '1' and zero volts represents binary '0'. Thus, the transmitted signal takes the form of a sequence of 6 V pulses separated by zero volt pulses. This is often known by the name of *pulse code modulation*. Such transmission in digital format provides very high immunity to noise because noise is typically much smaller than the amplitude of a pulse representing binary 1. At the receiving end of a transmitted signal, any pulse level between 0 and 3 volts can be interpreted as a binary '0' and anything greater than 3 V can be interpreted as a binary '1'. A further advantage of digital transmission is that other information, such as about plant status,

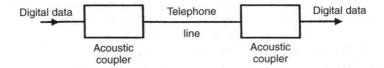

Fig. 8.10 Telephone transmission.

can be conveyed as well as parameter values. However, consideration must be given to the potential problems of aliasing and quantization, as discussed in section 6.4.3, and the sampling frequency must therefore be chosen carefully.

Many different mediums can be used to transmit digital signals. Electrical cable, in the form of a twisted pair or coaxial cable, is commonly used as the transmission path. However, in some industrial environments, the noise levels are so high that even digital data becomes corrupted when transmitted as electrical pulses. In such cases, alternative transmission mechanisms have to be used.

One alternative is to modulate the pulses onto a high-frequency carrier, with positive and zero pulses being represented as two distinct frequencies either side of a centre carrier frequency. Once in such a frequency modulated format, a normal mains electricity supply cable operating at mains frequency is often used to carry the data signal. The large frequency difference between the signal carrier and the mains frequency prevents any corruption of the data transmitted, and simple filtering and demodulation is able to extract the measurement signal after transmission. The public switched telephone network can also be used to transmit frequency modulated data at speeds up to 1200 bits/s, using acoustic couplers as shown in Figure 8.10. The transmitting coupler converts each binary '1' into a tone at 1.4 kHz and each binary '0' into a tone at 2.1 kHz, whilst the receiving coupler converts the tones back into binary digits.

Another solution is to apply the signal to a digital-to-current converter unit and then use current loop transmission, with 4 mA representing binary '0' and 20 mA representing binary '1'. This permits baud rates up to 9600 bit/s at transmission distances up to 3 km. Fibre-optic links and radio telemetry are also widely used to transmit digital data.

References and further reading

Grattan, K.T.V. (1989) New developments in sensor technology – fibre-optics and electro-optics, *Measurement and Control*, **22**(6), pp. 165–175.

9

Digital computation and intelligent devices

This chapter is concerned with introducing the principles of digital computation and its application in measurement systems. Digital computers have been used in conjunction with measurement systems for many years in the typical control system scenario where a computer uses data on process variables supplied by a measurement system to compute a control signal that is then applied to an actuator in order to modify some aspect of the controlled process. In this case, the computer is not actually part of the measurement system but merely works with it by taking data from the system. However, the rapid fall in the cost of computers has led to their widespread inclusion actually within measurement systems, performing various signal processing operations digitally that were previously carried out by analogue electronic circuits.

In early applications of digital signal processing, the computer remained as a distinctly separate component within the measurement system. However, the past few years have seen the development of measurement systems in the form of *intelligent devices* in which the computational element (usually called a microcomputer or microprocessor) is much more closely integrated into the measurement system. These devices are known by various names such as *intelligent instruments*, *smart sensors* and *smart transmitters*. However, before discussing these in detail, the basic principles of digital computation need to be covered first.

9.1 Principles of digital computation

9.1.1 Elements of a computer

The primary function of a digital computer is the manipulation of data. The three elements that are essential to the fulfilment of this task are the central processing unit, the memory and the input–output interface, as shown in Figure 9.1. These elements are collectively known as the computer hardware, and each element exists physically as one or more integrated circuit chips mounted on a printed circuit board. Where the central processing unit (CPU) consists of a single microprocessor, it is usual to regard the system as a microcomputer. The distinction between the terms 'microcomputer', 'minicomputer' and 'mainframe computer' is a very arbitrary division made according to relative computer power. However, this classification has become

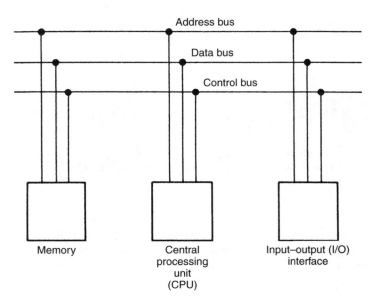

Fig. 9.1 Elements of a microcomputer.

somewhat meaningless, with present day 'microcomputers' being more powerful than mainframe computers of only a few years ago.

The *central processing unit* (CPU) part of a computer can be regarded as the brain of the system. A relatively small CPU is commonly called a *microprocessor*. The CPU determines what computational operations are carried out and the sequence in which the operations are executed. During such operation, the CPU makes use of one or more special storage locations within itself known as *registers*. Another part of the CPU is the *arithmetic and logic unit* (ALU), which is where all arithmetic operations are evaluated. The CPU operates according to a sequential list of required operations defined by a computer program, known as the computer software. This program is held in the second of the three system components known as the computer memory.

The *computer memory* also serves several other functions besides this role of holding the computer program. One of these is to provide temporary storage locations that the CPU uses to store variables during execution of the computer program. A further common use of memory is to store data tables that are used for scaling and variable conversion purposes during program execution.

Memory can be visualized as a consecutive sequence of boxes in which various items are stored, as shown in Figure 9.2 for a typical memory size of 65 536 storage units. If this storage mechanism is to be useful, then it is essential that a means be provided for giving a unique label to each storage box. This is achieved by labelling the first box as 0, the next one as 1 and so on for the rest of the storage locations. These numbers are known as the *memory addresses*. Whilst these can be labelled by decimal numbers, it is more usual to use hexadecimal notation (see section 9.1.2).

Two main types of computer memory exist and there are important differences between these. The two kinds are *random access memory* (RAM) and *read only memory* (ROM). The CPU can both read from and write to the former, but it can only read from

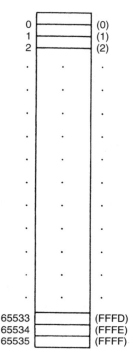

Fig. 9.2 Schematic representation of computer memory (numbers in parentheses are memory addresses in hexadecimal notation).

the latter. The importance of ROM becomes apparent if the behaviour of each kind of memory when the power supply is turned off is considered. At power-off time, RAM loses its contents but ROM maintains them, and this is the value of ROM. Intelligent devices normally use ROM for storage of the program and data tables and just have a small amount of RAM that is used by the CPU for temporary variable storage during program execution.

The third essential element of a computer system is the *input–output (I/O) interface*, which allows the computer to communicate with the outside world by reading in data values and outputting results after the appropriate computation has been executed. In the case of a microcomputer performing a signal processing function within an intelligent device, this means reading in the values obtained from one or more sensors and outputting a processed value for presentation at the instrument output. All such external peripherals are identified by a unique number, as for memory addresses.

Communication between these three computer elements is provided by three electronic highways known as the *data bus*, the *address bus* and the *control bus*. At each data transfer operation executed by the CPU, two items of information must be conveyed along the electronic highway, the item of data being transferred and the address where it is being sent. Whilst both of these items of information could be conveyed along a single bus, it is more usual to use two buses that are called the data bus and the address bus. The timing of data transfer operations is important, particularly when transfers take place to peripherals such as disk drives and keyboards where the

CPU often has to wait until the peripheral is free before it can initialize a data transfer. This timing information is carried by a third highway known as the control bus.

The latest trend made possible by advances in very large-scale integration (VLSI) technology is to incorporate all three functions of central processor unit, memory and I/O within a single chip (known as a computer on a chip or *microcomputer*). The term 'microprocessor' is often used to describe such an integrated unit, but this is strictly incorrect since the device contains more than just processing power.

9.1.2 Computer operation

As has already been mentioned, the fundamental role of a computer is the manipulation of data. Numbers are used both in quantifying items of data and also in the form of codes that define the computational operations that are to be executed. All numbers that are used for these two purposes must be stored within the computer memory and also transported along the communication buses. A detailed consideration of the conventions used for representing numbers within the computer is therefore required.

Number systems

The decimal system is the best known number system, but it is not very suitable for use by digital computers. It uses a base of ten, such that each digit in a number can have any one of ten values within the range 0–9. Items of electronic equipment such as the digital counter, which are often used as computer peripherals, have liquid crystal display elements that can each display any of the ten decimal digits, and therefore a four element display can directly represent decimal numbers in the range 0–9999. The decimal system is therefore perfectly suitable for use with such output devices.

The fundamental unit of data storage within a digital computer is a memory element known as a *bit*. This holds information by switching between one of two possible states. Each storage unit can therefore only represent two possible values and all data to be entered into memory must be organized into a format that recognizes this restriction. This means that numbers must be entered in binary format, where each digit in the number can only have one of two values, 0 or 1. The binary representation is particularly convenient for computers because bits can be represented very simply electronically as either zero or non-zero voltages. However, the conversion is tedious for humans. Starting from the right-hand side of a binary number, where the first digit represents 2^0 (i.e. 1), each successive binary digit represents progressively higher powers of 2. For example, in the binary number 1111, the first digit (starting from the right-hand side) represents 1, the next 2, the next 4 and the final, leftmost digit represents 8, Thus the decimal equivalent is $1 + 2 + 4 + 8 = 15$.

Example 9.1
Convert the following 8-bit binary number to its decimal equivalent: 10110011

Solution
Starting at the right-hand side, we have:

$$(1 \times 2^0) + (1 \times 2^1) + (0 \times 2^2) + (0 \times 2^3) + (1 \times 2^4) + (1 \times 2^5)$$
$$+ (0 \times 2^6) + (1 \times 2^7)$$
$$= 1 + 2 + 0 + 0 + 16 + 32 + 0 + 128 = 179$$

Measurement and Instrumentation Principles

For data storage purposes, memory elements are combined into larger units known as bytes, which are usually considered to consist of 8 bits each. Each bit holds one binary digit, and therefore a memory unit consisting of 8 bits can store 8-digit binary numbers in the range 00000000 to 11111111 (equivalent to decimal numbers in the range 0 to 255). A binary number in this system of 10010011 for instance would correspond with the decimal number 147.

This range is clearly inadequate for most purposes, including measurement systems, because even if all data could be conveniently scaled the maximum resolution obtainable is only 1 part in 128. Numbers are therefore normally stored in units of either 2 or 4 bytes, which allow the storage of integer (whole) numbers in the range 0–65 535 or 0–4 294 967 296.

No means have been suggested so far for expressing the sign of numbers, which is clearly necessary in the real world where negative as well as positive numbers occur. A simple way to do this is to reserve the most significant (left-hand bit) in a storage unit to define the sign of a number, with '0' representing a positive number and '1' a negative number. This alters the ranges of numbers representable in a 1-byte storage unit to -127 to $+127$, as only 7 bits are left to express the magnitude of the number, and also means that there are two representations of the value 0. In this system the binary number 10010011 translates to the decimal number -19 and 00010011 translates to $+19$. For reasons dictated by the mode of operation of the CPU, however, most computers use an alternative representation known as the two's complement form.

The *two's complement* of a number is most easily formed by going via an intermediate stage of the *one's complement*. The one's complement of a number is formed by reversing all digits in the binary representation of the magnitude of a number, changing 1s to 0s and 0s to 1s, and then changing the left-hand bit to a 1 if the original number was negative. The two's complement is then formed by adding 1 at the least significant (right-hand) end of the one's complement. As before for a 1-byte storage unit, only 7 bits are available for representing the magnitude of a number, but, because there is now only one representation of zero, the decimal range representable is -128 to $+127$.

Example 9.2
Find the one's and two's complement 8-bit binary representation of the following decimal numbers: 56 -56 73 119 27 -47

Method of Solution
Take first the decimal value of 56
Form 7-bit binary representation: 0111000
Reverse digits in this: 1000111
Add sign bit to left-hand end to form one's complement: 01000111
Form two's complement by adding one to one's complement: $01000111 + 1 = 01001000$
Take next the decimal value of -56
Form 7-bit binary representation: 0111000
Reverse digits in this: 1000111
Add sign bit to left-hand end to form one's complement: 11000111
Form two's complement by adding one to one's complement: $11000111 + 1 = 11001000$

Summary of solution for all values

Decimal number	Binary representation of magnitude (7 bit)	Digits reversed in 7-bit representation	One's complement (8 bit)	Two's complement (8 bit)
56	0111000	1000111	01000111	01001000
−56	0111000	1000111	11000111	11001000
73	1001001	0110110	00110110	00110111
119	1110111	0001000	00001000	00001001
27	0011011	1100100	01100100	01100101
−47	0101111	1010000	11010000	11010001

We have therefore established the binary code in which the computer stores positive and negative integers (whole numbers). However, it is frequently necessary also to handle real numbers (those with fractional parts). These are most commonly stored using the floating-point representation.

The *floating-point representation* divides each memory storage unit (notionally, not physically) into three fields, known as the sign field, the exponent field and the mantissa field. The sign field is always 1 bit wide but there is no formal definition for the relative sizes of the other fields. However, a common subdivision of a 32-bit (4-byte) storage unit is to have a 7-bit exponent field and a 24-bit mantissa field, as shown in Figure 9.3.

The value contained in the storage unit is evaluated by multiplying the number in the mantissa field by 2 raised to the power of the number in the exponent field. Negative as well as positive exponents are obtained by biasing the exponent field by 64 (for a 7-bit field), such that a value of 64 is interpreted as an exponent of 0, a value of 65 as an exponent of 1, a value of 63 as an exponent of −1 etc. Suppose therefore that the sign bit field has a zero, the exponent field has a value of 0111110 (decimal 62) and the mantissa field has a value of 000000000000000001110111 (decimal 119), i.e. the contents of the storage unit are 00111110000000000000000001110111. The number stored is $+119 \times 2^{-2}$. Changing the first (sign) bit to a 1 would change the number stored to -119×2^{-2}.

However, if a human being were asked to enter numbers in these binary forms, the procedure would be both highly tedious and also very prone to error. In consequence, simpler ways of entering binary numbers have been developed. Two such ways are to use octal and hexadecimal numbers, which are translated to binary numbers at the input–output interface to the computer.

Octal numbers use a base of 8 and consist of decimal digits in the range 0–7 that each represent 3 binary digits. Thus 8 octal digits represent a 24-bit binary number.

Fig. 9.3 Representation of memory storage unit.

Hexadecimal numbers have a base of 16 and are used much more commonly than octal numbers. They use decimal digits in the range 0–9 and letters in the range A–F that each represent 4 binary digits. The decimal digits 0–9 translate directly to the decimal values 0–9 and the letters A–F translate respectively to the decimal values 10–15. A 24-bit binary number requires 6 hexadecimal digits to represent it. The following table shows the octal, hexadecimal and binary equivalents of decimal numbers in the range 0–15.

Decimal	Octal	Hexadecimal	Binary
0	0	0	0
1	1	1	1
2	2	2	10
3	3	3	11
4	4	4	100
5	5	5	101
6	6	6	110
7	7	7	111
8	10	8	1000
9	11	9	1001
10	12	A	1010
11	13	B	1011
12	14	C	1100
13	15	D	1101
14	16	E	1110
15	17	F	1111

Octal/hexadecimal to binary conversion

Octal and hexadecimal conversion is very simple. Each octal/hexadecimal digit is taken in turn and converted to its binary representation according to the table above.

Example 9.3
Convert the octal number 7654 to binary.

Solution
Using the table above, write down the binary equivalent of each octal digit:

| 7 | 6 | 5 | 4 |
| 111 | 110 | 101 | 100 |

Thus, the binary code is 111110101100.

Example 9.4
Convert the hexadecimal number ABCD to binary.

Solution
Using the table above, write down the binary equivalent of each hexadecimal digit:

| A | B | C | D |
| 1010 | 1011 | 1100 | 1101 |

Thus, the binary code is 1010101111001101.

Binary to octal/hexadecimal conversion

Conversion from binary to octal or hexadecimal is also simple. The binary digits are taken in groups of three at a time (for octal) or four at a time (for hexadecimal), starting at the least significant end of the number (right-hand side) and writing down the appropriate octal or hexadecimal digit for each group.

Example 9.5
Convert the binary number 010111011001 into octal and hexadecimal.

Solution

$$| \ 010 \ | \ 111 \ | \ 011 \ | \ 001 \ |$$
$$| \ \ 2 \ \ | \ \ 7 \ \ | \ \ 3 \ \ | \ \ 1 \ \ |$$
$$= 2731 \text{ octal}$$

$$| \ 0101 \ | \ 1101 \ | \ 1001 \ |$$
$$| \ \ 5 \ \ | \ \ D \ \ | \ \ 9 \ \ |$$
$$= 5D9 \text{ hexadecimal}$$

Example 9.6
The 24-bit binary number 011111001001001101011010 is to be entered into a computer. How would it be entered using (a) octal code and (b) hexadecimal code?

Solution

(a) Divide the 24-bit number into groups of three, starting at the right-hand side:

$$011 \ | \ 111 \ | \ 001 \ | \ 001 \ | \ 001 \ | \ 101 \ | \ 011 \ | \ 010 \ | = 37111532 \text{ octal}$$

Thus, the number would be entered as 37111532 using octal code.
(b) Divide the 24-bit number into groups of four, starting at the right-hand side:

$$| \ 0111 \ | \ 1100 \ | \ 1001 \ | \ 0011 \ | \ 0101 \ | \ 1010 \ | = 7C935A \text{ hexadecimal}$$

In carrying out such conversions, it is essential that the groupings of binary digits start from the right-hand side. Groupings starting at the left-hand side give completely wrong values unless the number of binary digits happens to be an integer multiple of the grouping size. Consider a 10-digit binary number: 1011100011.

Grouping digits starting at the right-hand side gives the values 1343 octal and 2E3 hexadecimal.
Grouping digits starting at the left gives the (incorrect) values of 5611 octal and B83 hexadecimal.

When converting a binary number to octal or hexadecimal representation, a check must also be made that all of the binary digits represent data. In some systems, the first (left-hand) digit is used as a sign bit and the last (right-hand) digit is used as a parity bit.

Example 9.7
In a system that uses the first bit as a sign bit and the last bit as a parity bit, what is the octal and hexadecimal representation of the binary code: 110111000111?

Solution
The ten data bits are 1011100011. This converts to 1343 octal and 2E3 hexadecimal.

Programming and program execution

In most modes of usage, including use as part of intelligent devices, computers are involved in manipulating data. This requires data values to be input, processed and output according to a sequence of operations defined by the computer program. However, in practice, programming the microprocessor within an intelligent device is not normally the province of the instrument user, indeed, there is rarely any provision for the user to create or modify operating programs even if he/she wished to do so. There are several reasons for this. Firstly, the signal processing needed within an intelligent device is usually well defined, and therefore it is more efficient for a manufacturer to produce this rather than to have each individual user produce near identical programs separately. Secondly, better program integrity and instrument operation is achieved if a standard program produced by the instrument manufacturer is used. Finally, use of a standard program allows it to be burnt into ROM, thereby protecting it from any failure of the instrument power supply. This also facilitates software maintenance and updates, by the mechanism of the manufacturer providing a new ROM that simply plugs into the slot previously occupied by the old ROM.

However, even though it is not normally a task undertaken by the user, some appreciation of microprocessor programming for an intelligent device is useful background knowledge. To illustrate the techniques involved in programming, consider a very simple program that reads in a value from a sensor, adds a pre-stored value to it to compensate for a bias in the sensor measurement, and outputs a corrected reading to a display device.

Let us assume that the addresses of the sensor and output display device are 00C0 and 00C1 respectively, and that the required scaling value has already been stored in memory address 0100. The instructions below are formed from the instruction set for a Z80* microprocessor and make use of CPU registers A and B.

IN A,C0

IN B,100

ADD A,B

OUT C1,A

This list of four instructions constitutes the computer program that is necessary to execute the required task. The CPU normally executes the instructions one at a time, starting at the top of the list and working downwards (though jump and branch instructions change this order). The first instruction (IN A,C0) reads in a value from the sensor at address C0 and places the value in CPU register A (often called the accumulator). The mechanics of the execution of this instruction consist of the CPU putting the required address C0 on the address bus and then putting a command on the control bus that causes the contents of the target address (C0) to be copied onto the data bus

---

* The Z80 is now an obsolete 8-bit processor but its simplicity is well suited to illustrating programming techniques. Similar, but necessarily more complex, programming instructions are used with current 16- and 32-bit processors.

and subsequently transferred into the A register. The next instruction (IN B,100) reads in a value from address 100 (the pre-stored biasing value) and stores it in register B. The following instruction (ADD A,B) adds together the contents of registers A and B and stores the result in register A. Register A now contains the measurement read from the sensor but corrected for bias. The final instruction (OUT C1,A) transfers the contents of register A to the output device on address C1.

9.1.3 Interfacing

The input–output interface connects the computer to the outside world, and is therefore an essential part of the computer system. When the CPU puts the address of a peripheral onto the address bus, the input–output interface decodes the address and identifies the unique computer peripheral with which a data transfer operation is to be executed. The interface also has to interpret the command on the control bus so that the timing of the data transfer is correct. One further very important function of the input–output interface is to provide a physical electronic highway for the flow of data between the computer data bus and the external peripheral. In many computer applications, including their use within intelligent devices, the external peripheral requires signals to be in analogue form. Therefore the input–output interface must provide for conversion between these analogue signals and the digital signals required by a digital computer. This is satisfied by analogue-to-digital and digital-to-analogue conversion elements within the input–output interface.

A standard form of interface used to connect a computer to its peripheral devices is the UART (Universal Asynchronous Receiver/Transmitter). This has been used for around 30 years. A newer interface protocol that is particularly suitable for connecting a large number of devices and providing for communication between different computers is the PCI (Peripheral Component Interconnect) interface. Very recently, an alternative protocol called the Universal Serial Bus (USB) has been developed that is rapidly gaining in popularity.

The rest of this section presents some elementary concepts of interfacing in simple terms. A more detailed discussion follows later in Chapter 10, where the combination of intelligent devices into larger networks is discussed.

Address decoding

A typical address bus in a microcomputer is 16 bits wide*, allowing 65 536 separate addresses to be accessed in the range 0000–FFFF (in hexadecimal representation). Special commands on some computers are reserved for accessing the bottom end 256 of these addresses in the range 0000–00FF, and, if these commands are used, only 8 bits are needed to specify the required address. For the purpose of explaining address-decoding techniques, the scheme below shows how the lower 8 bits of the 16-bit address line are decoded to identify the unique address referenced by one of these special commands. Decoding of all 16 address lines follows a similar procedure but requires a substantially greater number of integrated circuit chips.

Address decoding is performed by a suitable combination of logic gates. Figure 9.4 shows a very simple hardware scheme for decoding 8 address lines. This consists

---

* Recently, 32-bit address fields have also become available in some devices.

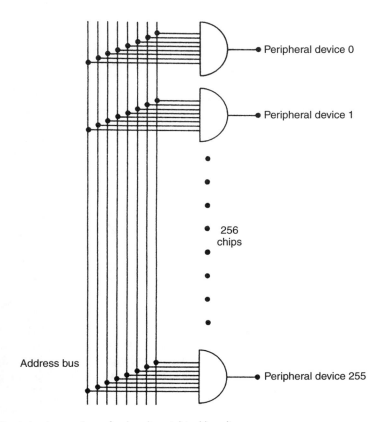

Fig. 9.4 Simple hardware scheme for decoding eight address lines.

of 256 8-input NAND gates, which each uniquely decode one of 256 addresses. A NAND gate is a logic element that only gives a logic level 1 output when all inputs are zero, and gives a logic level 0 output for any other combination of inputs. The inputs to the NAND gates are connected onto the lower 8 lines of the address bus and the computer peripherals are connected to the output of the particular gates that decode their unique addresses. There are two pins for each input to the NAND gates that respectively invert and do not invert the input signal. By connecting the 8 address lines appropriately to these two alternative pins at each input, the gate is made to decode a unique address. Consider for instance the pin connections shown in Figure 9.5. This NAND gate decodes address C5 (hexadecimal), which is 11000101 in binary. Because of the way in which the input pins to the chip are connected, the NAND gate will see all zeros at its input when 11000101 is on the lower 8 bits of the address bus and therefore will have an output of 1. Any other binary number on the address bus will cause this NAND gate to have a zero output.

Data transfer control

The transfer of data between the computer and peripherals is managed by control and status signals carried on the control bus that determine the exact sequencing and timing of I/O operations. Such management is necessary because of the different operating

176 Digital computation and intelligent devices

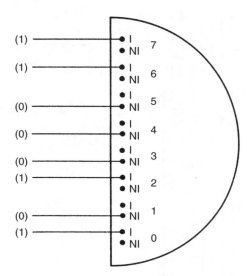

Fig. 9.5 Pin connections to NAND gate to decode address C5.

speeds of the computer and its peripherals and because of the multi-tasking operation of many computers. This means that, at any particular instant when a data transfer operation is requested, either the computer or the peripheral may not be ready to take part in the transfer. Typical control and status lines, and their meanings when set at a logic level of 1, are shown below.

- BUSY   Peripheral device busy
- READY   Peripheral device ready for data transfer
- ENABLE CPU ready for data transfer
- ERROR   Malfunction on peripheral device

Similar control signals are set up by both the computer and peripherals, but different conventions are often used to define the status of each device. Differing conventions occur particularly when the computer and peripherals come from different manufacturers, and might mean for instance that the computer interprets a logic level of 1 as defining a device to be busy but the peripheral device uses logic level 0 to define 'device busy' on the appropriate control line. Therefore, translation of the control lines between the computer and peripherals is required, which is achieved by a further series of logic gates within the I/O interface.

9.1.4 Practical considerations in adding computers to measurement systems

The foregoing discussion has presented some of the necessary elements in an input–output interface in a relatively simplistic manner that is just sufficient to give the reader the flavour of what is involved in an interface. Much fine detail has been omitted, and the amount of work involved in the practical design of a real interface should not be underestimated. One significant omission so far is discussion of the scaling that is generally required within the analogue–digital interface of a computer. The raw analogue input and output signals are generally either too large or too small for compatibility with the operating voltage levels of a digital computer and they have to be scaled upwards or downwards. This is normally achieved by operational amplifiers and/or potentiometers. The main features of an operational amplifier are its high gain (typically $\times 1\,000\,000$) and its large bandwidth (typically 1 MHz or better). However, when one is used at very high frequencies, the bandwidth becomes significant. The quality of an amplifier is often measured by a criterion called the gain–bandwidth product, which is the product of its gain and bandwidth. Other important attributes of the operational amplifier, particularly when used in a computer input–output interface or within intelligent devices, are its distortion level, overload recovery capacity and offset level. Special instrumentation amplifiers that are particularly good in these attributes have been developed for instrumentation applications, as described in section 5.5.1.

Suitable care must always be taken when introducing a computer into a measurement system to avoid creating sources of measurement noise. This applies particularly where one computer is used to process the output of several transducers and is connected to them by signal wires. In such circumstances, the connections and connecting wires can create noise through electrochemical potentials, thermoelectric potentials, offset voltages introduced by common mode impedances, and a.c. noise at power, audio and radio frequencies. Recognition of all these possible noise sources allows them to be eliminated in most cases by employing good practice when designing and constructing the measurement system.

9.2 Intelligent devices

The term 'intelligent device' is used to describe a package containing either a complete measurement system, or else a component within a measurement system, which incorporates a digital processor. Processing of the output of measurement sensors to correct for errors inherent in the measurement process brings about large improvements in measurement accuracy. Such intelligent devices are known by various names such as *intelligent instrument*, *smart sensor* and *smart transmitter*. There is no formal definition for any of these names, and there is considerable overlap between the characteristics of particular devices and the name given to them. The discussion below tries to lay out the historical development of intelligent devices, and it summarizes the general understanding of the sort of characteristics possessed by the various forms of intelligent device. Details of their application to measure particular physical variables will be covered in appropriate chapters in Part 2 of this book.

9.2.1 Intelligent instruments

The first intelligent instrument appeared over 20 years ago, although high prices when such devices first became available meant that their use within measurement systems grew very slowly initially. The processor within an intelligent instrument allows it to apply pre-programmed signal processing and data manipulation algorithms to measurements. One of the main functions performed by the first intelligent instruments to become available was compensation for environmental disturbances to measurements that cause systematic errors. Thus, apart from a primary sensor to measure the variable of interest, intelligent instruments usually have one or more secondary sensors to monitor the value of environmental disturbances. These extra measurements allow the output reading to be corrected for the effects of environmentally induced errors, subject to the following pre-conditions being satisfied:

(a) The physical mechanism by which a measurement sensor is affected by ambient condition changes must be fully understood and all physical quantities that affect the output must be identified.
(b) The effect of each ambient variable on the output characteristic of the primary sensor must be quantified.
(c) Suitable secondary sensors for monitoring the value of all relevant environmental variables must be available that will operate satisfactorily in the prevailing environmental conditions.

Condition (a) above means that the thermal expansion and contraction of all elements within a sensor must be considered in order to evaluate how it will respond to ambient temperature changes. Similarly, the sensor response, if any, to changes in ambient pressure, humidity, gravitational force or power supply level (active instruments) must be examined.

Quantification of the effect of each ambient variable on the characteristics of the measurement sensor is then necessary, as stated in condition (b). Analytic quantification of ambient condition changes from purely theoretical consideration of the construction of a sensor is usually extremely complex and so is normally avoided. Instead, the effect is quantified empirically in laboratory tests where the output characteristic of the sensor is observed as the ambient environmental conditions are changed in a controlled manner.

One early application of intelligent instruments was in volume flow rate measurement, where the flow rate is inferred by measuring the differential pressure across an orifice plate placed in a fluid-carrying pipe (see Chapter 16). The flow rate is proportional to the square root of the difference in pressure across the orifice plate. For a given flow rate, this relationship is affected both by the temperature and by the mean pressure in the pipe, and changes in the ambient value of either of these cause measurement errors. A typical intelligent flowmeter therefore contains three sensors, a primary one measuring pressure difference across the orifice plate and secondary ones measuring absolute pressure and temperature. The instrument is programmed to correct the output of the primary differential-pressure sensor according to the values measured by the secondary sensors, using appropriate physical laws that quantify the effect of ambient temperature and pressure changes on the fundamental relationship between flow and differential pressure. Even 20 years ago, such intelligent flow measuring instruments achieved typical inaccuracy levels of $\pm 0.1\%$, compared with $\pm 0.5\%$ for their non-intelligent equivalents.

Although automatic compensation for environmental disturbances is a very important attribute of intelligent instruments, many versions of such devices perform additional functions, and this was so even in the early days of their development. For example, the orifice-plate flowmeter just discussed usually converts the square root relationship between flow and signal output into a linear one, thus making the output much easier to interpret. Other examples of the sort of functions performed by intelligent instruments are:

- correction for the loading effect of measurement on the measured system
- signal damping with selectable time constants
- switchable ranges (using several primary sensors within the instrument that each measure over a different range)
- switchable output units (e.g. display in Imperial or SI units)
- linearization of the output
- self-diagnosis of faults
- remote adjustment and control of instrument parameters from up to 1500 metres away via 4-way, 20 mA signal lines.

These features will be discussed in greater detail under the later headings of smart sensors and smart transmitters.

Over the intervening years since their first introduction, the size of intelligent instruments has gradually reduced and the functions performed have steadily increased. One particular development has been the inclusion of a microprocessor within the sensor itself, in devices that are usually known as *smart sensors*. As further size reduction and device integration has taken place, such smart sensors have been incorporated into packages with other sensors and signal processing circuits etc. Whilst such a package conforms to the definition of an intelligent instrument given previously, most manufacturers now tend to call the package a *smart transmitter* rather than an intelligent instrument, although the latter term has continued in use in some cases.

9.2.2 Smart sensors

A smart sensor is a sensor with local processing power that enables it to react to local conditions without having to refer back to a central controller. Smart sensors are usually at least twice as accurate as non-smart devices, have reduced maintenance costs and require less wiring to the site where they are used. In addition, long-term stability is improved, reducing the required calibration frequency.

The functions possessed by smart sensors vary widely, but consist of at least some of the following:

- Remote calibration capability
- Self-diagnosis of faults
- Automatic calculation of measurement accuracy and compensation for random errors
- Adjustment for measurement of non-linearities to produce a linear output
- Compensation for the loading effect of the measuring process on the measured system.

Calibration capability

Self-calibration is very simple in some cases. Sensors with an electrical output can use a known reference voltage level to carry out self-calibration. Also, load-cell types of sensor, which are used in weighing systems, can adjust the output reading to zero when there is no applied mass. In the case of other sensors, two methods of self-calibration are possible, use of a look-up table and an interpolation technique. Unfortunately, a *look-up table* requires a large memory capacity to store correction points. Also, a large amount of data has to be gathered from the sensor during calibration. In consequence, the interpolation calibration technique is preferable. This uses an interpolation method to calculate the correction required to any particular measurement and only requires a small matrix of calibration points (van der Horn, 1996).

Self-diagnosis of faults

Smart sensors perform self-diagnosis by monitoring internal signals for evidence of faults. Whilst it is difficult to achieve a sensor that can carry out self-diagnosis of all possible faults that might arise, it is often possible to make simple checks that detect many of the more common faults. One example of self-diagnosis in a sensor is measuring the sheath capacitance and resistance in insulated thermocouples to detect breakdown of the insulation. Usually, a specific code is generated to indicate each type of possible fault (e.g. a failing of insulation in a device).

One difficulty that often arises in self-diagnosis is in differentiating between normal measurement deviations and sensor faults. Some smart sensors overcome this by storing multiple measured values around a set-point, calculating minimum and maximum expected values for the measured quantity.

Uncertainty techniques can be applied to measure the impact of a sensor fault on measurement quality. This makes it possible in certain circumstances to continue to use a sensor after it has developed a fault. A scheme for generating a validity index has been proposed that indicates the validity and quality of a measurement from a sensor (Henry, 1995).

Automatic calculation of measurement accuracy and compensation for random errors

Many smart sensors can calculate measurement accuracy on-line by computing the mean over a number of measurements and analysing all factors affecting accuracy. This averaging process also serves to greatly reduce the magnitude of random measurement errors.

Adjustment for measurement non-linearities

In the case of sensors that have a non-linear relationship between the measured quantity and the sensor output, digital processing can convert the output to a linear form, providing that the nature of the non-linearity is known so that an equation describing it can be programmed into the sensor.

9.2.3 Smart transmitters

In concept, a smart transmitter is almost identical to the intelligent instruments described earlier. The change in name has occurred over a number of years as intelligent instruments have become smaller and assumed a greater range of functions. Usage of the term 'smart transmitter' rather than 'intelligent instrument' is therefore mainly one of fashion. In some instances, smart transmitters are known alternatively as *intelligent transmitters*. The term *multivariable transmitter* is also sometimes used, particularly for a device like a smart flow-measuring instrument. This measures absolute pressure, differential pressure and process temperature, and computes both the mass flow rate and volume flow rate of the measured fluid.

There has been a dramatic reduction in the price of intelligent devices over the past few years and the cost differential between smart and conventional transmitters is now very small. Indeed, in a few cases, a smart transmitter is now cheaper than its non-smart equivalent because of the greater sales volume for the smart version. Thus, smart transmitters are now routinely bought instead of non-smart versions. However, in many cases, smart transmitters are only used at present in a conventional (non-smart) fashion to give a 4–20 mA analogue measurement signal on the two output wires. Where smart features are used at all, they are often only used during the commissioning phase of measurement systems. This is largely due to the past investment in analogue measurement systems, and the time and effort necessary to convert to measurement systems that can make proper use of intelligent features.

Almost all of the smart sensors that are presently available have an analogue output, because of the continuing popularity and investment in 4–20 mA current transmission systems. Whilst a small number of devices are now available with digital output, most users have to convert this to analogue form to maintain compatibility with existing instrumentation systems.

The capabilities of smart transmitters are perhaps best emphasized by comparing the attributes of the alternative forms of transmitter available. There are three types of transmitter, analogue, programmable and smart.

(a) Analogue transmitters:
- require one transmitter for every sensor type and every sensor range
- require additional transmitters to correct for environmental changes
- require frequent calibration.

(b) Programmable transmitters:
- include a microprocessor but do not have bi-directional communication (hence are not truly intelligent)
- require field calibration.

(c) Smart transmitters:
- include a microprocessor and have bi-directional communication
- include secondary sensors that can measure, and so compensate for, environmental disturbances
- usually incorporate signal conditioning and a–d conversion
- often incorporate multiple sensors covering different measurement ranges and allow automatic selection of the required range. The range can be readily altered if initially estimated incorrectly
- have a self-calibration capability that allows removal of zero drift and sensitivity drift errors
- have a self-diagnostic capability that allows them to report problems or requirements for maintenance
- can adjust for non-linearities to produce a linear output.

Smart transmitters are usually a little larger and heavier than non-smart equivalents. However, their advantages can be summarized as:

- Improved accuracy and repeatability
- Long-term stability is improved and required recalibration frequency is reduced

- Reduced maintenance costs
- Large range coverage, allowing interoperability and giving increased flexibility
- Remote adjustment of output range, on command from a portable keyboard or from a PC. This saves on technician time compared with carrying out adjustment manually
- Reduction in number of spare instruments required, since one spare transmitter can be configured to cover any range and so replace any faulty transmitter
- Possibility of including redundant sensors, which can be used to replace failed sensors and so improve device reliability
- Allowing remote recalibration or re-ranging by sending a digital signal to them
- Ability to store last calibration date and indicate when next calibration is required
- Single penetration into the measured process rather than the multiple penetration required by discrete devices, making installation easier and cheaper
- Ability to store data so that plant and instrument performance can be analysed. For example, data relating to the effects of environmental variations can be stored and used to correct output measurements over a large range.

Summary of smart transmitter features

Many of the features of smart transmitters are common with those of smart sensors, and the comments made earlier about smart sensors therefore apply equally. However, the use of multiple primary sensors and secondary sensors to measure environmental parameters mean that additional comments are necessary in respect of their self-calibration and self-diagnosis capabilities.

Self-calibration

Calibration techniques are very similar to those already described for smart sensors and the general principle is always to use simple calibration methods if these are available. Look-up tables in a smart transmitter have a particularly large memory requirement if correction for cross-sensitivity to another parameter (e.g. temperature) is required because a matrix of correction values has to be stored. Hence, interpolation calibration is even more preferable to look-up tables than it is in the case of calibrating smart sensors.

Self-diagnosis and fault detection

Fault diagnosis in sensors is often difficult because it is not easy to distinguish between measurement deviation due to a sensor fault and deviation due to a plant fault. The best theoretical approach to this difficulty is to apply mathematical modelling techniques to the sensor and plant in which it is working, with the aim of detecting inconsistencies in data from the sensor. However, there are very few industrial applications of this approach to fault detection in practice, firstly, because of the cost of implementation and, secondly, because of the difficulty of obtaining plant models that are robust to plant disturbances. Thus, it is usually necessary to resort to having multiple sensors and using a scheme such as two-out-of-three voting. Further advice on self-checking procedures can be found elsewhere (Brignell, 1996).

Effect of sensor errors

The effect of a sensor error on the quality of measurement varies according to the nature of the fault and the type of sensor. For example, a smart pressure sensor that loses temperature measurement will still give valid measurements but the uncertainty increases.

9.2.4 Communication with intelligent devices

The subject of instrumentation networks and digital communication with instruments is covered in detail in the next chapter. The aim over many years has been to use intelligent devices to their full potential by making all communications, including the measurement signal, entirely digital. A number of digital fieldbuses are now used for instrumentation systems, with protocols such as Profibus and WorldFIP being in widespread use. However, to date, despite international efforts over many years, no standard protocol for digital fieldbus communications has yet been established.

Partly because of this delay in developing an international digital fieldbus standard, and partly because of the need to maintain compatibility with the vast current investment in analogue instrumentation systems, a number of part analogue/part digital communication protocols have been developed as an interim measure. Prominent amongst these is a protocol called HART (**H**ighway **A**ddressable **R**emote **T**ransducer). This is a manufacturer-independent protocol that provides for analogue measurement signal transmission as well as sending command/status information digitally. The normal requirement for such dual analogue/digital communication with an intelligent device is six wires, two to convey the measurement signal, two to convey command/device status information and two to provide a power supply to the device. However, in order to economize on wiring and installation costs, HART allows this requirement to be reduced to four or even two wires by using the signal wires to convey device status/command signals or the power supply or both of these. HART has now achieved widespread use, even though it is not backed by an international standard.

Extended 4–20 mA current interface protocol

The 4–20 mA protocol is the most-used analogue transmission mechanism because of the protection against noise that it offers to the measurement values transmitted. This protocol has been extended for communication with intelligent devices to allow for the transmission, where necessary, of command/status information and the device power supply in analogue form on the signal wires. In this extended protocol, signals in the range 3.8 mA to 20.5 mA are regarded as 'normal' measurement signals, thus allowing for under- and over-range from the 4–20 mA measurement signal standard. The current bands immediately outside this in the range 3.6 mA to 3.8 mA and 20.5 mA to 21.0 mA are used for the conveyance of commands to the sensor/transmitter and the receipt of status information from it. This means that, if the signal wires are also used to carry the power supply to the sensor/transmitter, the power supply current must be limited to 3.5 mA or less to avoid the possibility of it being interpreted as a measurement signal or fault indicator. Signals greater than 21 mA (and less than 3.6 mA if the signal wires are not carrying a power supply) are normally taken to indicate either a short circuit or open circuit in the signal wiring.

Sending commands to sensor/transmitter

Commands can either be sent from a handheld keyboard or else communicated from a remote PC. Whilst a handheld keyboard is the cheaper option in terms of equipment requirement, it cannot store calibration data because it does not usually have any memory. It is therefore time consuming for a technician to enter the necessary calibration data manually. A PC makes communication easier because it can readily store calibration data. Also, its large screen allows more information to be viewed at one time. A PC is also able to receive status data from the sensor and store it for later use (e.g. to disclose trends in sensor status).

For hazardous environments, versions of smart transmitters are available that are made intrinsically safe by using reed-relay switches to alter transmitter parameters. In such cases, an LCD programming display is usually used to give commands to the transmitter, as this is also intrinsically safe.

9.2.5 Computation in intelligent devices

In the past, most computation in intelligent devices has been performed by software routines executed on a general-purpose microcomputer. However, there has been a trend in the last few years towards implementing digital signal processing, data conversion and communication interface functions in specially designed hardware elements. This achieves a large improvement in processing speed compared with the execution of software routines. The first implementations of this (Brignell, 1996) used ASICs (Application Specific Integrated Circuits). An ASIC is a gate array that is programmed by designing a mask that creates connections between elements in the device. Unfortunately, the mask design is a very costly process and therefore such devices are only cost effective in high-volume applications such as automobile systems.

More recently, alternative programming devices such as FPGAs (Field Programmable Gate Arrays) and CPLDs (Complex Programmable Logic Devices) have become available that offer a means of implementing digital signal processing and other functions that are cost effective in low-volume applications. Implementation cost is reduced because these devices are user-programmable and avoid the very expensive mask-design process required by ASICs. In fact, these alternative programming devices are now routinely used to build prototypes of designs before going into production of ASICs (which are still more cost effective than FPGAs and CPLDs in high-volume applications).

A typical programmable device consists of an array of configurable logic blocks, programmable input–output blocks and memory. However, FPGAs and CPLDs differ substantially in the way that elements are connected within the device, with the connections used in a CPLD allowing faster operation of the device. These differences mean that, in general, the CPLD is preferred for applications where there is a requirement for high processing speed and the FPGA is preferred where there is a need for high capacity (number of logic gates) in the device.

Some programmable devices contain both soft and hard cores, in which a hard core performing specific functions such as a PCI (Peripheral Component Interconnect) interface is embedded in a programmable soft core. In this hard-core/soft-core approach, the inclusion of hard-core elements increases computational speed and reduces size, but the increased specialization of the device reduces the number of potential applications and therefore increases unit cost.

Further information on these modern programmable devices can be found elsewhere (Amos, 1995; Brown, 1996).

9.2.6 Future trends in intelligent devices

The extent of application of smart transmitters is currently limited by:

• Lack of sufficient varieties of transmitter due to manufacturers' reluctance to invest in producing them ahead of agreement on an international fieldbus standard
• Limitations on the power of microprocessors available
• The large investment in conventional 4–20 mA signal transmission systems and cabling, thus inhibiting the use of transmitters to their full potential in fully digital transmission mode
• Limitations in the speed of bi-directional communication capabilities. If a common bus is used to transmit signals for several different transmitters, data transfer speed is slow because the bus can only service one transmitter at a time. This means that the time interval between measurements from a particular transmitter being read and responded to can become excessively large. In consequence, it is possible for dangerous conditions to develop in the controlled plant, such as high pressures.

Current research and international discussions are currently directed at solving all of these problems. Hence, a rapid growth in the application of intelligent devices, and their use in fully digital mode, is expected over the next few years. Size reductions will also continue, and indeed the first smart microsensors are now available. These are covered in greater detail in Chapter 13. The establishment of an international fieldbus standard (see Chapter 10) will also encourage greater use of intelligent devices in all-digital instrumentation and control schemes for industrial plant.

The use of programmable devices to perform signal processing functions within intelligent devices is likely to expand rapidly in the future. As well as further improvements to the processing capacity and computational speed of these devices, current research (Tempesti, 1999) is directed towards developing self-repairing capabilities in such devices.

Also, now that both sensors, processing elements and microcontrollers can all be constructed on silicon wafers, the next logical step is to extend the process of integration still further and include all of these elements on a single silicon chip. Apart from the reduction in system cost due to the reduction in the number of components, the requirement for fewer connections between components will lead to substantially improved system reliability, since most system faults can be traced to connection faults. However, whether, or how soon, this further integration will happen will depend on the relevant economics of separate and combined implementation of these system components.

9.3 Self-test questions

9.1 Describe briefly the three essential elements of a microcomputer system.
9.2 Write down the one's complement and two's complement 8-bit representations of the following decimal numbers: (a) 47 (b) −119 (c) −101 (d) 86 (e) 108
9.3 Write down the binary, octal and hexadecimal representations of the following decimal numbers: (a) 57 (b) 101 (c) 175 (d) 259 (e) 999 (f) 1234

9.4 The binary code representation of a number is 111101001101. If all 12 digits are data bits, what are (a) the octal and (b) the hexadecimal equivalents of this number?

9.5 The binary code representation of a number is 1011000110100101. If all 16 digits are data bits, what are (a) the octal and (b) the hexadecimal equivalents of this number?

9.6 The binary code representation of a number is 110100011000. The first and last binary digits are sign and parity bits respectively, so only the middle ten digits are data bits. What are (a) the octal, (b) the hexadecimal and (c) the decimal data values?

References and further reading

Amos, D. (1995) Interconnect trade-offs: CPLD versus FPGA, *Electronic Engineering*, March pp. 81–84.

Barney, G.C. (1988) *Intelligent Instruments*, Prentice-Hall, London.

Brignell, J. and White, N. (1996) *Intelligent Sensor Systems*, Institute of Physics Publishing.

Brown, S. and Rose, J. (1996) Architecture of FPGAs and CPLDs: A Tutorial, *IEEE Design and Test of Computers*, **13**, pp. 42–57.

Henry, M. (1995) Self-validation improves Coriolis flowmeter, *Control Engineering*, May pp. 81–86.

Tempesti, G. (1999) A biologically-inspired self-repairing FPGA, *Electronic Engineering*, August pp. 45–46.

Van der Horn, G *et al*. (1996) Calibration and linearisation method for microcontroller-based sensor systems, *Measurement and Control*, **29**, pp. 270–273.

10

Instrumentation/computer networks

10.1 Introduction

The inclusion of computer processing power in intelligent instruments and intelligent actuators creates the possibility of building an instrumentation system where several intelligent devices collaborate together, transmit information to one another and execute process control functions. Such an arrangement is often known as a *distributed control system*. Additional computer processors can also be added to the system as necessary to provide the necessary computational power when the computation of complex control algorithms is required. Such an instrumentation system is far more fault tolerant and reliable than older control schemes where data from several discrete instruments is carried to a centralized computer controller via long instrumentation cables. This improved reliability arises from the fact that the presence of computer processors in every unit injects a degree of redundancy into the system. Therefore, measurement and control action can still continue, albeit in a degraded form, if one unit fails.

In order to effect the necessary communication when two or more intelligent devices are to be connected together as nodes in a distributed system, some form of electronic highway must be provided between them that permits the exchange of information. Apart from data transfer, a certain amount of control information also has to be transferred. The main purpose of this control information is to make sure that the target device is ready to receive information before data transmission starts. This control information also prevents more than one device trying to send information at the same time.

In modern installations, all communication and data transmission between processing nodes in a distributed instrumentation and control system is carried out digitally along some form of electronic highway, although analogue data transmission (mainly current loop) is still widely used to transmit data from field devices into the processing nodes. If analogue transmission is used for measurement data, an analogue-to-digital converter must be provided at the interfaces between the measurement signal transmission cables and the processing nodes.

The electronic highway can either be a serial communication line, a parallel data bus, or a local area network. Serial data lines are very slow and are only used where a low data transmission speed is acceptable. Parallel data buses are limited to connecting a modest number of devices spread over a small geographical area, typically a single room, but provide reasonably fast data transmission. Local area networks are used to connect larger numbers of devices spread over larger geographical distances, typically a single building or site. They transmit data in digital format at high speed. Instrumentation networks that are geographically larger than a single building or site can also be built, but these generally require transmission systems that include telephone lines as well as local networks at particular sites within the large system.

The *input/output interface* of an intelligent device provides the necessary connection between the device and the electronic highway. The interface can be either serial or parallel.

A *serial interface* is used to connect a device onto a serial communication line. The connection is effected physically by a multi-pin plug that fits into a multi-pin socket on the casing of the device. The pins in this plug/socket match the signal lines used in the serial communication line exactly in number and function. Effectively, there is only one standard format for serial data transmission that enjoys international recognition. Whilst this is advantageous in avoiding compatibility problems when connecting together devices coming from different manufacturers, serial transmission is relatively slow.

A *parallel interface* is used to connect devices onto parallel instrument buses and also into all other types of network systems. Like the serial interface, the parallel interface exists physically as a multi-pin plug that fits into a multi-pin socket on the casing of the device. The pins in the plug/socket are matched exactly in number and function with the data and control lines used by a particular parallel instrument bus. Unfortunately, there are a number of different parallel instrument buses in use and thus a corresponding number of different parallel interface protocols, with little compatibility between them. Hence, whilst parallel data transmission is much faster than serial transmission, there are serious compatibility problems to be overcome when connecting together devices coming from different manufacturers because of the different parallel interface protocols used.

10.2 Serial communication lines

Serial communication only allows relatively slow data transfer rates, but it can operate over much larger distances than parallel communication. Transmission distances up to 3 km are possible with standard copper-wire links, and much greater distances can be achieved using either telephone lines or radio telemetry. Data are transferred down a single line on the electronic highway one bit at a time, and the start and finish of each item of data are denoted by special sequences of control characters that precede and follow the data bits.

Three alternative forms of serial communication exist, known as *simplex*, *half-duplex* and *full-duplex*. Simplex mode only allows transmission of data in one direction. For this reason, it is not widely used, since, although it permits a remote sensor to transmit information, the receiving station cannot send a message back to acknowledge receipt or request retransmission if the received data has been corrupted. In half-duplex mode, the same data wire is used by a device to both send and receive data, and thus the receiving station is able to acknowledge receipt of data from a sensor. However, sending and receiving of data simultaneously is not possible. In full-duplex operation, two separate data lines are used, one for send and one for receive, and so simultaneous sending and receiving of data is therefore possible. In addition to these three forms (simplex, half-duplex and full-duplex), two different transmission modes exist, known as *asynchronous transmission* and *synchronous transmission*.

10.2.1 Asynchronous transmission

The structure of the data and control characters used in asynchronous transmission are shown in Figure 10.1. The binary digits of '1' and '0' are represented by voltage logic levels of +V and zero. The start of transmission of each character is denoted by a binary '0' digit. The following seven digits represent a coded character. The next digit is known as a parity bit. Finally, the end of transmission of the character is denoted by either one or two stop bits, which are binary '1' digits.

The parity bit is provided as an error checking mechanism. It is set to make the total number of binary '1' digits in the character representation either odd or even, according to whether the odd or even parity system is being used (some manufacturers use odd parity and some use even parity). The seven character digits are usually coded using the ASCII system (American Standard Code for Information Interchange). This provides for the transmission of the full set of alphabetic, numerical, punctuation and control characters.

Asynchronous transmission allows the transmitter and receiver to use their own clock signals to put data on to the transmission line at one end and take it off at the other end. Receipt of the start bit causes the receiver to synchronize its clock with the incoming data, and this synchronization is maintained whilst the stream of character bits, parity bit and stop bit(s) are received. One particular disadvantage of asynchronous transmission

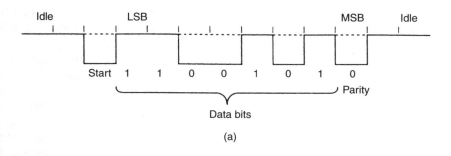

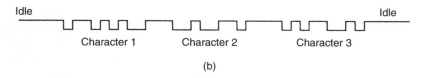

Fig. 10.1 Serial data transmission: (a) data and control bits for one character (even parity); (b) a string of serially transmitted characters.

is that transfer only occurs one bit at a time, and so the data transfer efficiency is low. Also, only seven out of each ten or 11 bits transmitted represent a character, with the other three or four bits being for synchronization. Maximum transmission speed is 19 200 bit/s, which is frequently referred to as a speed of 19.2 kbaud (where 1 baud = 1 bit/s).

However, although the data transfer rate is slow, asynchronous serial transmission does have the advantage that only two different standard formats exist, and these are very similar. These two formats, which have achieved international recognition, are the RS232 standard (USA) and the CCITT V24 standard (European). The only significant difference between these is the logic voltage level used, 3 V for RS232 and 6 V for V24, and this incompatibility can be handled very easily. Within either of these standards, there are options about the type of parity (odd or even), the number of stop bits and whether data transmission is in full- or half-duplex mode. The various equipment manufacturers use the options differently but it is a relatively simple matter to accommodate these differences when connecting together devices coming from different manufacturers.

When asynchronous transmission is not fast enough, local area networks are used that transmit data synchronously. This is covered in detail in section 10.4.

10.3 Parallel data bus

Parallel data buses allow much higher data rates than serial communication lines because data is transmitted in parallel, i.e. several bits are transmitted simultaneously. Control signals are transmitted on separate control lines within the bus. This means that the data lines are used solely for data transmission, thereby optimizing the data transmission rate capability.

There are a number of different parallel data buses in existence, but there is little compatibility between them. Differences exist in the number of data lines used, the number of control lines used, the interrupt structure used, the data timing system and the logic levels used for operation. Equipment manufacturers tend to keep to the same parallel interface protocol for all their range of devices, but different manufacturers use different protocols. Thus, while it will normally be easy to connect together a number of intelligent devices that all come from the same manufacturer, interfacing difficulties are likely to be experienced if devices from different manufacturers are connected together.

In practice, limiting the choice of devices to those available from the same manufacturer is unlikely to be acceptable. Even if all the units required can be obtained from the same manufacturer, this limitation is likely to mean having to use devices with a lower performance specification than desirable and at a cost penalty. Therefore, it is necessary to find a way of converting the different interface protocols into a common communication format at the interface between the device and the transmission medium. This is normally done by using some type of network as the transmission medium rather than using a single instrument bus. Many different systems exist for dealing with the different communication protocols used by different equipment manufacturers and thus allowing all the devices to be connected onto one particular type of instrument bus. Two examples of such systems are CAMAC and MEDIA, although these are rarely used nowadays as it is more common to solve the problem by using local area networks.

Some examples of parallel interface buses are *Multibus*, the *S100 Bus* and the *IEEE 488 bus* (also known by the alternative names of *HP-IB bus*, *GBIB bus*, *Plus bus* and *ANSI standard MC1-1 bus*). The IEEE 488 bus has now gained prominence for instrumentation networks, and it has been adopted by the International Electrotechnical Commission (IEC) as a standard, *IEC625*. It provides a parallel interface that facilitates the connection of intelligent instruments, actuators and controllers within a single room. The maximum length of bus allowable is about 20 m, and no more than 15 instruments should be distributed along its length. The maximum distance between two particular units on the bus should not exceed about 2 m. The maximum data transfer rate permitted by the bus is 1 Mbit/s in theory, though the maximum data rate achieved in practice over a full 20 m length of bus is more likely to be in the range 250–500 Kbit/s.

Physically, the bus consists of a shielded, 24-conductor cable. Sixteen of the conductors are used as signal lines, eight carrying data and eight carrying control signals. The remaining eight conductors are used as ground wires for the control signals, each control wire being twisted together along its full length with one of the ground wires. This minimizes cross-talk between the control wires. Normal practice is to route the eight twisted pairs carrying control signals in the centre of the cable and place the eight data wires round the periphery. This bundle of wires is surrounded by shielding and an outer insulated coating. Each end of the cable is connected to a standard 24-pin metal connector, with generally a female connector at one end and a male connector at the other. This facilitates several cables being chained from one device to another.

Figure 10.2 shows three devices connected onto an IEEE 488 bus. The bus can only carry one lot of information at a time, and which unit is sending data and which is receiving it is controlled by a supervisory computer connected to the bus. This supervisor ensures that only one unit can put data on the bus at a time, and thus prevents the corruption of data that would occur if several instruments had simultaneous access.

Having eight data lines means that the bus can transmit 8 bits of data in parallel at the same time. This was originally designed so that 8-bit computer words could be transmitted as whole words. This does not prevent the bus being used with computers of a different wordlength, for example 16 or 32 bits. However, if the wordlength is longer than 8 bits, whole words cannot be transmitted in one go: they have to be transmitted 8 bits at a time. The eight status lines provide the necessary control to ensure that when data transmission takes place between two units, three conditions are satisfied simultaneously. These three conditions are (a) that the sender unit is ready to transmit data, (b) that the receiver unit is ready to receive data and (c) that the bus does not currently have any data on it. The functioning of the eight status lines is as follows:

- DAV (Data valid)   This goes to a logic zero when the data on the eight data lines is valid.
- NFRD (Not ready for data)   This goes to logic zero when the receiver unit is ready to accept data.
- NDAC (Not data accepted)   This goes to logic zero when the receiver unit has finished receiving data.

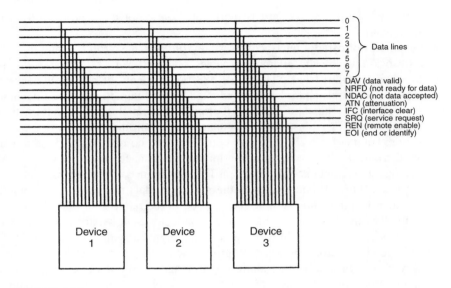

Fig. 10.2 IEEE 488 bus.

- ATN (Attention)   This is a general control signal that is used for various purposes to control the use of data lines and specify the send and receive devices to be used.
- IFC (Interface clear)   The controller uses this status line to put the interface into a wait state.
- SRQ (Service request)   This is an interrupt status line that allows high priority devices such as alarms to interrupt current bus traffic and get immediate access to the bus.
- REN (Remote enable)   This status line is used to specify which of two alternative sets of device programming data are to be used.
- EOI (End of output or identify)   This status line is used by the sending unit to indicate that it has finished transmitting data.

The IEEE 488 bus protocol uses a logic level of less than 0.8 V to represent a logic 0 signal and a voltage level greater than 2.0 V to represent a logic 1 signal.

10.4 Local area networks (LANs)

Local area networks transmit data in digital format along serial transmission lines. Synchronous transmission is normally used because this allows relatively high transmission speeds by transmitting blocks of characters at a time. A typical data block consists of 80 characters: this is preceded by a synchronization sequence and followed by a stop sequence. The synchronization sequence causes the receiver to synchronize its clock with that of the transmitter. The two main standards for synchronous, serial transmission are RS422 and RS485. A useful comparison between the performance and characteristics of each of these and RS232 (asynchronous serial transmission) can be found in Brook, (1996).

LANs have particular value in the monitoring and control of plants that are large and/or widely dispersed over a large area. Indeed, for such large instrumentation systems, a local area network is the only viable transmission medium in terms of performance and cost. Parallel data buses, which transmit data in analogue form, suffer from signal attenuation and noise pickup over large distances, and the high cost of the long, multi-core cables that they need is prohibitive.

The development of instrumentation networks is not without problems, however. Careful design of the network is required to prevent corruption of data when two or more devices on the network try to access it simultaneously and perhaps put information onto the data bus at the same time. This problem is solved by designing a suitable network protocol that ensures that network devices do not access the network simultaneously, thus preventing data corruption.

In a local area network, the electronic highway can take the form of either copper conductors or fibre-optic cable. Copper conductors are the cheapest option and allow transmission speeds up to 10 Mbit/s, using either a simple pair of twisted wires or a coaxial cable. However, fibre-optic cables are preferred in many networks for a number of reasons. The virtues of fibre-optic cables as a data transmission medium have been expounded in Chapter 8. Apart from the high immunity of the signals to noise, a fibre-optic transmission system can transfer data at speeds up to 240 Mbit/s. The reduction in signal attenuation during transmission also means that much longer transmission distances are possible without repeaters being necessary. For instance, the allowable distances between repeaters for a fibre-optic network are quoted as 1 km for half-duplex operation and up to 3.5 km for full-duplex operation. In addition, the bandwidth of fibre-optic transmission is higher than for electrical transmission. Some cost saving can be achieved by using plastic fibre-optic cables, but these cannot generally be used over distances greater than about 30 m because signal attenuation is too high.

There are many different protocols for local area networks but these are all based on one of three network structures known as star networks, bus networks and ring networks, as shown in Figure 10.3. A local area network operates within a single building or site and can transmit data over distances up to about 500 m without signal attenuation being a problem. For transmission over greater distances, telephone lines are used in the network. Intelligent devices are interfaced to the telephone line used for data transmission via a modem. The *modem* converts the signal into a frequency-modulated analogue form. In this form, it can be transmitted over either the public switched telephone network or over private lines rented from telephone companies. The latter, being dedicated lines, allow higher data transmission rates.

10.4.1 Star networks

In a *star network*, each instrument and actuator is connected directly to the supervisory computer by its own signal cable. One apparent advantage of a star network is that data can be transferred if necessary using a serial communication protocol such as

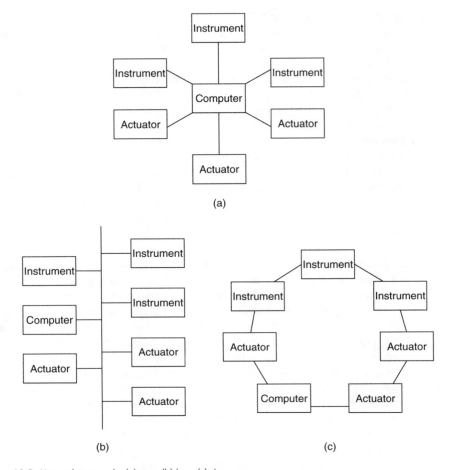

Fig. 10.3 Network protocols: (a) star; (b) bus; (c) ring.

RS232. This is an industry standard protocol and so compatibility problems do not arise, but of course data transfer is very slow. Because of this speed problem, parallel communication is usually preferred even for star networks.

Whilst star networks are simple in structure, the central supervisory computer node is a critical point in the system and failure of this means total failure of the whole system. When any device in the network needs to communicate with another device, a request has to be made to the central supervisory computer and all data transferred is routed through this central node. If the central node is inoperational for any reason then data communication in the network is stopped.

10.4.2 Ring and bus networks

In contrast, both ring and bus networks have a high degree of resilience in the face of one node breaking down. Hence, they are generally preferred to star networks. If the processor in any node breaks down, the data transmission paths in the network are still maintained. Thus, the network can continue to operate, albeit at a degraded performance level, using the remaining computational power in the other processors. Most computer and intelligent instrument/actuator manufacturers provide standard conversion modules that allow their equipment to interface to one of these standard networks.

In a *ring network*, all the intelligent devices are connected to a bus that is formed into a continuous ring. Ring protocol sends a special packet (or token) continuously round the ring to control access to the network. A station can only send data when it receives the token. During data transmission, the token is attached to the back of the message sent so that, once the information has been safely received, the token can continue on its journey round the network. A typical data transmission speed is 10 Mbit/s. Cambridge Ring, Arcnet and the IEEE 802.5 bus are examples of token ring protocols.

A *bus network* is similar to a ring network but the bus that the devices are connected onto is not continuous. Bus networks are also resilient towards the breakdown of one node in the network. A *contention protocol* is normally used. This allows any station to have immediate access to the network unless another station is using it simultaneously, in which case the protocol manages the situation and prevents data loss/corruption. They have a similar data transmission speed to ring networks of 10 Mbit/s. Ethernet and the IEEE 802.3 standard bus are examples of bus networks.

10.5 Gateways

Gateways, such as P1451 produced by the IEEE, are interfaces between intelligent devices and local area networks that overcome the non-compatibility problem between buses using different protocols. As a different gateway is required for each different LAN that a device may be connected to, this theoretically adds cost to the system and imposes a time delay that reduces performance. However, the availability of fast processing power at low cost means that the use of a gateway is a feasible solution to the problem of using devices from different suppliers that are designed for different buses. Alternative forms of gateway also provide a means of connecting analogue devices into a digital network, particularly those using 4–20 mA current loop transmission standards. In many cases, gateways provide a means of retaining existing equipment in a new digital network and thus avoid the expense of buying new devices throughout a plant.

10.6 HART

As intelligent devices developed over the years, the need arose for network protocols that could provide for the necessary digital communications to and from such devices. HART (**H**ighway **A**ddressable **R**emote **T**ransducer) is a well-known bus-based networking protocol that satisfies this need. Over the years, this has gained widespread international use, and has now become a *de facto* standard, with HART-compatible devices being available from all major instrument manufacturers. Recent surveys have predicted that HART will continue in widespread use for the next 15 to 20 years, irrespective of the timing of the long-promised, internationally accepted, all-digital fieldbus standard.

HART was always intended to be an interim network protocol to satisfy communication needs in the transitional period between the use of analogue communication with non-intelligent devices and fully digital communication with intelligent devices according to an international standard digital fieldbus protocol. Because of this need to support both old and new systems, HART supports two modes of use, a hybrid mode and a fully digital mode.

In *hybrid mode*, status/command signals are digital but data transmission takes place in analogue form (usually in 4–20 mA format). One serious limitation of this mode is that it is not possible to transmit multiple measurement signals on a single bus, since the analogue signals would corrupt each other. Hence, when HART is used in hybrid mode, the network must be arranged in a star configuration, using a separate line for each field device rather than a common bus.

In *fully digital mode*, data transmission is digital as well as status/command signals. This enables one cable to carry signals for up to 15 intelligent devices. In practice, the fully digital mode of HART is rarely used, since the data transmission speed is very limited compared with alternative fieldbus protocols such as Profibus. Therefore, the main application of the HART protocol has been to provide a communication capability with intelligent devices when existing analogue measurement signal transmission has to be retained because conversion to fully digital operation would be too expensive.

10.7 Digital fieldbuses

'Fieldbus' is a generic word that describes a range of high-speed, bus-based, network protocols that support two-way communication in digital format between a number of intelligent devices in a local area network. All forms of transmission are supported including twisted pair, coaxial cable, fibre optic and radio links. Compared with analogue networks that use 4–20 mA current loop data transmission, fieldbus-based systems have many advantages including faster system design, faster commissioning, reduced cabling costs, easier maintenance, facilities for automatic fault diagnosis (which also improves safety), the flexibility to interchange components derived from different suppliers, and estimated reductions of 40% in installation and maintenance costs. However, it should be noted that cost savings alone would not be sufficient justification for replacing an analogue system with a fieldbus-based one if the analogue system was operating satisfactorily.

Intelligent devices in an automated system comprise of a range of control elements, actuators, information processing devices, storage systems and operator displays as well as measurement devices. Hence, any fieldbus protocol must include provision for the needs of all system elements, and the communication requirements of field measurement devices cannot be viewed in isolation from these other elements. The design of a network protocol also has to cater for implementation in both large and small plants. A large plant may contain a number of processors in a distributed control system and have a large number of sensors and actuators. On the other hand, a small plant may be controlled by a single personal computer that provides an operator display on its monitor as well as communicating with plant sensors and actuators.

Many different digital fieldbus protocols now exist, and names of some of the more prominent ones include Profibus (Germany), WorldFIP (France), P-net (Denmark), Lonworks (USA), Devicenet (USA), IEEE 1118 (USA), Milbus (UK), Canbus, Interbus-S and SDS. These differ in many major respects such as message format, access protocols and rules for performance prediction. In recognition of the difficulties inherent in attempting to connect devices from different manufacturers that use a variety of incompatible interface standards and network protocols, the International Electrotechnical Commission (IEC) set up a working part in 1985 that was charged with defining a standard interface protocol, which was to be called the IEC Fieldbus. However, at the time of the IEC initiative, a number of companies were already developing their own fieldbus standards, and commercial interests have continually blocked agreement on a common, internationally recognized standard. In the meantime, some countries have adopted their own national fieldbus standard. Also, the European Union established a European standard (EN50170) in 1996 as an interim measure until the appearance of the promised IEC standard. EN50170 has adopted Profibus, WorldFIP and P-net as the three standards authorized for use, and the intention is to allow their use until six years after the IEC standard has been published. The inclusion of three protocols rather than one in EN50170 is unfortunate, but three protocols are much better than the 50 protocols that were in use prior to the adoption of EN50170.

In the meantime, efforts to achieve a single fieldbus standard have continued, with development now being carried out by a body called the Fieldbus Foundation, which is a consortium of instrument system manufacturers and users supported by the IEC. The present approach is to define a protocol whose basic architecture is in two levels, known as upper and lower. The lower level provides for communication between field devices and field input–output devices whilst the upper level enables field input–output devices to communicate with controllers. These two levels have quite different characteristics. The lower level generally requires few connections, only needs a slow data-transfer rate and must support intrinsically safe working. On the other hand, the upper level requires numerous connections and fast data transfer, but does not have to satisfy intrinsic safety requirements. In the fieldbus standard proposed, the lower level will conform with the specifications for the Application Layer in ISO-7* and the upper layer will satisfy the specifications for the Physical and Data Link Layers in ISO-7. Three standard bus speeds are currently specified for the Foundation Fieldbus lower level of 31.25 kbit/s, 1 Mbit/s and 2.5 Mbit/s. Maximum cable lengths allowed are 1900 m at 31.25 kbit/s, 750 m at 1 Mbit/s and 500 m at 2.5 Mbit/s. For the upper Foundation Fieldbus layer, a high-speed ethernet is currently being developed that will provide a data transfer rate up to 100 Mbit/s.

At the time of writing, it appears that we may now be very close to achieving an international fieldbus standard. The Fieldbus Foundation published a draft standard in 1998 known as IEC61158. This was agreed by a majority vote, but with dissent from a number of major instrument manufacturers. In the confirmed standard, which is expected in 2000, there are likely to be supplements that will allow devices operating under other protocols such as Profibus to be interfaced with the IEC61158 system.

---

* ISO-7 is an abbreviation for the International Standards Organization Open System Interconnection seven-layer model which covers all aspects of connections between computers and intelligent devices. It is discussed further in section 10.8.

10.8 Communication protocols for very large systems

Once a system gets too large to be covered by a local area network, it is generally necessary to use telephone lines. These provide communication over large distances within a protocol that is often called a *wide area network*. Public telephone lines are readily available, but there is a fundamental problem about their use for a wide area network. Whilst instrumentation networks need high bandwidths, public networks operate at the low bandwidth required to satisfy speech-based telephony. High bandwidths can be obtained by leasing private telephone lines but this solution is expensive and often uneconomic.

The solution that is emerging is to extend LAN technology into public telephone networks. A LAN extended in this way is renamed a metropolitan area network (MAN). An IEEE standard for MAN (IEEE 802.6) was first published in 1990. Messages between nodes are organized in packets. MANs cover areas that are typically up to 50 km in diameter, but in some cases links can be several hundred kilometres long. Use of the public switched telephone network for transmission is most common although private lines are sometimes used.

Both ring and bus networks lose efficiency as the number of nodes increases and are unsuitable for adoption by MAN. Instead, a protocol known as distributed queue dual bus (DQDB) is used. DQDB is a hybrid bus that carries isochronous data for the public switched telephone network as well as providing the data bus for a MAN. For handling data on a MAN, DQDB has a pair of buses on which data, preceded by the target address, circulates in fixed size packets in opposite directions, i.e. there is a clockwise bus and an anticlockwise bus. All stations have access to both buses and the protocol establishes a distributed queue. This ensures that all stations have access to the bus on a fair basis. Thus, the stations have their access demands satisfied in the order in which they arise (i.e. a first-come, first-served basis) but commensurate with ensuring that the bus is used efficiently. Fibre-optic cables are commonly used for the buses, allowing data transmission at speeds up to 140 Mbit/s.

10.8.1 Protocol standardization

Many years ago, the International Standards Organization recognized the enormous problems that would ensue, as the size of networks increased, if a diversity of communication protocols developed. In response, it published the *Open Systems Interconnection seven-layer model (ISO-7)* in 1978. This provides standard protocols for all aspects of computer communications required in a large-scale system, that is, management and stock control information etc. as well as instrumentation/process control networks.

The *ISO seven-layer model* defines a set of standard message formats, and rules for their interchange. The model can be applied both within local area networks and in much larger global networks that involve data transmission via telephone lines. Whilst the standards and protocols involved in ISO-7 are highly complex, the network builder does not need to have a detailed understanding of them as long as all devices used in the network are certified by their manufacturer as conforming to the standard. The main functions of each of the seven layers are summarized below:

Layer 1: Physical protocol: Defines how data are physically transported between two devices, including specification of cabling, connectors, I/O ports, modems, voltage levels, signal format and transfer speed.

Layer 2: Data protocol: Establishes paths to ensure data can be exchanged between two devices, and provides error detection and correction (by retransmitting corrupted data).

Layer 3: Network protocol: Controls the flow of data in packets between all devices in a network.

Layer 4: Transport protocol: Allows a user-task on one computer to communicate with another user-task on a different computer transparently of network characteristics, thus ensuring high reliability in data exchange.

Layer 5: Session protocol: Synchronizes communication activities during a session, and maintains a communication path between active user-tasks (a session is defined as the period of time during which two user-tasks remain connected).

Layer 6: Presentation protocol: Provides for code conversion as necessary, so that user-tasks using different data formats can communicate with each other.

Layer 7: Application protocol: Controls functions specific to user-tasks, such as resource sharing, file transfer and remote file access.

It should be noted that only layers 1, 2 and 7 are usually relevant to instrumentation and plant control systems.

The *Manufacturing Automation Protocol* (MAP) was conceived by General Motors in 1980 in order to support computer integrated manufacturing. It conforms with ISO-7, and is a similar attempt at providing a standard computer communications protocol for large systems.

10.9 Future development of networks

Network design and protocol are changing at a similar rapid rate to that of computer systems as a whole. Hence, it would be impossible in a text of this nature to cover all current developments, and, in any case, any such coverage would rapidly become out of date. The past few pages have covered some aspects of the general concepts and design of networks, and this will prove useful in helping the reader to understand the mode of operation of existing networks. However, network specialists should always be consulted to obtain up-to-date information about the current situation whenever a new network is being planned.

References and further reading

Brook, N. and Herklot, T. (1996) Choosing and implementing a serial interface, *Electronic Engineering*, June pp. 46–54.

11

Display, recording and presentation of measurement data

The earlier chapters in this book have been essentially concerned with describing ways of producing high-quality, error-free data at the output of a measurement system. Having got the data, the next consideration is how to present it in a form where it can be readily used and analysed. This chapter therefore starts by covering the techniques available to either display measurement data for current use or record it for future use. Following this, standards of good practice for presenting data in either graphical or tabular form are covered, using either paper or a computer monitor screen as the display medium. This leads on to a discussion of mathematical regression techniques for fitting the best lines through data points on a graph. Confidence tests to assess the correctness of the line fitted are also described. Finally, correlation tests are described that determine the degree of association between two sets of data when they are both subject to random fluctuations.

11.1 Display of measurement signals

Measurement signals in the form of a varying electrical voltage can be displayed either by an *oscilloscope* or else by any of the *electrical meters* described earlier in Chapter 6. However, if signals are converted to digital form, other display options apart from meters become possible, such as electronic output displays or using a computer monitor.

11.1.1 Electronic output displays

Electronic displays enable a parameter value to be read immediately, thus allowing for any necessary response to be made immediately. The main requirement for displays is that they should be clear and unambiguous. Two common types of character format used in displays, seven-segment and 7×5 dot matrix, are shown in Figure 11.1. Both types of display have the advantage of being able to display alphabetic as well as numeric information, although the seven-segment format can only display a limited nine-letter subset of the full 26-letter alphabet. This allows added meaning to be given to the number displayed by including a word or letter code. It also allows a single

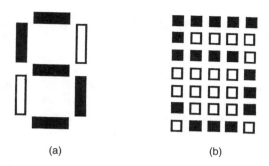

Fig. 11.1 Character formats used in electronic displays: (a) seven-segment; (b) 7 × 5 dot matrix.

display unit to send information about several parameter values, cycling through each in turn and including alphabetic information to indicate the nature of the variable currently displayed.

Electronic output units usually consist of a number of side-by-side cells, where each cell displays one character. Generally, these accept either serial or parallel digital input signals, and the input format can be either binary-coded decimal (BCD) or ASCII. Technologies used for the individual elements in the display are either light-emitting diodes (LEDs) or liquid-crystal elements.

11.1.2 Computer monitor displays

Now that computers are part of the furniture in most homes, the ability of computers to display information is widely understood and appreciated. Computers are now both cheap and highly reliable, and they provide an excellent mechanism for both displaying and storing information. As well as alphanumeric displays of industrial plant variable and status data, for which the plant operator can vary the size of font used to display the information at will, it is also relatively easy to display other information such as plant layout diagrams, process flow layouts etc. This allows not only the value of parameters that go outside control limits to be displayed, but also their location on a schematic map of the plant. Graphical displays of the behaviour of a measured variable are also possible. However, this poses a difficulty when there is a requirement to display the variable's behaviour over a long period of time since the length of the time axis is constrained by the size of the monitor's screen. To overcome this, the display resolution has to decrease as the time period of the display increases.

Touch screens are the very latest development in computer displays. Apart from having the ability to display the same sort of information as a conventional computer monitor, they also provide a command-input facility in which the operator simply has to touch the screen at points where images of keys or boxes are displayed. A full 'qwerty' keyboard is often provided as part of the display. The sensing elements behind the screen are protected by the glass and continue to function even if the glass gets scratched. Touch screens are usually totally sealed, and thus provide intrinsically safe operation in hazardous environments.

11.2 Recording of measurement data

Many techniques now exist for recording measurement data in a form that permits subsequent analysis, particularly for looking at the historical behaviour of measured parameters in fault diagnosis procedures. The earliest recording instruments used were various forms of mechanical chart recorder. Whilst many of these remain in use, most modern forms of chart recorder exist in hybrid forms in which microprocessors are incorporated to improve performance. The sections below discuss these, along with other methods of recording signals including digital recorders, magnetic tape recorders, digital (storage) oscilloscopes and hard-copy devices such as dot-matrix, inkjet and laser printers.

11.2.1 Mechanical chart recorders

Mechanical chart recorders are a long-established means of making permanent records of electrical signals in a simple, cheap and reliable way, even though they have poor dynamic characteristics which means that they are unable to record signals at frequencies greater than about 30 Hz. They have particular advantages in providing a non-corruptible record that has the merit of instant 'viewability', thereby satisfying regulations in many industries that require variables to be monitored and recorded continuously with hard-copy output. ISO 9000 quality assurance procedures and ISO 14000 environmental protection systems set similar requirements, and special regulations in the defence industry go even further by requiring hard-copy output to be kept for ten years. Hence, whilst many people have been predicting the demise of chart recorders, the reality of the situation is that they are likely to be needed in many industries for many years to come. This comment applies particularly to the more modern, hybrid form of chart recorder, which contains a microprocessor to improve performance. Mechanical chart recorders are either of the galvanometric type or potentiometric type. Both of these work on the same principle of driving chart paper at a constant speed past a pen whose deflection is a function of the magnitude of the measured signal. This produces a time history of the measured signal.

Galvanometric recorders

These work on the same principle as a moving-coil meter except that the pointer draws an ink trace on paper, as illustrated in Figure 11.2, instead of merely moving against a scale. The measured signal is applied to the coil, and the angular deflection of this and its attached pointer is proportional to the magnitude of the signal applied. Inspection of Figure 11.3(a) shows that the displacement y of the pen across the chart recorder is given by $y = R \sin \theta$. This sine relationship between the input signal and the displacement y is non-linear, and results in an error of 0.7% for deflections of $\pm 10°$. A more serious problem arising from the pen moving in an arc is that it is difficult to relate the magnitude of deflection with the time axis. One way of overcoming this is to print a grid on the chart paper in the form of circular arcs, as illustrated in Figure 11.3(b). Unfortunately, measurement errors often occur in reading this type of chart, as interpolation for points drawn between the curved grid lines is difficult. An alternative solution is to use heat-sensitive chart paper directed over a knife-edge, and

Measurement and Instrumentation Principles 203

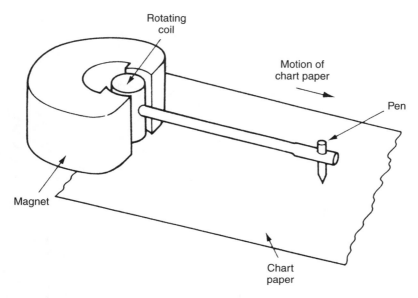

Fig. 11.2 Simple galvanometric recorder.

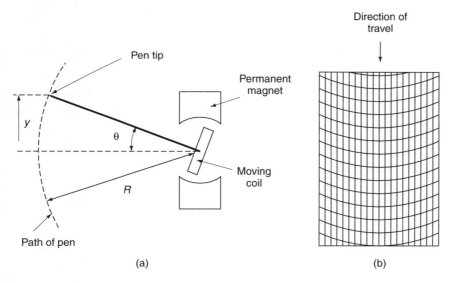

Fig. 11.3 Output of simple chart recorder: (a) y versus θ relationship; (b) curvilinear chart paper.

to replace the pen by a heated stylus, as illustrated in Figure 11.4. The input–output relationship is still non-linear, with the deflection y being proportional to $\tan\theta$ as shown in Figure 11.5(a), and the reading error for excursions of $\pm 10°$ is still 0.7%. However, the rectilinearly scaled chart paper now required, as shown in Figure 11.5(b), allows much easier interpolation between grid lines.

204 Display, recording and presentation of measurement data

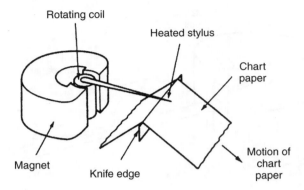

Fig. 11.4 Knife-edge galvanometric recorder.

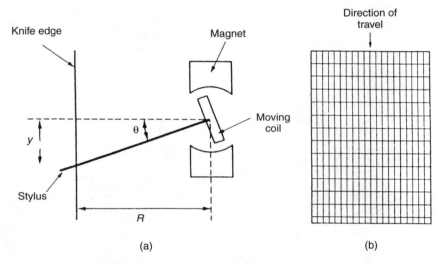

Fig. 11.5 Knife-edge recorder output: (a) y versus θ relationship; (b) rectilinear chart paper.

Neglecting friction, the torque equation for a galvanometric recorder in steady state can be expressed as:

Torque due to current in coil = Torque due to spring

Following a step input, we can write:

Torque due to current in coil = Torque due to spring + Accelerating torque

or:

$$K_i I = K_s \theta + J\ddot{\theta} \qquad (11.1)$$

where I is the coil current, θ is the angular displacement, J is the moment of inertia and K_i and K_s are constants. Consider now what happens if a recorder with resistance R_r is connected to a transducer with resistance R_t and output voltage V_t, as shown in

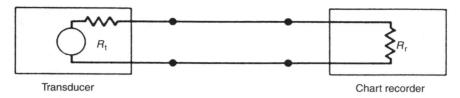

Fig. 11.6 Connection of transducer to chart recorder.

Figure 11.6. The current flowing in steady state is given by: $I = V_t/(R_t + R_r)$. When the transducer voltage V_t is first applied to the recorder coil, the coil will accelerate and, because the coil is moving in a magnetic field, a backward voltage will be induced in it given by

$$V_i = -K_i \dot{\theta}$$

Hence, the coil current is now given by:

$$I = \frac{V_t - K_i \dot{\theta}}{R_t + R_r}$$

Now substituting for I in the system equation (11.1):

$$K_i \left(\frac{V_t - K_i \dot{\theta}}{R_t + R_r} \right) = K_s \theta + J \ddot{\theta}$$

or, rearranging:

$$\ddot{\theta} + \frac{K_i^2 \dot{\theta}}{J(R_t + R_r)} + \frac{K_s \theta}{J} = \frac{K_i V_t}{J(R_t + R_r)} \qquad (11.2)$$

This is the standard equation of a second order dynamic system, with natural frequency ω and damping factor ξ given by:

$$\omega_n = \sqrt{\frac{K_s}{J}}; \quad \xi = \frac{K_i^2}{2K_s J(R_t + R_r)}$$

In steady-state, $\ddot{\theta} = \dot{\theta} = 0$ and equation (11.2) reduces to:

$$\frac{\theta}{V_t} = \frac{K_i}{K_s(R_t + R_r)} \qquad (11.3)$$

which is an expression describing the measurement sensitivity of the system.

The dynamic characteristics of a galvanometric chart recorder can therefore be represented by one of the output-reading/time characteristics shown in Figure 2.12. Which particular characteristic applies depends on the damping factor of the instrument. At the design stage, the usual aim is to give the instrument a damping factor of about 0.7. Achieving this is not straightforward, since the damping factor depends not only on the coil and spring constants (K_i and K_s) but also on the total circuit resistance ($R_t + R_r$). Adding a series or parallel resistance between the transducer and recorder,

206 Display, recording and presentation of measurement data

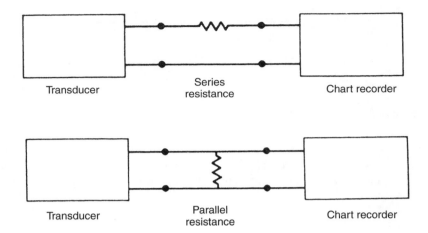

Fig. 11.7 Addition of series and parallel resistances between transducer and chart recorder.

as illustrated in Figure 11.7, respectively reduces or increases the damping factor. However, consideration of the sensitivity expression of (11.3) shows that any reduction in the damping factor takes place at the expense of a reduction in measurement sensitivity. Other methods to alter the damping factor are therefore usually necessary, and these techniques include decreasing the spring constant and system moment of inertia. The second order nature of the instrument's characteristics also means that the maximum frequency of signal that it can record is about 30 Hz. If there is a need to record signals at higher frequencies than this, other instruments such as ultra-violet recorders have to be used.

Galvanometric recorders have a typical quoted measurement inaccuracy of ±2% and a resolution of 1%. However, their accuracy is liable to decrease over time as dirt affects performance, particularly because it increases friction in the bearings carrying the suspended coil. In consequence, potentiometric types of recorder are usually preferred in modern instrumentation systems.

Potentiometric recorders

Potentiometric recorders have much better specifications than galvanometric recorders, with a typical inaccuracy of ±0.1% of full scale and measurement resolution of 0.2% f.s. being achievable. Such instruments employ a servo system, as shown in Figure 11.8, in which the pen is driven by a servomotor, and a potentiometer on the pen feeds back a signal proportional to pen position. This position signal is compared with the measured signal, and the difference is applied as an error signal that drives the motor. However, a consequence of this electromechanical balancing mechanism is to give the instrument a slow response time in the range 0.2–2.0 seconds. This means that potentiometric recorders are only suitable for measuring d.c. and slowly time-varying signals. In addition, this type of recorder is susceptible to commutator problems when a standard d.c. motor is used in the servo system. However, the use of brushless servo motors in many recent models overcomes this problem. Newer models also often use a non-contacting ultrasonic sensor to provide feedback on pen position in place of a

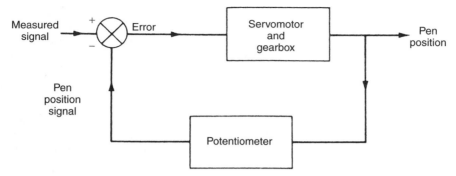

Fig. 11.8 Servo system of potentiometric chart recorder.

potentiometer. Another recent trend is to include a microprocessor controller (this is discussed under hybrid chart recorders).

Circular chart recorders

Before leaving the subject of standard mechanical chart recorders, mention must also be made of circular chart recorders. These consist of a rotating circular paper chart, as shown in Figure 11.9, which typically turns through one full revolution in 24 hours, allowing charts to be removed once per day and stored. The pen in such instruments is often driven pneumatically to record 200–1000 mbar (3–15 psi) pneumatic process signals, although versions with electrically driven pens also exist. This type of chart recorder was one of the earliest recording instruments to be used and, whilst they have now largely been superseded by other types of recorder, new ones continue to be bought for some applications. Apart from single channel versions, models recording up to six channels, with traces in six different colours, can be obtained.

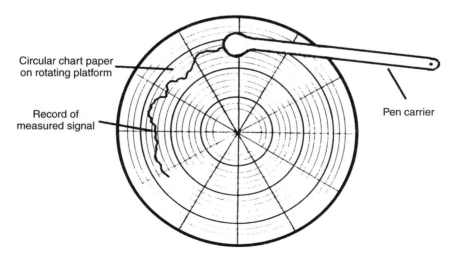

Fig. 11.9 Circular chart recorder.

208 Display, recording and presentation of measurement data

11.2.2 Ultra-violet recorders

The earlier discussion about galvanometric recorders concluded that restrictions on how far the system moment of inertia and spring constants can be reduced limited the maximum bandwidth to about 100 Hz. Ultra-violet recorders work on very similar principles to standard galvanometric chart recorders, but achieve a very significant reduction in system inertia and spring constants by mounting a narrow mirror rather than a pen system on the moving coil. This mirror reflects a beam of ultra-violet light onto ultra-violet sensitive paper. It is usual to find several of these mirror-galvanometer systems mounted in parallel within one instrument to provide a multi-channel recording capability, as illustrated in Figure 11.10. This arrangement enables signals at frequencies up to 13 kHz to be recorded with a typical inaccuracy of ±2% f.s. Whilst it is possible to obtain satisfactory permanent signal recordings by this method, special precautions are necessary to protect the ultra-violet-sensitive paper from light before use and to spray a fixing lacquer on it after recording. Such instruments must also be handled with extreme care, because the mirror galvanometers and their delicate

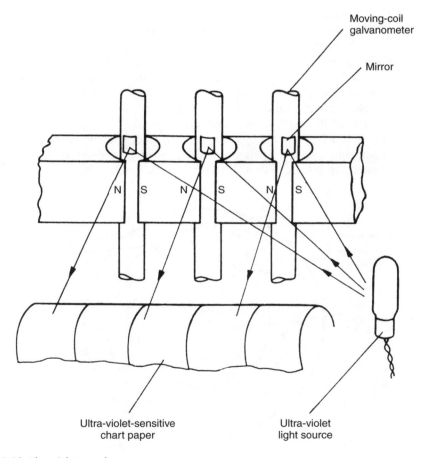

Fig. 11.10 Ultra-violet recorder.

mounting systems are easily damaged by relatively small shocks. In addition, ultra-violet recorders are significantly more expensive than standard chart recorders.

11.2.3 Fibre-optic recorders (recording oscilloscopes)

The *fibre optic recorder* uses a fibre-optic system to direct light onto light-sensitive paper. Fibre-optic recorders are similar to oscilloscopes in construction, insofar as they have an electron gun and focusing system that directs a stream of electrons onto one point on a fluorescent screen, and for this reason they are alternatively known as *recording oscilloscopes*. The screen is usually a long thin one instead of the square type found in an oscilloscope and only one set of deflection plates is provided. The signal to be recorded is applied to the deflection plates and the movement of the focused spot of electrons on the screen is proportional to the signal amplitude. A narrow strip of fibre optics in contact with the fluorescent screen transmits the motion of the spot to photosensitive paper held in close proximity to the other end of the fibre-optic strip. By driving the photosensitive paper at a constant speed past the fibre-optic strip, a time history of the measured signal is obtained. Such recorders are much more expensive than ultra-violet recorders but have an even higher bandwidth up to 1 MHz.

Whilst the construction above is the more common in fibre-optic recorders, a second type also exists that uses a conventional square screen instead of a long thin one. This has a square faceplate attached to the screen housing a square array of fibre-optics. The other side of the fibre-optic system is in contact with chart paper. The effect of this is to provide a hard copy of the typical form of display obtainable on a cathode ray oscilloscope.

11.2.4 Hybrid chart recorders

Hybrid chart recorders represent the latest generation of chart recorder and basically consist of a potentiometric chart recorder with an added microprocessor. The microprocessor provides for selection of range and chart speed, and also allows specification of alarm modes and levels to detect when measured variables go outside acceptable limits. Additional information can also be printed on charts, such as names, times and dates of variables recorded. Microprocessor-based, hybrid versions of circular chart recorders also now exist.

11.2.5 Magnetic tape recorders

Magnetic tape recorders can record analogue signals up to 80 kHz in frequency. As the speed of the tape transport can be switched between several values, signals can be recorded at high speed and replayed at a lower speed. Such time scaling of the recorded information allows a hard copy of the signal behaviour to be obtained from instruments such as ultra-violet and galvanometric recorders whose bandwidth is insufficient to allow direct signal recording. A 200 Hz signal cannot be recorded directly on a chart recorder, but if it is recorded on a magnetic tape recorder running at high speed and then replayed at a speed ten times lower, its frequency will be time scaled to 20 Hz which can be recorded on a chart recorder. Instrumentation tape recorders typically have between four and ten channels, allowing many signals to be recorded simultaneously.

The two basic types of analogue tape recording technique are direct recording and frequency-modulated recording. Direct recording offers the best data bandwidth but the accuracy of signal amplitude recording is quite poor, and this seriously limits the usefulness of this technique in most applications. The frequency-modulated technique offers better amplitude-recording accuracy, with a typical inaccuracy of $\pm 5\%$ at signal frequencies of 80 kHz. In consequence, this technique is very much more common than direct recording.

11.2.6 Digital recorders

For some time, the only technique available for recording signals at frequencies higher than 80 kHz has been to use a digital processor. As the signals to be recorded are usually in analogue form, a prerequisite for this is an analogue-to-digital (A–D) converter board to sample the analogue signals and convert them to digital form. The relevant aspects of computer hardware necessary to achieve this were covered in Chapter 9. Correct choice of the sampling interval is always necessary to ensure that an accurate digital record of the signal is obtained and problems of aliasing etc. are not encountered, as explained in Chapter 5. Some prior analogue signal conditioning may also be required in some circumstances, again as mentioned in Chapter 5.

Until a few years ago, the process of recording data digitally was carried out by standard computer equipment equipped with the necessary analogue interface boards etc., and the process was known as *data-logging*. More recently, purpose-designed digital recorders have become available for this purpose. These are usually multi-channel, and are available from many suppliers. Typically, a 10-bit A–D converter is used, which gives a 0.1% measurement resolution. Alternatively, a 12-bit converter gives 0.025% resolution. Specifications typically quoted for digital recorders are frequency response of 25 kHz, maximum sampling frequency of 200 MHz and data storage up to 4000 data points per channel.

When there is a requirement to view recorded data, for instance to look at the behaviour of parameters in a production process immediately before a fault has occurred in the process, it is usually necessary to use the digital recorder in conjunction with a chart recorder*, applying speed scaling as appropriate to allow for the difference in frequency-response capability between a digital recorder and a chart recorder. However, in these circumstances, it is only necessary to use the chart recorder to display the process parameters for the time period of interest. This saves a large amount of paper compared with the alternative of running the chart recorder continuously if a digital recorder is not used as the main data-capture mechanism.

As an alternative to chart recorders when hard-copy records are required, numerical data can be readily output from digital recorders onto alphanumeric digital printers in the form of dot-matrix, inkjet or laser printing devices. However, when there are trends in data, the graphical display of the time history of a variable provided by a chart recorder shows up the trends much more readily.

---

* Note that some digital recorders actually incorporate a recording oscilloscope to provide a hard copy of recorded data, thus obviating the need for a chart recorder.

As an alternative to hard-copy displays of measured variables when there is a need to view their behaviour over a particular time period, there is an increasing trend to use a computer monitor to display the variables graphically. Digital recorders with this kind of display facility are frequently known as *paperless recorders*.

11.2.7 Storage oscilloscopes

Storage oscilloscopes exist in both analogue and digital forms, although the latter is now much more common. An *analogue storage oscilloscope* is a conventional oscilloscope that has a special phosphorescent coating on its screen that allows it to 'hold' a trace for up to one hour. This can be photographed if a permanent record of the measured signal is required.

The *digital storage oscilloscope*, commonly referred to simply as a *digital oscilloscope*, is merely a conventional analogue oscilloscope that has digital storage capabilities. The components of a digital oscilloscope are illustrated in Figure 11.11. The input analogue measurement signal is sampled at a high frequency, converted to digital form by an analogue-to-digital converter, and stored in memory as a record of the amplitude/time history of the measured signal. Subsequently, the digital signal is converted back into an analogue signal that has the same amplitude/time characteristics as the original signal, and this is applied to the *xy* deflector plates of the analogue part of the oscilloscope at a frequency that is sufficient to ensure that the display on the screen is refreshed without inducing 'flicker'. One difference compared with a normal analogue oscilloscope is that the output display consists of a sequence of dots rather than a continuous trace. The density of the dots depends partly on the sampling frequency of the input signal and partly on the rate at which the digitized signal is converted back into analogue form. However, when used to measure signals in the medium-frequency range, the dot density is high enough to give the display a pseudo-continuous appearance.

Digital oscilloscopes generally offer a higher level of performance than analogue versions, as well as having the ability to either display a measurement signal in real time or else store it for future display. However, there are also additional advantages. The digitization of the measured signal means that it is possible for the instrument to compute many waveform parameters such as minimum and maximum values, r.m.s. values, mean values, rise time and signal frequency. Such parameter values can be presented on the oscilloscope screen on demand. Also, digital oscilloscopes can record transient signals when used in single-sweep mode. This task is very difficult when using analogue oscilloscopes because of the difficulties in achieving the necessary synchronization. If permanent, hard-copy records of signals are required, digital oscilloscopes usually have analogue output terminals that permit stored signals to be transferred into a chart recorder.

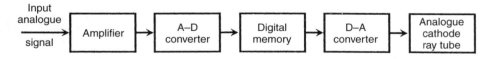

Fig. 11.11 Components of a digital oscilloscope.

11.3 Presentation of data

The two formats available for presenting data on paper are tabular and graphical ones and the relative merits of these are compared below. In some circumstances, it is clearly best to use only one or other of these two alternatives alone. However, in many data collection exercises, part of the measurements and calculations are expressed in tabular form and part graphically, so making best use of the merits of each technique. Very similar arguments apply to the relative merits of graphical and tabular presentations if a computer screen is used for the presentation instead of paper.

11.3.1 Tabular data presentation

A tabular presentation allows data values to be recorded in a precise way that exactly maintains the accuracy to which the data values were measured. In other words, the data values are written down exactly as measured. Besides recording the raw data values as measured, tables often also contain further values calculated from the raw data. An example of a tabular data presentation is given in Table 11.1. This records the results of an experiment to determine the strain induced in a bar of material that is subjected to a range of stresses. Data were obtained by applying a sequence of forces to the end of the bar and using an extensometer to measure the change in length. Values of the stress and strain in the bar are calculated from these measurements and are also included in the table. The final row, which is of crucial importance in any tabular presentation, is the estimate of possible error in each calculated result.

A table of measurements and calculations should conform to several rules as illustrated in Table 11.1:

(i) The table should have a title that explains what data are being presented within the table.

Table 11.1 Sample tabular presentation of data

Table of measured applied forces and extensometer readings and calculations of stress and strain

	Force applied (KN)	Extensometer reading (divisions)	Stress (N/m^2)	Strain
	0	0	0	0
	2	4.0	15.5	19.8×10^{-5}
	4	5.8	31.0	28.6×10^{-5}
	6	7.4	46.5	36.6×10^{-5}
	8	9.0	62.0	44.4×10^{-5}
	10	10.6	77.5	52.4×10^{-5}
	12	12.2	93.0	60.2×10^{-5}
	14	13.7	108.5	67.6×10^{-5}
Possible error in measurements (%)	±0.2	±0.2	±1.5	$±1.0 \times 10^{-5}$

(ii) Each column of figures in the table should refer to the measurements or calculations associated with one quantity only.
(iii) Each column of figures should be headed by a title that identifies the data values contained in the column.
(iv) The units in which quantities in each column are measured should be stated at the top of the column.
(v) All headings and columns should be separated by bold horizontal (and sometimes vertical) lines.
(vi) The errors associated with each data value quoted in the table should be given. The form shown in Table 11.1 is a suitable way to do this when the error level is the same for all data values in a particular column. However, if error levels vary, then it is preferable to write the error boundaries alongside each entry in the table.

11.3.2 Graphical presentation of data

Presentation of data in graphical form involves some compromise in the accuracy to which the data are recorded, as the exact values of measurements are lost. However, graphical presentation has important advantages over tabular presentation.

(i) Graphs provide a pictorial representation of results that is more readily comprehended than a set of tabular results.
(ii) Graphs are particularly useful for expressing the quantitative significance of results and showing whether a linear relationship exists between two variables. Figure 11.12 shows a graph drawn from the stress and strain values given in the Table 11.1. Construction of the graph involves first of all marking the points corresponding to the stress and strain values. The next step is to draw some lines through these data points that best represents the relationship between the two variables. This line will normally be either a straight one or a smooth curve. The data points will not usually lie exactly on this line but instead will lie on either side of it. The magnitude of the excursions of the data points from the line drawn will depend on the magnitude of the random measurement errors associated with the data.
(iii) Graphs can sometimes show up a data point that is clearly outside the straight line or curve that seems to fit the rest of the data points. Such a data point is probably due either to a human mistake in reading an instrument or else to a momentary malfunction in the measuring instrument itself. If the graph shows such a data point where a human mistake or instrument malfunction is suspected, the proper course of action is to repeat that particular measurement and then discard the original data point if the mistake or malfunction is confirmed.

Like tables, the proper representation of data in graphical form has to conform to certain rules:

(i) The graph should have a title or caption that explains what data are being presented in the graph.
(ii) Both axes of the graph should be labelled to express clearly what variable is associated with each axis and to define the units in which the variables are expressed.

214 Display, recording and presentation of measurement data

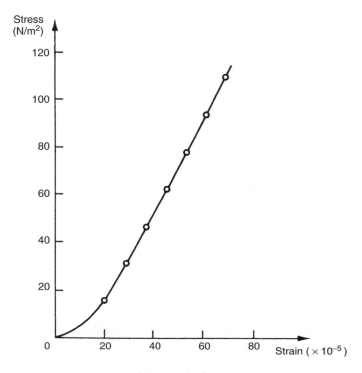

Fig. 11.12 Sample graphical presentation of data: graph of stress against strain.

(iii) The number of points marked along each axis should be kept reasonably small – about five divisions is often a suitable number.
(iv) No attempt should be made to draw the graph outside the boundaries corresponding to the maximum and minimum data values measured, i.e. in Figure 11.12, the graph stops at a point corresponding to the highest measured stress value of 108.5.

Fitting curves to data points on a graph

The procedure of drawing a straight line or smooth curve as appropriate that passes close to all data points on a graph, rather than joining the data points by a jagged line that passes through each data point, is justified on account of the random errors that are known to affect measurements. Any line between the data points is mathematically acceptable as a graphical representation of the data if the maximum deviation of any data point from the line is within the boundaries of the identified level of possible measurement errors. However, within the range of possible lines that could be drawn, only one will be the optimum one. This optimum line is where the sum of negative errors in data points on one side of the line is balanced by the sum of positive errors in data points on the other side of the line. The nature of the data points is often such that a perfectly acceptable approximation to the optimum can be obtained by drawing a line through the data points by eye. In other cases, however, it is necessary to fit a line mathematically, using regression techniques.

Regression techniques

Regression techniques consist of finding a mathematical relationship between measurements of two variables y and x, such that the value of variable y can be predicted from a measurement of the other variable x. However, regression techniques should not be regarded as a magic formula that can fit a good relationship to measurement data in all circumstances, as the characteristics of the data must satisfy certain conditions. In determining the suitability of measurement data for the application of regression techniques, it is recommended practice to draw an approximate graph of the measured data points, as this is often the best means of detecting aspects of the data that make it unsuitable for regression analysis. Drawing a graph of the data will indicate, for example, whether there are any data points that appear to be erroneous. This may indicate that human mistakes or instrument malfunctions have affected the erroneous data points, and it is assumed that any such data points will be checked for correctness.

Regression techniques cannot be successfully applied if the deviation of any particular data point from the line to be fitted is greater than the maximum possible error that is calculated for the measured variable (i.e. the predicted sum of all systematic and random errors). The nature of some measurement data sets is such that this criterion cannot be satisfied, and any attempt to apply regression techniques is doomed to failure. In that event, the only valid course of action is to express the measurements in tabular form. This can then be used as a $x-y$ look-up table, from which values of the variable y corresponding to particular values of x can be read off. In many cases, this problem of large errors in some data points only becomes apparent during the process of attempting to fit a relationship by regression.

A further check that must be made before attempting to fit a line or curve to measurements of two variables x and y is to examine the data and look for any evidence that both variables are subject to random errors. It is a clear condition for the validity of regression techniques that only one of the measured variables is subject to random errors, with no error in the other variable. If random errors do exist in both measured variables, regression techniques cannot be applied and recourse must be made instead to correlation analysis (covered later in this chapter). A simple example of a situation where both variables in a measurement data set are subject to random errors are measurements of human height and weight, and no attempt should be made to fit a relationship between them by regression.

Having determined that the technique is valid, the regression procedure is simplest if a straight-line relationship exists between the variables, which allows a relationship of the form $y = a + bx$ to be estimated by linear least squares regression. Unfortunately, in many cases, a straight-line relationship between the points does not exist, which is readily shown by plotting the raw data points on a graph. However, knowledge of physical laws governing the data can often suggest a suitable alternative form of relationship between the two sets of variable measurements, such as a quadratic relationship or a higher order polynomial relationship. Also, in some cases, the measured variables can be transformed into a form where a linear relationship exists. For example, suppose that two variables y and x are related according to $y = ax^c$. A linear relationship from this can be derived, using a logarithmic transformation, as $\log(y) = \log(a) + c \log(x)$.

Thus, if a graph is constructed of $\log(y)$ plotted against $\log(x)$, the parameters of a straight-line relationship can be estimated by linear least squares regression.

All quadratic and higher order relationships relating one variable y to another variable x can be represented by a power series of the form:

$$y = a_0 + a_1 x + a_2 x^2 + \cdots + a_p x^p$$

Estimation of the parameters $a_0 \ldots a_p$ is very difficult if p has a large value. Fortunately, a relationship where p only has a small value can be fitted to most data sets. Quadratic least squares regression is used to estimate parameters where p has a value of two, and for larger values of p, polynomial least squares regression is used for parameter estimation.

Where the appropriate form of relationship between variables in measurement data sets is not obvious either from visual inspection or from consideration of physical laws, a method that is effectively a trial and error one has to be applied. This consists of estimating the parameters of successively higher order relationships between y and x until a curve is found that fits the data sufficiently closely. What level of closeness is acceptable is considered in the later section on confidence tests.

Linear least squares regression

If a linear relationship between y and x exists for a set of n measurements $y_1 \ldots y_n$, $x_1 \ldots x_n$, then this relationship can be expressed as $y = a + bx$, where the coefficients a and b are constants. The purpose of least squares regression is to select the optimum values for a and b such that the line gives the best fit to the measurement data.

The deviation of each point (x_i, y_i) from the line can be expressed as d_i, where $d_i = y_i - (a + bx_i)$.

The best-fit line is obtained when the sum of the squared deviations, S, is a minimum, i.e. when

$$S = \sum_{i=1}^{n}(d_i^2) = \sum_{i=1}^{n}(y_i - a - bx_i)^2$$

is a minimum.

The minimum can be found by setting the partial derivatives $\partial S/\partial a$ and $\partial S/\partial b$ to zero and solving the resulting two simultaneous (normal) equations:

$$\partial S/\partial a = \sum 2(y_i - a - bx_i)(-1) = 0 \qquad (11.4)$$

$$\partial S/\partial b = \sum 2(y_i - a - bx_i)(-x_i) = 0 \qquad (11.5)$$

The values of the coefficients a and b at the minimum point can be represented by $\hat{a}$ and $\hat{b}$, which are known as the least squares estimates of a and b. These can be calculated as follows:

From (11.4),

$$\sum y_i = \sum \hat{a} + \hat{b} \sum x_i = n\hat{a} + \hat{b} \sum x_i \text{ and thus, } \hat{a} = \frac{\sum y_i - \hat{b} \sum x_i}{n} \qquad (11.6)$$

From (11.5),

$$\sum (x_i y_i) = \hat{a} \sum x_i + \hat{b} \sum x_i^2 \qquad (11.7)$$

Now substitute for $\hat{a}$ in (11.7) using (11.6):

$$\sum(x_i y_i) = \frac{\left(\sum y_i - \hat{b} \sum x_i\right)}{n} \sum x_i + \hat{b} \sum x_i^2$$

Collecting terms in $\hat{b}$,

$$\hat{b}\left[\sum x_i^2 - \frac{\left(\sum x_i\right)^2}{n}\right] = \sum(x_i y_i) - \frac{\sum x_i \sum y_i}{n}$$

Rearranging gives:

$$\hat{b}\left[\sum x_i^2 - n\left\{\left(\sum x_i/n\right)\right\}^2\right] = \sum(x_i y_i) - n\sum(x_i/n)\sum(y_i/n).$$

This can be expressed as:

$$\hat{b}\left[\sum x_i^2 - n x_m^2\right] = \sum(x_i y_i) - n x_m y_m,$$

where x_m and y_m are the mean values of x and y.
Thus:

$$\hat{b} = \frac{\sum(x_i y_i) - n x_m y_m}{\sum x_i^2 - n x_m^2} \quad (11.8)$$

And, from (11.6):

$$\hat{a} = y_m - \hat{b} x_m \quad (11.9)$$

Example 11.1
In an experiment to determine the characteristics of a displacement sensor with a voltage output, the following output voltage values were recorded when a set of standard displacements was measured:

Displacement (cm)	1.0	2.0	3.0	4.0	5.0	6.0	7.0	8.0	9.0	10.0
Voltage (V)		2.1	4.3	6.2	8.5	10.7	12.6	14.5	16.3	18.3	21.2

Fit a straight line to this set of data using least squares regression and estimate the output voltage when a displacement of 4.5 cm is measured.

Solution
Let y represent the output voltage and x represent the displacement. Then a suitable straight line is given by $y = a + bx$. We can now proceed to calculate estimates for

the coefficients a and b using equations (11.8) and (11.9) above. The first step is to calculate the mean values of x and y. These are found to be $x_m = 5.5$ and $y_m = 11.47$. Next, we need to tabulate $x_i y_i$ and x_i^2 for each pair of data values:

x_i	y_i	$x_i y_i$	x_i^2
1.0	2.1	2.1	1
2.0	4.3	8.6	4
3.0	6.2	18.6	9
⋮	⋮	⋮	⋮
⋮	⋮	⋮	⋮
10.0	21.2	212.0	100

Now calculate the values needed from this table: $n = 10$; $\sum(x_i y_i) = 801.0$; $\sum(x_i^2) = 385$ and enter these values into (11.8) and (11.9).

$$\hat{b} = \frac{801.0 - (10 \times 5.5 \times 11.47)}{385 - (10 \times 5.5^2)} = 2.067; \quad \hat{a} = 11.47 - (2.067 \times 5.5) = 0.1033;$$

i.e. $y = 0.1033 + 2.067x$

Hence, for $x = 4.5$, $y = 0.1033 + (2.067 \times 4.5) = 9.40$ volts. Note that in this solution, we have only specified the answer to an accuracy of three figures, which is the same accuracy as the measurements. Any greater number of figures in the answer would be meaningless.

Least squares regression is often appropriate for situations where a straight-line relationship is not immediately obvious, for example where $y \propto x^2$ or $y \propto \exp(x)$.

Example 11.2
From theoretical considerations, it is known that the voltage (V) across a charged capacitor decays with time (t) according to the relationship: $V = K \exp(-t/\tau)$. Estimate values for K and τ if the following values of V and t are measured.

V	8.67	6.55	4.53	3.29	2.56	1.95	1.43	1.04	0.76
t	0	1	2	3	4	5	6	7	8

Solution
If $V = K \exp(-T/\tau)$ then, $\log_e(V) = \log_e(K) - t/\tau$. Now let $y = \log_e(V)$, $a = \log(K)$, $b = -1/\tau$ and $x = t$. Hence, $y = a + bx$, which is the equation of a straight line whose coefficients can be estimated by applying equations (11.8) and (11.9). Therefore, proceed in the same way as example 11.1 and tabulate the values required:

V	$\log_e V$ (y_i)	t (x_i)	($x_i y_i$)	(x_i^2)
8.67	2.16	0	0	0
6.55	1.88	1	1.88	1
4.53	1.51	2	3.02	4
⋮	⋮	⋮	⋮	⋮
⋮	⋮	⋮	⋮	⋮
0.76	−0.27	8	−2.16	64

Now calculate the values needed from this table: $n = 9$; $\sum(x_i y_i) = 15.86$; $\sum(x_i^2) = 204$; $x_m = 4.0$; $y_m = 0.9422$, and enter these values into (11.8) and (11.9).

$$\hat{b} = \frac{15.86 - (9 \times 4.0 \times 0.9422)}{204 - (9 \times 4.0^2)} = -0.301; \quad \hat{a} = 0.9422 + (0.301 \times 4.0) = 2.15$$

$$K = \exp(a) = \exp(2.15) = 8.58; \quad \tau = -1/b = -1/(-0.301) = 3.32$$

Quadratic least squares regression

Quadratic least squares regression is used to estimate the parameters of a relationship $y = a + bx + cx^2$ between two sets of measurements $y_1 \ldots y_n$, $x_1 \ldots x_n$.

The deviation of each point (x_i, y_i) from the line can be expressed as d_i, where $d_i = y_i - (a + bx_i + cx_i^2)$.

The best-fit line is obtained when the sum of the squared deviations, S, is a minimum, i.e. when

$$S = \sum_{i=1}^{n}(d_i^2) = \sum_{i=1}^{n}(y_i - a - bx_i + cx_i^2)^2$$

is a minimum.

The minimum can be found by setting the partial derivatives $\partial S / \partial a$, $\partial S / \partial b$ and $\partial S / \partial c$ to zero and solving the resulting simultaneous equations, as for the linear least squares regression case above. Standard computer programs to estimate the parameters a, b and c by numerical methods are widely available and therefore a detailed solution is not presented here.

Polynomial least squares regression

Polynomial least squares regression is used to estimate the parameters of the pth order relationship $y = a_0 + a_1 x + a_2 x^2 + \cdots + a_p x^p$ between two sets of measurements $y_1 \ldots y_n$, $x_1 \ldots x_n$.

The deviation of each point (x_i, y_i) from the line can be expressed as d_i, where:

$$d_i = y_i - (a_0 + a_1 x_i + a_2 x_i^2 + \cdots + a_p x_i^p)$$

The best-fit line is obtained when the sum of the squared deviations given by

$$S = \sum_{i=1}^{n}(d_i^2)$$

is a minimum.

The minimum can be found as before by setting the p partial derivatives $\partial S/\partial a_0 \ldots \partial S/\partial a_p$ to zero and solving the resulting simultaneous equations. Again, as for the quadratic least squares regression case, standard computer programs to estimate the parameters $a_0 \ldots a_p$ by numerical methods are widely available and therefore a detailed solution is not presented here.

Confidence tests in curve fitting by least squares regression

Once data has been collected and a mathematical relationship that fits the data points has been determined by regression, the level of confidence that the mathematical relationship fitted is correct must be expressed in some way. The first check that must be made is whether the fundamental requirement for the validity of regression techniques is satisfied, i.e. whether the deviations of data points from the fitted line are all less than the maximum error level predicted for the measured variable. If this condition is violated by any data point that a line or curve has been fitted to, then use of the fitted relationship is unsafe and recourse must be made to tabular data presentation, as described earlier.

The second check concerns whether or not random errors affect both measured variables. If attempts are made to fit relationships by regression to data where both measured variables contain random errors, any relationship fitted will only be approximate and it is likely that one or more data points will have a deviation from the fitted line or curve that is greater than the maximum error level predicted for the measured variable. This will show up when the appropriate checks are made.

Having carried out the above checks to show that there are no aspects of the data which suggest that regression analysis is not appropriate, the next step is to apply least squares regression to estimate the parameters of the chosen relationship (linear, quadratic etc.). After this, some form of follow-up procedure is clearly required to assess how well the estimated relationship fits the data points. A simple curve-fitting confidence test is to calculate the sum of squared deviations S for the chosen y/x relationship and compare it with the value of S calculated for the next higher order regression curve that could be fitted to the data. Thus if a straight-line relationship is chosen, the value of S calculated should be of a similar magnitude to that obtained by fitting a quadratic relationship. If the value of S were substantially lower for a quadratic relationship, this would indicate that a quadratic relationship was a better fit to the data than a straight-line one and further tests would be needed to examine whether a cubic or higher order relationship was a better fit still.

Other more sophisticated confidence tests exist such as the *F-ratio test*. However, these are outside the scope of this book.

Correlation tests

Where both variables in a measurement data set are subject to random fluctuations, correlation analysis is applied to determine the degree of association between the variables. For example, in the case already quoted of a data set containing measurements of human height and weight, we certainly expect some relationship between the variables of height and weight because a tall person is heavier *on average* than a short person. Correlation tests determine the strength of the relationship (or interdependence) between the measured variables, which is expressed in the form of a correlation coefficient.

For two sets of measurements $y_1 \ldots y_n$, $x_1 \ldots x_n$ with means x_m and y_m, the correlation coefficient Φ is given by:

$$\Phi = \frac{\sum (x_i - x_m)(y_i - y_m)}{\sqrt{\left[\sum (x_i - x_m)^2\right]\left[\sum (y_i - y_m)^2\right]}}$$

The value of $|\Phi|$ always lies between 0 and 1, with 0 representing the case where the variables are completely independent of one another and 1 the case where they are totally related to one another. For $0 < |\Phi| < 1$, linear least squares regression can be applied to find relationships between the variables, which allows x to be predicted from a measurement of y, and y to be predicted from a measurement of x. This involves finding two separate regression lines of the form:

$$y = a + bx \quad \text{and} \quad x = c + dy$$

These two lines are not normally coincident as shown in Figure 11.13. Both lines pass through the centroid of the data points but their slopes are different.

As $|\Phi| \rightarrow 1$, the lines tend to coincidence, representing the case where the two variables are totally dependent upon one another.

As $|\Phi| \rightarrow 0$, the lines tend to orthogonal ones parallel to the x and y axes. In this case, the two sets of variables are totally independent. The best estimate of x given any measurement of y is x_m and the best estimate of y given any measurement of x is y_m.

For the general case, the best fit to the data is the line that bisects the angle between the lines on Figure 11.13.

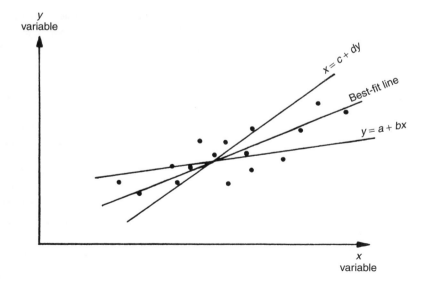

Fig. 11.13 Relationship between two variables with random fluctuations.

11.4 Self-test questions

11.1 (a) Explain the derivation of the expression $\ddot{\theta} + \frac{K_i^2 \dot{\theta}}{JR} + \frac{K_s \theta}{J} = \frac{K_i V_t}{JR}$ describing the dynamic response of a chart recorder following a step change in the electrical voltage output of a transducer connected to its input. Explain also what all the terms in the expression stand for. (Assume that the impedances of both the transducer and recorder have a resistive component only and that there is negligible friction in the system.)
  (b) Derive expressions for the measuring system natural frequency, ω_n, the damping factor, ξ, and the steady-state sensitivity.
  (c) Explain simple ways of increasing and decreasing the damping factor and describe the corresponding effect on measurement sensitivity.
  (d) What damping factor gives the best system bandwidth?
  (e) What aspects of the design of a chart recorder would you modify in order to improve the system bandwidth? What is the maximum bandwidth typically attainable in chart recorders, and if such a maximum-bandwidth instrument is available, what is the highest-frequency signal that such an instrument would be generally regarded as being suitable for measuring if the accuracy of the signal amplitude measurement is important?

11.2 Theoretical considerations show that quantities x and y are related in a linear fashion such that $y = ax + b$. Show that the best estimate of the constants a and b are given by:

$$\hat{a} = \frac{\sum(x_i y_i) - n x_m y_m}{\sum x_i^2 - n x_m^2}; \quad \hat{b} = y_m - \hat{a} x_m$$

Explain carefully the meaning of all the terms in the above two equations.

11.3 The characteristics of a chromel-constantan thermocouple is known to be approximately linear over the temperature range 300°C–800°C. The output e.m.f. was measured practically at a range of temperatures and the following table of results obtained. Using least squares regression, calculate the coefficients a and b for the relationship $T = aE + b$ that best describes the temperature–e.m.f. characteristic.

Temp (°C)	300	325	350	375	400	425	450	475	500	525	550
e.m.f. (mV)	21.0	23.2	25.0	26.9	28.6	31.3	32.8	35.0	37.2	38.5	40.7

Temp (°C)	575	600	625	650	675	700	725	750	775	800
e.m.f. (mV)	43.0	45.2	47.6	49.5	51.1	53.0	55.5	57.2	59.0	61.0

11.4 Measurements of the current (I) flowing through a resistor and the corresponding voltage drop (V) are shown below:

I	1	2	3	4	5
V	10.8	20.4	30.7	40.5	50.0

The instruments used to measure voltage and current were accurate in all respects except that they each had a zero error that the observer failed to take account of or to correct at the time of measurement. Determine the value of the resistor from the data measured.

11.5 A measured quantity y is known from theoretical considerations to depend on a variable x according to the relationship $y = a + bx^2$. For the following set of measurements of x and y, use linear least squares regression to determine the estimates of the parameters a and b that fit the data best.

x	0	1	2	3	4	5
y	0.9	9.2	33.4	72.5	130.1	200.8

11.6 The mean-time-to-failure ($MTTF$) of an integrated circuit is known to obey a law of the following form: $MTTF = C \exp T_0/T$, where T is the operating temperature and C and T_0 are constants.

The following values of $MTTF$ at various temperatures were obtained from accelerated-life tests.

$MTTF$ (hours)	54	105	206	411	941	2145
Temperature (K)	600	580	560	540	520	500

(a) Estimate the values of C and T_0. (Hint – $\log_e(MTTF) = \log_e(C) + T_0/T$. This equation is now a straight-line relationship between $\log(MTTF)$ and $1/T$, where $\log(C)$ and T_0 are constants.)
(b) For a $MTTF$ of 10 years, calculate the maximum allowable temperature.

References and further reading

Chatfield, C. (1983) *Statistics for Technology*, Chapman and Hall, London.
Topping, J. (1972) *Errors of Observation and their Treatment*, Chapman and Hall, London.

12

Measurement reliability and safety systems

12.1 Reliability

The reliability of measurement systems can be quantified as the mean time between faults occurring in the system. In this context, a fault means the occurrence of an unexpected condition in the system that causes the measurement output to either be incorrect or not to exist at all. The following sections summarize the principles of reliability theory that are relevant to measurement systems. A fuller account of reliability theory, and particularly its application in manufacturing systems, can be found elsewhere (Morris, 1997).

12.1.1 Principles of reliability

The reliability of a measurement system is defined as the ability of the system to perform its required function within specified working conditions for a stated period of time. Unfortunately, factors such as manufacturing tolerances in an instrument and varying operating conditions conspire to make the faultless operating life of a system impossible to predict. Such factors are subject to random variation and chance, and therefore reliability cannot be defined in absolute terms. The nearest one can get to an absolute quantification of reliability are quasi-absolute terms like the mean-time-between-failures, which expresses the average time that the measurement system works without failure. Otherwise, reliability has to be expressed as a statistical parameter that defines the probability that no faults will develop over a specified interval of time.

In quantifying reliability for a measurement system, an immediate difficulty that arises is defining what counts as a fault. Total loss of a measurement output is an obvious fault but a fault that causes a finite but incorrect measurement is more difficult to identify. The usual approach is to identify such faults by applying statistical process control techniques (Morris, 1997).

Reliability quantification in quasi-absolute terms
Whilst reliability is essentially probabilistic in nature, it can be quantified in quasi-absolute terms by the mean-time-between-failures and the mean-time-to-failure parameters. It

must be emphasized that these two quantities are usually average values calculated over a number of identical instruments, and therefore the actual values for any particular instrument may vary substantially from the average value.

The *mean-time-between-failures* (MTBF) is a parameter that expresses the average time between faults occurring in an instrument, calculated over a given period of time. For example, suppose that the history of an instrument is logged over a 360 day period and the time intervals in days between faults occurring was as follows:

$$11\ 23\ 27\ 16\ 19\ 32\ 6\ 24\ 13\ 21\ 26\ 15\ 14\ 33\ 29\ 12\ 17\ 22$$

The mean interval is 20 days, which is therefore the mean-time-between-failures. An alternative way of calculating MTBF is simply to count the number of faults occurring over a given period. In the above example, there were 18 faults recorded over a period of 360 days and so the MTBF can be calculated as:

$$MTBF = 360/18 = 20 \text{ days}$$

Unfortunately, in the case of instruments that have a high reliability, such in-service calculation of reliability in terms of the number of faults occurring over a given period of time becomes grossly inaccurate because faults occur too infrequently. In this case, MTBF predictions provided by the instrument manufacturer can be used, since manufacturers have the opportunity to monitor the performance of a number of identical instruments installed in different companies. If there are a total of F faults recorded for N identical instruments in time T, the MTBF can be calculated as $MTBF = TN/F$. One drawback of this approach is that it does not take the conditions of use, such as the operating environment, into account.

The mean-time-to-failure (MTTF) is an alternative way of quantifying reliability that is normally used for devices like thermocouples that are discarded when they fail. MTTF expresses the average time before failure occurs, calculated over a number of identical devices.

The final reliability-associated term of importance in measurement systems is the *mean-time-to-repair* (MTTR). This expresses the average time needed for repair of an instrument. MTTR can also be interpreted as the *mean-time-to-replace*, since replacement of a faulty instrument by a spare one is usually preferable in manufacturing systems to losing production whilst an instrument is repaired.

The MTBF and MTTR parameters are often expressed in terms of a combined quantity known as the availability figure. This measures the proportion of the total time that an instrument is working, i.e. the proportion of the total time that it is in an unfailed state. The availability is defined as the ratio:

$$\text{Availability} = \frac{MTBF}{MTBF + MTTR}$$

In measurement systems, the aim must always be to maximize the MTBF figure and minimize the MTTR figure, thereby maximizing the availability. As far as the MTBF and MTTF figures are concerned, good design and high-quality control standards during manufacture are the appropriate means of maximizing these figures. Design procedures that mean that faults are easy to repair are also an important factor in reducing the MTTR figure.

Failure patterns

The pattern of failure in an instrument may increase, stay the same or decrease over its life. In the case of *electronic components*, the failure rate typically changes with time in the manner shown in Figure 12.1(a). This form of characteristic is frequently known as a *bathtub curve*. Early in their life, electronic components can have quite a high rate of fault incidence up to time t_1 (see Figure 12.1(a)). After this initial working period, the fault rate decreases to a low level and remains at this low level until time t_2 when ageing effects cause the fault rate to start to increase again. Instrument manufacturers often 'burn in' electronic components for a length of time corresponding to the time t_1. This means that the components have reached the high-reliability phase of their life before they are supplied to customers.

Mechanical components usually have different failure characteristics as shown in Figure 12.1(b). Material fatigue is a typical reason for the failure rate to increase over the life of a mechanical component. In the early part of their life, when all components are relatively new, many instruments exhibit a low incidence of faults. Then, at a later stage, when fatigue and other ageing processes start to have a significant effect, the rate of faults increases and continues to increase thereafter.

Complex systems containing many different components often exhibit a constant pattern of failure over their lifetime. The various components within such systems each have their own failure pattern where the failure rate is increasing or decreasing with time. The greater the number of such components within a system, the greater is

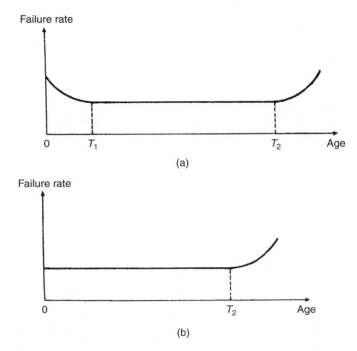

Fig. 12.1 Typical variation of reliability with component age: (a) electronic components ('bathtub' curve); (b) mechanical components.

the tendency for the failure patterns in the individual components to cancel out and for the rate of fault-incidence to assume a constant value.

Reliability quantification in probabilistic terms

In probabilistic terms, the reliability $R(T)$ of an instrument X is defined as the probability that the instrument will not fail within a certain period of time T. The unreliability or likelihood of failure $F(T)$ is a corresponding term which expresses the probability that the instrument will fail within the specified time interval. $R(T)$ and $F(T)$ are related by the expression:

$$F(T) = 1 - R(T) \qquad (12.1)$$

To calculate $R(T)$, accelerated lifetime testing* is carried out for a number (N) of identical instruments. Providing all instruments have similar conditions of use, the times of failure, $t_1, t_2 \ldots t_n$ will be distributed about the mean time to failure t_m. If the probability density of the time-to-failure is represented by $f(t)$, the probability that a particular instrument will fail in a time interval δt is given by $f(t)\delta t$, and the probability that the instrument will fail before a time T is given by:

$$F(T) = \int_0^T f(t)\,dt$$

The probability that the instrument will fail in a time interval ΔT following T, assuming that it has survived to time T, is given by:

$$\frac{F(T + \Delta T) - F(T)}{R(T)}$$

where $R(T)$ is the probability that the instrument will survive to time T. Dividing this expression by ΔT gives the average failure rate in the interval from T to $T + \Delta T$ as:

$$\frac{F(T + \Delta T) - F(T)}{\Delta T R(T)}$$

In the limit as $\Delta T \to 0$, the instantaneous failure rate at time T is given by:

$$\theta_f = \frac{d[F(T)]}{dt}\frac{1}{R(T)} = \frac{F'(T)}{R(T)} \qquad (12.2)$$

If it is assumed that the instrument is in the constant-failure-rate phase of its life, denoted by the interval between times t_1 and t_2 in Figure 12.1, then the instantaneous failure rate at T is also the mean failure rate which can be expressed as the reciprocal of the MTBF, i.e. mean failure rate $= \theta_f = 1/t_m$.

Differentiating (12.1) with respect to time gives $F'(T) = -R'(T)$. Hence, substituting for $F'(T)$ in (12.2) gives:

$$\theta_f = -\frac{R'(T)}{R(T)}$$

---

* Accelerated lifetime testing means subjecting an instrument to a much greater frequency of use than would normally be expected. If an instrument is normally used ten times per day, then one hundred days of normal use can be simulated by using it one thousand times in a single day.

This can be solved (Miller, 1990) to give the following expression:

$$R(T) = \exp(-\theta_f T) \qquad (12.3)$$

Examination of equation (12.3) shows that, at time $t = 0$, the unreliability is zero. Also, as t tends to ∞, the unreliability tends to a value of 1. This agrees with intuitive expectations that the value of unreliability should lie between values of 0 and 1. Another point of interest in equation (12.3) is to consider the unreliability when $T = MTBF$, i.e. when $T = t_m$. Then: $F(T) = 1 - \exp(-1) = 0.63$, i.e. the probability of a product failing after it has been operating for a length of time equal to the MTBF is 63%.

Further analysis of equation (12.3) shows that, for $T/t_m \leq 0.1$:

$$F(T) \approx T/t_m \qquad (12.4)$$

This is a useful formula for calculating (approximately) the reliability of a critical product which is only used for a time that is a small proportion of its MTBF.

Example 12.1
If the mean-time-to-failure of an instrument is 50 000 hours, calculate the probability that it will not fail during the first 10 000 hours of operation.

Solution
From (12.3), $R(T) = \exp(-\theta_f T) = \exp(-10\,000/50\,000) = 0.8187$

Example 12.2
If the mean-time-to-failure of an instrument is 80 000 hours, calculate the probability that it will not fail during the first 8000 hours of operation.

Solution
In this case, $T/t_m = 80\,000/8000 = 0.1$ and so equation (12.4) can be applied, giving $R(T) = 1 - F(T) \approx 1 - T/t_m \approx 0.9$. To illustrate the small level of inaccuracy involved in using the approximate expression (12.4), if we calculate the probability according to (12.3) we get $R(T) = \exp(-0.1) = 0.905$. Thus, there is a small but finite error in applying (12.4) instead of (12.3).

12.1.2 Laws of reliability in complex systems

Measurement systems usually comprise a number of components that are connected together in series, and hence it is necessary to know how the reliabilities of individual components are aggregated into a reliability figure for the whole system. In some cases, identical measurement components are put in parallel to improve reliability, because the measurement system then only fails if all of the parallel components fail. These two cases are covered by particular laws of reliability.

Reliability of components in series

A measurement system consisting of several components in series fails when any one of the separate components develops a fault. The reliability of such a system can be quantified as the probability that none of the components will fail within a given interval of time. For a system of n series components, the reliability R_S is the product of the separate reliabilities of the individual components according to the joint probability rule (Morris, 1997):

$$R_S = R_1 R_2 \ldots R_n \qquad (12.5)$$

Example 12.3
A measurement system consists of a sensor, a variable conversion element and a signal processing circuit, for which the reliability figures are 0.9, 0.95 and 0.99 respectively. Calculate the reliability of the whole measurement system.

Solution
Applying (12.5), $R_S = 0.9 \times 0.95 \times 0.99 = 0.85$.

Reliability of components in parallel

One way of improving the reliability of a measurement system is to connect two or more instruments in parallel. This means that the system only fails if every parallel instrument fails. For such systems, the system reliability R_S is given by:

$$R_S = 1 - F_S \qquad (12.6)$$

where F_S is the unreliability of the system. The equation for calculating F_S is similar to (12.5). Thus, for n instruments in parallel, the unreliability is given by:

$$F_S = F_1 F_2 \ldots F_n \qquad (12.7)$$

If all the instruments in parallel are identical then (12.7) can be written in the simpler form:

$$F_S = (F_X)^n \qquad (12.8)$$

where F_X is the unreliability of each instrument.

Example 12.4
In a particular safety critical measurement system, three identical instruments are connected in parallel. If the reliability of each instrument is 0.95, calculate the reliability of the measurement system.

Solution
From (12.1), the unreliability of each instrument F_X is given by $F_X = 1 - R_X = 1 - 0.95 = 0.05$.
Applying (12.8), $F_S = (F_X)^3 = (0.05)^3 = 0.000125$.
Thus, from (12.6), $R_S = 1 - F_S = 1 - 0.000125 = 0.999875$.

12.1.3 Improving measurement system reliability

When designing a measurement system, the aim is always to reduce the probability of the system failing to as low a level as possible. An essential requirement in achieving this is to ensure that the system is replaced at or before the time t_2 in its life shown in Figure 12.1 when the statistical frequency of failures starts to increase. Therefore, the initial aim should be to set the lifetime T equal to t_2 and minimize the probability $F(T)$

of the system failing within this specified lifetime. Once all measures to reduce $F(T)$ have been applied, the acceptability of the reliability $R(T)$ has to be assessed against the requirements of the measurement system. Inevitably, cost enters into this, as efforts to increase $R(T)$ usually increase the cost of buying and maintaining the system. Lower reliability is acceptable in some measurement systems where the cost of failure is low, such as in manufacturing systems where the cost of lost production, or the loss due to making out-of-specification products, is not serious. However, in other applications, such as where failure of the measurement system incurs high costs or causes safety problems, high reliability is essential. Some special applications where human access is very difficult or impossible, such as measurements in unmanned spacecraft, satellites and nuclear power plants, demand especially high reliability because repair of faulty measurement systems is impossible.

The various means of increasing $R(T)$ are considered below. However, once all efforts to increase $R(T)$ have been exhausted, the only solution available if the reliability specified for a working period T is still not high enough is to reduce the period T over which the reliability is calculated by replacing the measurement system earlier than time t_2.

Choice of instrument

The type of components and instruments used within measuring systems has a large effect on the system reliability. Of particular importance in choosing instruments is to have regard to the type of operating environment in which they will be used. In parallel with this, appropriate protection must be given (for example, enclosing thermocouples in sheaths) if it is anticipated that the environment may cause premature failure of an instrument. Some instruments are more affected than others, and thus more likely to fail, in certain environments. The necessary knowledge to make informed choices about the suitability of instruments for particular environments, and the correct protection to give them, requires many years of experience, although instrument manufacturers can give useful advice in most cases.

Instrument protection

Adequate protection of instruments and sensors from the effects of the operating environment is necessary. For example, thermocouples and resistance thermometers should be protected by a sheath in adverse operating conditions.

Regular calibration

The most common reason for faults occurring in a measurement system, whereby the error in the measurement goes outside acceptable limits, is drift in the performance of the instrument away from its specified characteristics. Such faults can usually be avoided by ensuring that the instrument is recalibrated at the recommended intervals of time. Types of intelligent instrument and sensor that perform self-calibration have clear advantages in this respect.

Redundancy

Redundancy means the use of two or more measuring instruments or measurement system components in parallel such that any one can provide the required measurement.

Example 12.4 showed the use of three identical instruments in parallel to make a particular measurement instead of a single instrument. This increased the reliability from 95% to 99.99%. Redundancy can also be applied in larger measurement systems where particular components within it seriously degrade the overall reliability of the system. Consider the five-component measurement system shown in Figure 12.2(a) in which the reliabilities of the individual system components are $R_1 = R_3 = R_5 = 0.99$ and $R_2 = R_4 = 0.95$.

Using (12.5), the system reliability is given by $R_S = 0.99 \times 0.95 \times 0.99 \times 0.95 \times 0.99 = 0.876$.

Now, consider what happens if redundant instruments are put in parallel with the second and fourth system components, as shown in Figure 12.2(b). The reliabilities of these sections of the measurement system are now modified to new values R'_2 and R'_4, which can be calculated using (12.1), (12.6) and (12.8) as follows: $F_2 = 1 - R_2 = 0.05$. Hence, $F'_2 = (0.05)^2 = 0.0025$ and $R'_2 = 1 - F'_2 = 0.9975$. $R'_4 = R'_2$ since $R_4 = R_2$. Using (12.5) again, the system reliability is now $R_S = 0.99 \times 0.9975 \times 0.99 \times 0.9975 \times 0.99 = 0.965$.

Thus, the redundant instruments have improved the system reliability by a large amount. However, this improvement in reliability is only achieved at the cost of buying and maintaining the redundant components that have been added to the measurement system. If this practice of using redundant instruments to improve reliability is followed, provision must be provided for replacing failed components by the standby units. The most efficient way of doing this is to use an automatic switching system, but manual methods of replacement can also work reasonably well in many circumstances.

The principle of increasing reliability by placing components in parallel is often extended to other aspects of measurement systems such as the connectors in electrical circuits, as bad connections are a frequent cause of malfunction. For example, two separate pairs of plugs and sockets are frequently used to make the same connection. The second pair is redundant, i.e. the system can usually function at 100% efficiency without it, but it becomes useful if the first pair of connectors fails.

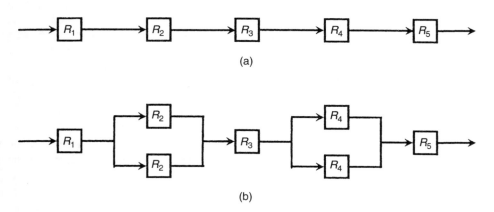

Fig. 12.2 Improving measurement system reliability: (a) original system; (b) duplicating components that have poor reliability.

12.1.4 Software reliability

As computer processors, and the software within them, are increasingly found in most measurement systems, the issue of the reliability of such components has become very important. Computer hardware behaves very much like electronic components in general, and the rules for calculating reliability given earlier can be applied. However, the factors affecting reliability in software are fundamentally different. Application of the general engineering definition of reliability to software is not appropriate because the characteristics of the error mechanisms in software and in engineering hardware are fundamentally different. Hardware systems that work correctly when first introduced can develop faults at any time in the future, and so the MTBF is a sensible measure of reliability. However, software does not change with time: if it starts off being error free, then it will remain so. Therefore, what we need to know, in advance of its use, is whether or not faults are going to be found in the software after it has been put into use. Thus, for software, a MTBF reliability figure is of little value. Instead, we must somehow express the probability that errors will not occur in it.

Quantifying software reliability

A fundamental problem in predicting that errors will not occur in software is that, however exhaustive the testing, it is impossible to say with certainty that all errors have been found and eliminated. Errors can be quantified by three parameters, D, U and T, where D is the number of errors detected by testing the software, U is the number of undetected errors and T is the total number of errors (both detected and undetected).

Hence:

$$U = T - D \qquad (12.9)$$

Good program testing can detect most errors and so make D approach T so that U tends towards zero. However, as the value of T can never be predicted with certainty, it is very difficult to predict that software is error free, whatever degree of diligence is applied during testing procedures.

Whatever approach is taken to quantifying reliability, software testing is an essential prerequisite to the quantification methods available. Whilst it is never possible to detect all the errors that might exist, the aim must always be to find and correct as many errors as possible by applying a rigorous testing procedure. Software testing is a particularly important aspect of the wider field of software engineering. However, as it is a subject of considerable complexity, the detailed procedures available are outside the scope of this book. A large number of books now cover good software engineering in general and software testing procedures in particular, and the reader requiring further information is referred to the referenced texts such as Pfleeger (1987) and Shooman (1983).

One approach to quantifying software reliability (Fenton, 1991) is to monitor the rate of error discovery during testing and then extrapolate this into an estimate of the mean-time-between-failures for the software once it has been put into use. Testing can then be extended until the predicted MTBF is greater than the projected time-horizon of usage of the software. This approach is rather unsatisfactory because it accepts that errors in the software exist and only predicts that errors will not emerge very frequently.

Confidence in the measurement system is much greater if we can say, 'There is a high probability that there are zero errors in the software' rather than 'There are a finite number of errors in the software but they are unlikely to emerge within the expected lifetime of its usage.' One way of achieving this is to estimate the value of T (total number of errors) from initial testing and then carry out further software testing until the predicted value of T is zero, in a procedure known as *error seeding* (Mills, 1972). In this method, the programmer responsible for producing the software deliberately puts a number of errors E into the program, such that the total number of errors in the program increases from T to T', where $T' = T + E$. Testing is then carried out by a different programmer who will identify a number of errors given by D', where $D' = D + E'$ and E' is the number of deliberately inserted errors that are detected by this second programmer. Normally, the real errors detected (D) will be less than T and the seeded errors detected (E') will be less than E. However, on the assumption that the ratio of seeded errors detected to the total number of seeded errors will be the same as the ratio of the real errors detected to the total number of real errors, the following expression can be written:

$$\frac{D}{T} = \frac{E'}{E} \qquad (12.10)$$

As E' is measured, E is known and D can be calculated from the number of errors D' detected by the second programmer according to $D = D' - E'$, the value of T can then be calculated as:

$$T = DE/E' \qquad (12.11)$$

Example 12.5
The author of a digital signal-processing algorithm that forms a software component within a measurement system adds 12 deliberate faults to the program. The program is then tested by a second programmer, who finds 34 errors. Of these detected errors, the program author recognizes 10 of them as being seeded errors. Estimate the original number of errors present in the software (i.e. excluding the seeded errors).

Solution
The total number of errors detected (D') is 34 and the program author confirms that the number of seeded errors (E') within these is 10 and that the total number of seeded errors (E) was 12. Because $D' = D + E'$ (see earlier), $D = D' - E' = 24$. Hence, from (12.11), $T = DE/E' = 24 \times 12/10 = 28.8$.

One flaw in expression (12.11) is the assumption that the seeded errors are representative of all the real (unseeded) errors in the software, both in proportion and character. This assumption is never entirely valid in practice because, if errors are unknown, then their characteristics are also unknown. Thus, whilst this approach may be able to give an approximate indication of the value of T, it can never predict its actual value with certainty.

An alternative to error seeding is the *double-testing* approach, where two independent programmers test the same program (Pfleeger, 1987). Suppose that the number of errors detected by each programmer is D_1 and D_2 respectively. Normally, the errors detected by the two programmers will be in part common and in part different. Let C be the number of common errors that both programmers find. The error-detection success of each programmer can be quantified as:

$$S_1 = D_1/T; \quad S_2 = D_2/T \qquad (12.12)$$

It is reasonable to assume that the proportion of errors D_1 that programmer 1 finds out of the total number of errors T is the same proportion as the number of errors C that he/she finds out of the number D_2 found by programmer 2, i.e.:

$$\frac{D_1}{T} = \frac{C}{D_2} = S_1, \quad \text{and hence } D_2 = \frac{C}{S_1} \qquad (12.13)$$

From (12.12), $T = D_2/S_2$, and substituting in the value of D_2 obtained from (12.13), the following expression for T is obtained:

$$T = C/S_1 S_2 \qquad (12.14)$$

From (12.13), $S_1 = C/D_2$ and from (12.12), $S_2 = D_2 S_1/D_1 = C/D_1$. Thus, substituting for S_1 and S_2 in (12.14):

$$T = D_1 D_2 / C \qquad (12.15)$$

Example 12.6
A piece of software is tested independently by two programmers, and the number of errors found is 24 and 26 respectively. Of the errors found by programmer 1, 21 are the same as errors found by programmer 2.

Solution
$D_1 = 24$, $D_2 = 26$ and $C = 21$. Hence, applying (12.15), $T = D_1 D_2/C = 24 \times 26/21 = 29.7$.

Program testing should continue until the number of errors that have been found is equal to the predicted total number of errors T. In the case of example 12.6, this means continuing testing until 30 errors have been found. However, the problem with doing this is that T is only an estimated quantity and there may actually be only 28 or 29 errors in the program. Thus, to continue testing until 30 errors have been found would mean testing forever! Hence, once 28 or 29 errors have been found and continued testing for a significant time after this has detected no more errors, the testing procedure should be terminated, even though the program could still contain one or two errors. The approximate nature of the calculated value of T also means that its true value could be 31 or 32, and therefore the software may still contain errors if testing is stopped once 30 errors have been found. Thus, the fact that T is only an estimated value means the statement that a program is error free once the number of errors detected is equal to T can only be expressed in probabilistic terms.

To quantify this probability, further testing of the program is necessary (Pfleeger, 1987). The starting point for this further testing is the stage when the total number of errors T that are predicted have been found (or when the number found is slightly less than T but further testing does not seem to be finding any more errors). The next step is to seed the program with W new errors and then test it until all W seeded errors have been found. Provided that no new errors have been found during this further testing phase, the probability that the program is error free can then be expressed as:

$$P = W/(W+1) \qquad (12.16)$$

However, if any new error is found during this further testing phase, the error must be corrected and then the seeding and testing procedure must be repeated. Assuming that no new errors are detected, a value of $W = 10$ gives $P = 0.91$ (probability 91% that program is error free). To get to 99% error-free probability, W has to be 99.

Improving software reliability

The *a priori* requirement in achieving high reliability in software is to ensure that it is produced according to sound software engineering principles. Formal standards for achieving high quality in software are set out in BS 7165 (1991) and ISO 9000-3 (1991). Libraries and bookshops, especially academic ones, offer a number of texts on good software design procedures. These differ significantly in their style of approach, but all have the common aim of encouraging the production of error-free software that conforms to the design specification. It is not within the scope of this book to enter into arguments about which software design approach is best, as choice between the different software design techniques largely depends on personal preferences. However, it is essential that software contributing to a measurement system is produced according to good software engineering principles.

The second stage of reliability enhancement is the application of a rigorous testing procedure as described in the last section. This is a very time-consuming and hence expensive business, and so testing should only continue until the calculated level of reliability is the minimum needed for the requirements of the measurement system. However, if a very high level of reliability is demanded, such rigorous testing becomes extremely expensive and an alternative approach known as N-version programming is often used. *N-version programming* requires N different programmers to produce N different versions of the same software according to a common specification. Then, assuming that there are no errors in the specification itself, any difference in the output of one program compared with the others indicates an error in that program. Commonly, $N = 3$ is used, that is, three different versions of the program are produced, but $N = 5$ is used for measurement systems that are very critical. In this latter case, a 'voting' system is used, which means that up to two out of the five versions can be faulty without incorrect outputs being generated.

Unfortunately, whilst this approach reduces the chance of software errors in measurement systems, it is not foolproof because the degree of independence between programs cannot be guaranteed. Different programmers, who may be trained in the same place and use the same design techniques, may generate different programs that have the same errors. Thus, this method has the best chance of success if the programmers are trained independently and use different design techniques.

Languages such as ADA also improve the safety of software because they contain special features that are designed to detect the kind of programming errors that are commonly made. Such languages have been specifically developed with safety critical applications in mind.

12.2 Safety systems

12.2.1 Introduction to safety systems

Measurement system reliability is usually inexorably linked with safety issues, since measuring instruments to detect the onset of dangerous situations that may potentially compromise safety are a necessary part of all safety systems implemented. Statutory safety legislation now exists in all countries around the world. Whilst the exact content of legislation varies from country to country, a common theme is to set out responsibilities for all personnel whose actions may affect the safety of themselves or others. Penalties are prescribed for contravention of the legislation, which can include fines or custodial sentences or both. Legislation normally sets out duties for both employers and employees.

Duties of employers include:

- To ensure that process plant is operated and maintained in a safe way so that the health and safety of all employees is protected
- To provide such training and supervision as is necessary to ensure the health and safety of all employees
- To provide a monitoring and shutdown system (safety system) for any process plant or other equipment that may cause danger if certain conditions arise
- To ensure the health and safety, as far as is reasonably practical, of all persons who are not employees but who may reasonably be expected to be at risk from operations carried out by a company.

Duties of employees include:

- To take reasonable care for their own safety
- To take reasonable care for the safety of others
- To avoid misusing or damaging any equipment or system that is designed to protect people's safety.

The primary concern of measurement and instrumentation technologists with regard to safety legislation is, firstly, to ensure that all measurement systems are installed and operated in a safe way and, secondly, to ensure that instruments and alarms installed as part of safety protection systems operate reliably and effectively.

Intrinsic safety

Intrinsic safety describes the ability of measuring instruments and other systems to operate in explosive or flammable environments without any risk of sparks or arcs causing an explosion or fire. The detailed design of systems to make them intrinsically safe is outside the scope of this book. However, the general principles are either to design electrical systems in a way that avoids any possibility of parts that may spark coming into contact with the operating environment or else to avoid using electrical components altogether. The latter point means that pneumatic sensors and actuators continue to find favour in some applications despite the advantages of electrical devices in most other respects.

Installation practice

Good installation practice is necessary to prevent any possibility of people getting electrical shocks from measurement systems. Instruments that have a mains power supply must be subject to the normal rules about the condition of supply cables, clamping of wires and earthing of all metal parts. However, most measurement systems operate at low voltages and so pose no direct threat unless parts of the system come into contact with mains conductors. This should be prevented by applying codes of practice that require that all cabling for measurement systems be kept physically separate to that used for carrying mains voltages to equipment. Normally, this prohibits the use of the same trunking to house both signal wires and mains cables, although some special forms of trunking are available that have two separate channels separated by a metal barrier, thus allowing them to be used for both mains cables and signal wires. This subject is covered in depth in the many texts on electrical installation practice.

12.2.2 Operation of safety systems

The purpose of safety systems is to monitor parameter values in manufacturing plant and other systems and to make an effective response when plant parameters vary away from normal operating values and cause a potentially dangerous situation to develop. The response can either be to generate an alarm for the plant operator to take action or else to take more direct action to shut down the plant automatically. The design and operation of safety systems is now subject to guidelines set by the international standard IEC61508.

IEC61508

IEC61508 (1999) sets out a code of practice that is designed to ensure that safety systems work effectively and reliably. Although primarily concerned with electrical, electronic and programmable-electronic safety systems, the principles embodied by the standard can be applied as well to systems with other technologies, such as mechanical, pneumatic and hydraulic devices.

The IEC61508 standard is subdivided into three sets of requirements:

- Proper management of design, implementation and maintenance of safety systems
- Competence and training of personnel involved in designing, implementing or maintaining safety systems
- Technical requirements for the safety system itself.

A full analysis of these various requirements can be found elsewhere (Dean, 1999).

A key feature of IEC61508 is the *safety integrity level* (SIL), which is expressed as the degree of confidence that a safety system will operate correctly and ensure that there is an adequate response to any malfunctions in manufacturing plant etc. that may cause a hazard and put human beings at risk. The SIL value is set according to what the tolerable risk is in terms of the rate of failure for a process. The procedure for defining the required SIL value is known as *risk analysis*. What is 'tolerable' depends on what the consequences of a dangerous failure are in terms of injury to one or more people or death to one or more people. The acceptable level of tolerance for particular industries and processes is set according to guidelines defined by safety regulatory authorities, expert advice and legal requirements. Table 12.1 gives the SIL value corresponding to various levels of tolerable risk for continuous operating plant.

The safety system is required to have sufficient reliability to match the rate of dangerous failures in a plant to the SIL value set. This reliability level is known as the *safety integrity* of the system. *Plant reliability* is calculated by identical principles to those set out in section 12.1 for measurement systems, and is based on a count of the number of faults that occur over a certain interval of time. However, it must be emphasized that the frequency of potentially dangerous failures is usually less than the rate of occurrence of faults in general. Thus, the reliability value for a plant cannot be used directly as a prediction of the rate of occurrence of dangerous failures. Hence, the total failures over a period of time must be analysed and divided between faults that are potentially dangerous and those that are not.

Once risk analysis has been carried out to determine the appropriate SIL value, the required performance of the safety protection system can be calculated. For example, if the maximum allowable probability of dangerous failures per hour is specified as 10^{-8} and the actual probability of dangerous failures in a plant is calculated as 10^{-3} per hour, then the safety system must have a minimum reliability of $10^{-8}/10^{-3}$, i.e. 10^{-5} failures for a one-hour period. A fuller account of calculating safety system requirements is given elsewhere (Simpson, 1999).

12.2.3 Design of a safety system

A typical safety system consists of a sensor, a trip amplifier and either an actuator or alarm generator, as shown in Figure 12.3. For example, in a safety system designed to protect against abnormally high pressures in a process, the sensor would be some form of pressure transducer, and the trip amplifier would be a device that amplifies the measured pressure signal and generates an output that activates either an actuator or an alarm if the measured pressure signal exceeded a preset threshold value. A typical actuator in this case would be a relief valve.

Table 12.1

SIL	Probability of dangerous failure per hour	Probability of dangerous failure per year
4	10^{-9} to 10^{-8}	10^{-5} to 10^{-4}
3	10^{-8} to 10^{-7}	10^{-4} to 10^{-3}
2	10^{-7} to 10^{-6}	10^{-3} to 10^{-2}
1	10^{-6} to 10^{-5}	10^{-2} to 10^{-1}

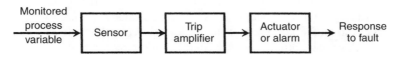

Fig. 12.3 Elements of a safety system.

Software is increasingly embedded within safety systems to provide intelligent interpretation of sensor outputs, such as identifying trends in measurements. Safety systems that incorporate software and a computer processor are commonly known as *microprocessor-based protection systems*. In any system containing software, the reliability of the software is crucial to the overall reliability of the safety system, and the reliability–quantification techniques described in section 12.2 assume great importance.

To achieve the very high levels of reliability normally specified for safety systems, it is usual to guard against system failure by either triplicating the safety system and implementing two-out-of-three voting or, alternatively, by providing a switchable, standby safety system. These techniques are considered below.

Two-out-of-three voting system

This system involves triplicating the safety system, as shown in Figure 12.4. Shutdown action is taken, or an alarm is generated, if two out of the three systems indicate the requirement for action. This allows the safety system to operate reliably if any one of the triplicated systems fails and is often known as a two-out-of-three voting system. The reliability R_S is given by:

$$R_S = \text{Probability of all three systems operating correctly}$$
$$+ \text{Probability of any two systems operating correctly}$$
$$= R_1 R_2 R_3 + (R_1 R_2 F_3 + R_1 F_2 R_3 + F_1 R_2 R_3) \quad (12.17)$$

where R_1, R_2, R_3 and F_1, F_2 and F_3 are the reliabilities and unreliabilities of the three systems respectively. If all of the systems are identical (such that $R_1 = R_2 = R_3 = R$ etc.):

$$R_S = R^3 + 3R^2 F = R^3 + 3R^2(1 - R) \quad (12.18)$$

Example 12.7
In a particular protection system, three safety systems are connected in parallel and a two-out-of-three voting strategy is applied. If the reliability of each of the three systems is 0.95, calculate the overall reliability of the whole protection system.

Solution
Applying (12.18), $R_S = 0.95^3 + [3 \times 0.95^2 \times (1 - 0.95)] = 0.993$.

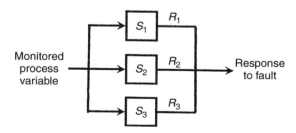

Fig. 12.4 Two-out-of-three voting system.

Standby system

A standby system avoids the cost of providing and running three separate safety systems in parallel. Use of a standby system means that only two safety systems have to be provided. The first system is in continuous use but the second system is normally not operating and is only switched into operation if the first system develops a fault. The flaws in this approach are the necessity for faults in the primary system to be reliably detected and the requirement that the switch must always work correctly. The probability of failure F_S of a standby system of the form shown in Figure 12.5, assuming no switch failures during normal operation, can be expressed as:

F_S = Probability of systems S_1 and S_2 both failing, given successful switching
+ Probability of S_1 and the switching system both failing at the same time

$= F_1 F_2 R_D R_W + F_1(1 - R_D R_W)$

System reliability is given by:

$$R_S = 1 - F_S = 1 - F_1(1 + F_2 R_D R_W - R_D R_W) \tag{12.19}$$

where R_D is the reliability of the fault detector and R_W is the reliability of the switch.

The derivation of (12.19) assumes that there are no switch failures during normal operation of the system, that is, there are no switch failures during the time that the controlled process is operating satisfactorily and there is no need to switch over to the standby system. However, because the switch is subject to a continuous flow of current, its reliability cannot be assumed to be 100%. If the reliability of the switch in normal operation is represented by R_N, the expression in (12.19) must be multiplied by R_N and the reliability of the system becomes:

$$R_S = R_N[1 - F_1(1 + F_2 R_D R_W - R_D R_W)] \tag{12.20}$$

The problem of detecting faults in the primary safety system reliably can be solved by operating both safety systems in parallel. This enables faults in the safety system to be distinguished from faults in the monitored process. If only one of the two safety systems indicates a failure, this can be taken to indicate a failure of one of the safety systems rather than a failure of the monitored process. However, if both safety systems indicate a fault, this almost certainly means that the monitored process has developed a potentially dangerous fault. This scheme is known as *one-out-of-two voting*, but it is obviously inferior in reliability to the two-out-of-three scheme described earlier.

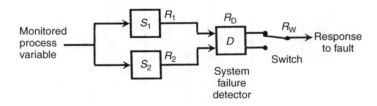

Fig. 12.5 Standby system.

Example 12.8
In a particular protection system, a switchable standby safety system is used to increase reliability. If the reliability of the main system is 0.95, that of the standby system is 0.96*, that of the switching system is 0.95 and the reliability of the switch in normal operation is 0.98, calculate the reliability of the protection system.

Solution
Applying (12.19), the parameter values are $F_1 = 0.05$, $F_2 = 0.04$, $R_D R_W = 0.95$. Hence:
$$R_S = 0.98[1 - 0.05(1 + \{0.04 \times 0.95\} - 0.95)] = 0.976$$

Actuators and alarms

The final element in a safety system is either an automatic actuator or an alarm that requires a human response. The reliability of the actuator can be calculated in the same way as all other elements in the system and incorporated into the calculation of the overall system reliability as expressed in equations (12.17)–(12.20). However, the reliability of alarms cannot be quantified in the same manner. Therefore, safety system reliability calculations have to exclude the alarm element. In consequence, the system designer needs to take steps to maximize the probability that the human operator will take the necessary response to alarms that indicate a dangerous plant condition.

Some useful guidelines for measurement technologists involved in designing alarm systems are provided in a paper by Bransby (1999). A very important criterion in system design is that alarms about dangerous conditions in plant must be much more prominent than alarms about conditions that are not dangerous. Care should also be taken to ensure that the operator of a plant is not bombarded by too many alarms, as this leads the operator to get into the habit of ignoring alarms. Ignoring an alarm indicating that a fault is starting to occur may cause dangerous conditions in the plant to develop. Thus, alarms should be uncommon rather than routine, so that they attract the attention of the plant operator. This ensures, as far as possible, that the operator will take the proper action in response to an alarm about a potentially dangerous situation.

12.3 Self-test questions

1. The performance of a measuring instrument is monitored over a one-year (365-day) period and the intervals between faults being recorded are as follows:

   27 6 18 41 54 29 46 14 49 38 17 26

   Calculate the mean-time-between-failures.

2. The days on which an instrument failed were recorded over a 12-month period as follows (such that day 1 = 1 Jan, day 32 = 1 Feb etc.):

   Day number of faults: 18 72 111 173 184 227 286 309 356

   Calculate the mean-time-between-failures.

---

* Because the standby system is not subject to normal use, its reliability tends to be higher than the primary system even if the two systems consist of nominally identical components.

3. A manufacturer monitors the performance of a new type of instrument that is installed at 20 different locations. If a total of nine faults are recorded over a 100-day period, calculate the mean-time-between-failure that should be specified for any one instrument.
4. If the mean-time-between-failure for an instrument is 247 days and the mean-time-to-repair is 3 days, calculate its availability.
5. If the mean-time-to-failure of an instrument is 100 000 hours, calculate the probability that it will not fail in the first 50 000 hours.
6. If the mean-time-to-failure of an instrument is 100 000 hours, calculate the probability that it will not fail in the first 5000 hours.
7. Four measurement components connected in series have the following reliabilities: 0.98 0.93 0.95 0.99. Calculate the reliability of the whole measurement system.
8. In a particular measurement system, two instruments with individual reliability of 0.95 are connected together in parallel. Calculate the reliability of the measurement system if it can continue to function as long as both of the instruments do not fail at the same time.
9. Calculate the reliability of the measurement system shown in Figure 12.2(b) if the reliabilities of the individual components are $R_1 = R_3 = R_5 = 0.98$; $R_2 = R_4 = 0.90$.
10. In order to estimate the number of errors in a new piece of software by applying the error-seeding approach, a programmer puts ten deliberate (seeded) faults into the program. A second programmer then tests the program and finds 27 errors, of which eight are confirmed by the first programmer to be seeded errors. Estimate the original number of faults in the program (i.e. excluding the seeded errors).
11. The double-testing approach is applied to test a new computer program and the two programmers who do the testing find 31 and 34 errors respectively. If 27 of the errors found by programmer 1 are the same as errors in the list produced by programmer 2, estimate the actual number of errors in the program.
12. A program is tested and the total number of errors is estimated as 16. The program is then seeded with 20 deliberate errors and further testing is then carried out until all 20 seeded errors have been found.
    (a) If no new (previously undetected) errors are found during this further testing to find all the seeded errors, calculate the probability that the program is error free after this further testing.
    (b) How many seeded errors would have to be put into the program and then detected to achieve a 98% probability that the program is error free?
13. Three safety systems are connected in parallel in a protection system and a two-out-of-three voting strategy is applied. If the reliability of each of the three systems is 0.90, calculate the overall reliability of the whole protection system.
14. A switchable standby safety system is used to increase reliability in a protection system. If the reliability of the main system is 0.90, that of the standby system is 0.91, that of the switching system is 0.90 and the reliability of the switch in normal operation is 0.96, calculate the reliability of the protection system.

References and further reading

Bahr, M.J. (1997) *System Safety: A Practical Approach*, Taylor and Francis.
Barnand, M.J. (1998) *Health and Safety for Engineers*, Thomas Telford.

Bransby, M. (1999) The human contribution to safety – designing alarm systems, *Measurement and Control*, **32**, pp. 209–213.

BS 7165 (1991) Recommendations for the achievement of quality in software, British Standards Institute.

Cluley, J.C. (1993) *Reliability in Instrumentation and Control*, Butterworth-Heinemann.

Dean, S. (1999) IEC61508 – understanding functional safety assessment, *Measurement and Control*, **32**, pp. 201–204.

Dhillon, B.S. and Singh, C. (1981) *Engineering Reliability: New Techniques and Applications*, Wiley.

Fenton, N.E. (1991) *Software Metrics: A Rigorous Approach*, Chapman and Hall.

IEC61508 (1991) Functional safety of electrical, electronic and programmable-electronic safety related systems, International Electrotechnical Commission, Geneva.

ISO 9000-3 (1991) International Organization for Standards, Geneva.

Mather, R.N. (1987) Methodology for business system development, *IEEE Trans Software Eng.*, **13**(5), pp. 593–601.

Miller, I.R., Freund, J.E. and Johnson, R. (1990) *Probability and Statistics for Engineers*, Prentice Hall.

Mills, H.D. (1972) *On the Statistical Validation of Computer Programs*, IBM Federal Systems Division, Maryland.

Morris, A.S. (1997) *Measurement and Calibration Requirements for Quality Assurance to ISO9000*, John Wiley.

Pfleeger, S.L. (1987) *Software Engineering: The Production of Quality Software*, Macmillan.

Shooman, M.L. (1983) *Software Engineering: Design, Reliability and Management*, McGraw-Hill.

Simpson, K. and Smith, D.J. (1999) Assessing safety related systems and architectures, *Measurement and Control*, **32**, pp. 205–208.

Thomson, J.R. (1987) *Engineering Safety Assessment – An Introduction*, Longman.

Part 2   Measurement Sensors and Instruments

The second part of the book starts by discussing some of the main physical principles used in measurement sensors, and then it goes on to discuss the range of sensors and instruments that are available for measuring various physical quantities. In presenting each range of measurement devices, the aim has been to be as comprehensive as possible, so that the book can be used as a source of reference when choosing an instrument for any particular measurement situation. This method of treatment means that some quite rare and expensive instruments are included as well as very common and cheap ones, and therefore it is necessary to choose from the instrument ranges presented with care.

The first step in choosing an instrument for a particular measurement situation is to specify the static and dynamic characteristics required. These must always be defined carefully, as a high degree of accuracy, sensitivity etc. in the instrument specification inevitably involves a high cost. The characteristics specified should therefore be the minimum necessary to achieve the required level of performance in the system that the instrument is providing a measurement for. Once the static and dynamic characteristics required have been defined, the range of instruments presented in each of the following chapters can be reduced to a subset containing those instruments that satisfy the defined specifications.

The second step in choosing an instrument is to consider the working conditions in which the instrument will operate. Conditions demanding special consideration are those where the instrument will be exposed to mechanical shocks, vibration, fumes, dust or fluids. Such considerations further subdivide the subset of possible instruments already identified. If a choice still exists at this stage, then the final criterion is one of cost.

The relevant criteria in instrument choice for measuring particular physical quantities are considered further in the following chapters.

13

Sensor technologies

13.1 Capacitive and resistive sensors

Capacitive sensors consist of two parallel metal plates in which the dielectric between the plates is either air or some other medium. The capacitance C is given by $C = \varepsilon_0 \varepsilon_r A/d$, where ε_0 is the absolute permittivity, ε_r is the relative permittivity of the dielectric medium between the plates, A is the area of the plates and d is the distance between them. Capacitive devices are often used as displacement sensors, in which motion of a moveable capacitive plate relative to a fixed one changes the capacitance. Often, the measured displacement is part of instruments measuring pressure, sound or acceleration. Alternatively, fixed plate capacitors can also be used as sensors, in which the capacitance value is changed by causing the measured variable to change the dielectric constant of the material between the plates in some way. This principle is used in devices to measure moisture content, humidity values and liquid level, as discussed in later chapters.

Resistive sensors rely on the variation of the resistance of a material when the measured variable is applied to it. This principle is most commonly applied in temperature measurement using resistance thermometers or thermistors, and in displacement measurement using strain gauges or piezoresistive sensors. In addition, some moisture meters work on the resistance-variation principle.

13.2 Magnetic sensors

Magnetic sensors utilize the magnetic phenomena of inductance, reluctance and eddy currents to indicate the value of the measured quantity, which is usually some form of displacement.

Inductive sensors translate movement into a change in the mutual inductance between magnetically coupled parts. One example of this is the inductive displacement transducer shown in Figure 13.1. In this, the single winding on the central limb of an 'E'-shaped ferromagnetic body is excited with an alternating voltage. The displacement to be measured is applied to a ferromagnetic plate in close proximity to the 'E' piece. Movements of the plate alter the flux paths and hence cause a change in the current flowing in the winding. By Ohm's law, the current flowing in the winding is

248 Sensor technologies

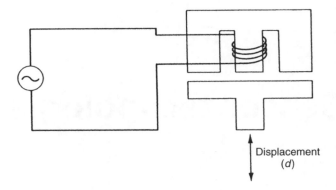

Fig. 13.1 Inductive displacement sensor.

given by $I = V/\omega L$. For fixed values of w and V, this equation becomes $I = 1/KL$, where K is a constant. The relationship between L and the displacement, d, applied to the plate is a non-linear one, and hence the output-current/displacement characteristic has to be calibrated.

The inductance principle is also used in differential transformers to measure translational and rotational displacements.

In *variable reluctance sensors*, a coil is wound on a permanent magnet rather than on an iron core as in variable inductance sensors. Such devices are commonly used to measure rotational velocities. Figure 13.2 shows a typical instrument in which a ferromagnetic gearwheel is placed next to the sensor. As the tip of each tooth on the gearwheel moves towards and away from the pick-up unit, the changing magnetic flux in the pick-up coil causes a voltage to be induced in the coil whose magnitude is proportional to the rate of change of flux. Thus, the output is a sequence of positive and negative pulses whose frequency is proportional to the rotational velocity of the gearwheel.

Eddy current sensors consist of a probe containing a coil, as shown in Figure 13.3, that is excited at a high frequency, which is typically 1 MHz. This is used to measure the displacement of the probe relative to a moving metal target. Because of the high frequency of excitation, eddy currents are induced only in the surface of the target,

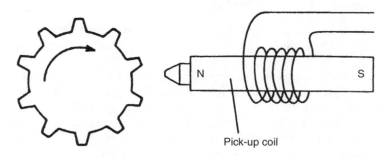

Fig. 13.2 Variable reluctance sensor.

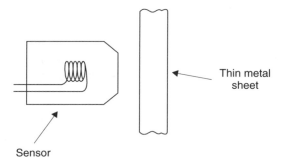

Fig. 13.3 Eddy current sensor.

and the current magnitude reduces to almost zero a short distance inside the target. This allows the sensor to work with very thin targets, such as the steel diaphragm of a pressure sensor. The eddy currents alter the inductance of the probe coil, and this change can be translated into a d.c. voltage output that is proportional to the distance between the probe and the target. Measurement resolution as high as $0.1\,\mu m$ can be achieved. The sensor can also work with a non-conductive target if a piece of aluminium tape is fastened to it.

13.3 Hall-effect sensors

Basically, a Hall-effect sensor is a device that is used to measure the magnitude of a magnetic field. It consists of a conductor carrying a current that is aligned orthogonally with the magnetic field, as shown in Figure 13.4. This produces a transverse voltage difference across the device that is directly proportional to the magnetic field strength. For an excitation current I and magnetic field strength B, the output voltage is given by $V = KIB$, where K is known as the Hall constant.

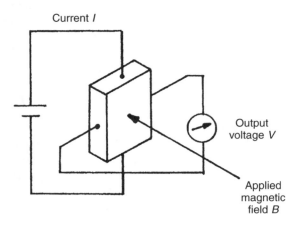

Fig. 13.4 Principles of Hall-effect sensor.

The conductor in Hall-effect sensors is usually made from a semiconductor material as opposed to a metal, because a larger voltage output is produced for a magnetic field of a given size. In one common use of the device as a proximity sensor, the magnetic field is provided by a permanent magnet that is built into the device. The magnitude of this field changes when the device becomes close to any ferrous metal object or boundary. The Hall effect is also commonly used in keyboard pushbuttons, in which a magnet is attached underneath the button. When the button is depressed, the magnet moves past a Hall-effect sensor. The induced voltage is then converted by a trigger circuit into a digital output. Such pushbutton switches can operate at high frequencies without contact bounce.

13.4 Piezoelectric transducers

Piezoelectric transducers produce an output voltage when a force is applied to them. They are frequently used as ultrasonic receivers and also as displacement transducers, particularly as part of devices measuring acceleration, force and pressure. In ultrasonic receivers, the sinusoidal amplitude variations in the ultrasound wave received are translated into sinusoidal changes in the amplitude of the force applied to the piezoelectric transducer. In a similar way, the translational movement in a displacement transducer is caused by mechanical means to apply a force to the piezoelectric transducer. Piezoelectric transducers are made from piezoelectric materials. These have an asymmetrical lattice of molecules that distorts when a mechanical force is applied to it. This distortion causes a reorientation of electric charges within the material, resulting in a relative displacement of positive and negative charges. The charge displacement induces surface charges on the material of opposite polarity between the two sides. By implanting electrodes into the surface of the material, these surface charges can be measured as an output voltage. For a rectangular block of material, the induced voltage is given by:

$$V = \frac{kFd}{A} \tag{13.1}$$

where F is the applied force in g, A is the area of the material in mm, d is the thickness of the material and k is the piezoelectric constant. The polarity of the induced voltage depends on whether the material is compressed or stretched.

The input impedance of the instrument used to measure the induced voltage must be chosen carefully. Connection of the measuring instrument provides a path for the induced charge to leak away. Hence, the input impedance of the instrument must be very high, particularly where static or slowly varying displacements are being measured.

Materials exhibiting piezoelectric behaviour include natural ones such as quartz, synthetic ones such as lithium sulphate and ferroelectric ceramics such as barium titanate. The piezoelectric constant varies widely between different materials. Typical values of k are 2.3 for quartz and 140 for barium titanate. Applying equation (13.1) for a force of 1 g applied to a crystal of area 100 mm^2 and thickness 1 mm gives an output of 23 µV for quartz and 1.4 mV for barium titanate.

Certain polymeric films such as polyvinylidine also exhibit piezoelectric properties. These have a higher voltage output than most crystals and are very useful in many applications where displacement needs to be translated into a voltage. However, they have very limited mechanical strength and are unsuitable for applications where resonance might be generated in the material.

The piezoelectric principle is invertible, and therefore distortion in a piezoelectric material can be caused by applying a voltage to it. This is commonly used in ultrasonic transmitters, where the application of a sinusoidal voltage at a frequency in the ultra-sound range causes a sinusoidal variation in the thickness of the material and results in a sound wave being emitted at the chosen frequency. This is considered further in the section below on ultrasonic transducers.

13.5 Strain gauges

Strain gauges are devices that experience a change in resistance when they are stretched or strained. They are able to detect very small displacements, usually in the range 0–50 µm, and are typically used as part of other transducers, for example diaphragm pressure sensors that convert pressure changes into small displacements of the diaphragm. Measurement inaccuracies as low as ±0.15% of full-scale reading are achievable and the quoted life expectancy is usually three million reversals. Strain gauges are manufactured to various nominal values of resistance, of which 120 Ω, 350 Ω and 1000 Ω are very common. The typical maximum change of resistance in a 120 Ω device would be 5 Ω at maximum deflection.

The traditional type of strain gauge consists of a length of metal resistance wire formed into a zigzag pattern and mounted onto a flexible backing sheet, as shown in Figure 13.5(a). The wire is nominally of circular cross-section. As strain is applied to the gauge, the shape of the cross-section of the resistance wire distorts, changing the cross-sectional area. As the resistance of the wire per unit length is inversely proportional to the cross-sectional area, there is a consequential change in resistance. The input–output relationship of a strain gauge is expressed by the *gauge factor*, which is defined as the change in resistance (R) for a given value of strain (S), i.e.

$$\text{gauge factor} = \delta R/\delta S$$

In recent years, wire-type gauges have largely been replaced, either by metal-foil types as shown in Figure 13.5(b), or by semiconductor types. Metal-foil types are very

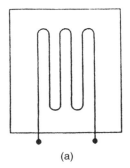

(a)   (b)

Fig. 13.5 Strain gauges: (a) wire type; (b) foil type.

similar to metal-wire types except the active element consists of a piece of metal foil cut into a zigzag pattern. Cutting a foil into the required shape is much easier than forming a piece of resistance wire into the required shape, and this makes the devices cheaper to manufacture. A popular material in metal strain gauge manufacture is a copper–nickel–manganese alloy, which is known by the trade name of 'Advance'. Semiconductor types have piezoresistive elements, which are considered in greater detail in the next section. Compared with metal gauges, semiconductor types have a much superior gauge factor (up to 100 times better) but they are more expensive. Also, whilst metal gauges have an almost zero temperature coefficient, semiconductor types have a relatively high temperature coefficient.

In use, strain gauges are bonded to the object whose displacement is to be measured. The process of bonding presents a certain amount of difficulty, particularly for semiconductor types. The resistance of the gauge is usually measured by a d.c. bridge circuit and the displacement is inferred from the bridge output measured. The maximum current that can be allowed to flow in a strain gauge is in the region of 5 to 50 mA depending on the type. Thus, the maximum voltage that can be applied is limited and consequently, as the resistance change in a strain gauge is typically small, the bridge output voltage is also small and amplification has to be carried out. This adds to the cost of using strain gauges.

13.6 Piezoresistive sensors

A piezoresistive sensor is made from semiconductor material in which a p-type region has been diffused into an n-type base. The resistance of this varies greatly when the sensor is compressed or stretched. This is frequently used as a strain gauge, where it produces a significantly higher gauge factor than that given by metal wire or foil gauges. Also, measurement uncertainty can be reduced to ±0.1%. It is also used in semiconductor-diaphragm pressure sensors and in semiconductor accelerometers.

It should also be mentioned that the term piezoresistive sensor is sometimes used to describe all types of strain gauge, including metal types. However, this is incorrect since only about 10% of the output from a metal strain gauge is generated by piezoresistive effects, with the remainder arising out of the dimensional cross-section change in the wire or foil. Proper piezoelectric strain gauges, which are alternatively known as *semiconductor strain gauges*, produce most (about 90%) of their output through piezoresistive effects, and only a small proportion of the output is due to dimensional changes in the sensor.

13.7 Optical sensors (air path)

Optical sensors are based on the modulation of light travelling between a light source and a light detector, as shown in Figure 13.6. The transmitted light can travel along either an air path or a fibre-optic cable. Either form of transmission gives immunity to electromagnetically induced noise, and also provides greater safety than electrical sensors when used in hazardous environments.

Measurement and Instrumentation Principles

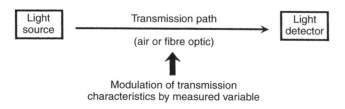

Fig. 13.6 Operating principles of optical sensors.

Light sources suitable for transmission across an air path include tungsten-filament lamps, laser diodes and light-emitting diodes (LEDs). However, as the light from tungsten lamps is usually in the visible part of the light frequency spectrum, it is prone to interference from the sun and other sources. Hence, infrared LEDs or infrared laser diodes are usually preferred. These emit light in a narrow frequency band in the infrared region and are not affected by sunlight.

The main forms of light detector used with optical systems are photocells (cadmium sulphide or cadmium selenide being the most common type of photocell), phototransistors and photodiodes. These are all photoconductive devices, whose resistance is reduced according to the intensity of light to which they are exposed. Photocells and phototransistors are particularly sensitive in the infrared region, and so are ideal partners for infrared LED and laser diode sources.

Air-path optical sensors are commonly used to measure proximity, translational motion, rotational motion and gas concentration. These uses are discussed in more detail in later chapters.

13.8 Optical sensors (fibre-optic)

As an alternative to using air as the transmission medium, optical sensors can use fibre-optic cable instead to transmit light between a source and a detector. In such sensors, the variable being measured causes some measurable change in the characteristics of the light transmitted by the cable. However, the problems and solutions that were described in Chapter 8 for fibre-optic signal transmission, in ensuring that the proportion of light entering the cable is maximized, apply equally when optical fibres are used as sensors.

The basis of operation of fibre-optic sensors is the translation of the physical quantity measured into a change in one or more parameters of a light beam. The light parameters that can be modulated are one or more of the following:

- intensity
- phase
- polarization
- wavelength
- transmission time.

Fibre-optic sensors usually incorporate either glass/plastic cables or all plastic cables. All glass types are rarely used because of their fragility. Plastic cables have particular advantages for sensor applications because they are cheap and have a relatively large diameter of 0.5–1.0 mm, making connection to the transmitter and receiver easy. However, plastic cables should not be used in certain hostile environments where they may be severely damaged. The cost of the fibre-optic cable itself is insignificant for sensing applications, as the total cost of the sensor is dominated by the cost of the transmitter and receiver.

Fibre-optic sensors characteristically enjoy long life. For example, the life expectancy of reflective fibre-optic switches is quoted at ten million operations. Their accuracy is also good, with, for instance, $\pm 1\%$ of full-scale reading being quoted as a typical inaccuracy level for a fibre-optic pressure sensor. Further advantages are their simplicity, low cost, small size, high reliability and capability of working in many kinds of hostile environment.

Two major classes of fibre-optic sensor exist, intrinsic sensors and extrinsic sensors. In *intrinsic sensors*, the fibre-optic cable itself is the sensor, whereas in *extrinsic sensors*, the fibre-optic cable is only used to guide light to/from a conventional sensor.

13.8.1 Intrinsic sensors

Intrinsic sensors can modulate either the intensity, phase, polarization, wavelength or transit time of light. Sensors that modulate light intensity tend to use mainly multimode fibres, but only monomode cables are used to modulate other light parameters. A particularly useful feature of intrinsic fibre-optic sensors is that they can, if required, provide distributed sensing over distances of up to 1 metre.

Light intensity is the simplest parameter to manipulate in intrinsic sensors because only a simple source and detector are required. The various forms of switches shown in Figure 13.7 are perhaps the simplest form of these, as the light path is simply blocked and unblocked as the switch changes state.

Modulation of the intensity of transmitted light takes place in various simple forms of proximity, displacement, pressure, pH and smoke sensors. Some of these are sketched in Figure 13.8. In proximity and displacement sensors (the latter are often given the special name *fotonic sensors*), the amount of reflected light varies with the distance between the fibre ends and a boundary. In pressure sensors, the refractive index of the fibre, and hence the intensity of light transmitted, varies according to the mechanical deformation of the fibres caused by pressure. In the pH probe, the amount of light reflected back into the fibres depends on the pH-dependent colour of the chemical indicator in the solution around the probe tip. Finally, in a form of smoke detector, two fibre-optic cables placed either side of a space detect any reduction in the intensity of light transmission between them caused by the presence of smoke.

A simple form of accelerometer can be made by placing a mass subject to the acceleration on a multimode fibre. The force exerted by the mass on the fibre causes a change in the intensity of light transmitted, hence allowing the acceleration to be determined. The typical inaccuracy quoted for this device is $\pm 0.02\,g$ in the measurement range $\pm 5\,g$ and $\pm 2\%$ in the measurement range up to $100\,g$.

A similar principle is used in probes that measure the internal diameter of tubes. The probe consists of eight strain-gauged cantilever beams that track changes in diameter, giving a measurement resolution of $20\,\mu m$.

A slightly more complicated method of effecting light intensity modulation is the variable shutter sensor shown in Figure 13.9. This consists of two fixed fibres with

Measurement and Instrumentation Principles 255

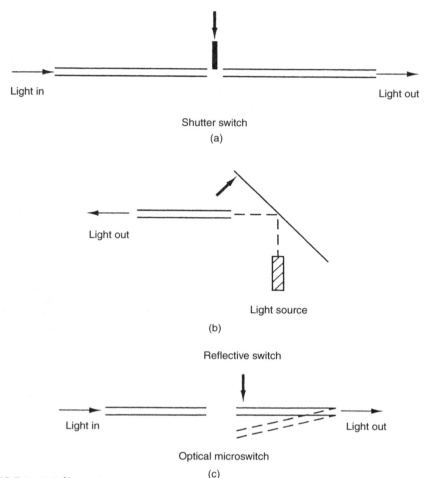

Fig. 13.7 Intrinsic fibre-optic sensors.

two collimating lenses and a variable shutter between them. Movement of the shutter changes the intensity of light transmitted between the fibres. This is used to measure the displacement of various devices such as Bourdon tubes, diaphragms and bimetallic thermometers.

Yet another type of intrinsic sensor uses cable where the core and cladding have similar refractive indices but different temperature coefficients. This is used as a temperature sensor. Temperature rises cause the refractive indices to become even closer together and losses from the core to increase, thus reducing the quantity of light transmitted.

Refractive index variation is also used in a form of intrinsic sensor used for cryogenic leak detection. The fibre used for this has a cladding whose refractive index becomes greater than that of the core when it is cooled to cryogenic temperatures. The fibre-optic cable is laid in the location where cryogenic leaks might occur. If any leaks do occur, light travelling in the core is transferred to the cladding, where it is attenuated.

256 Sensor technologies

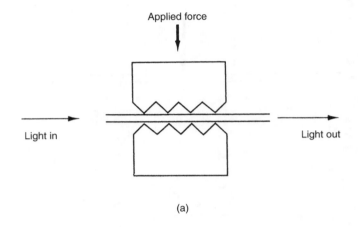

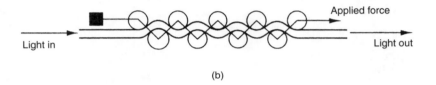

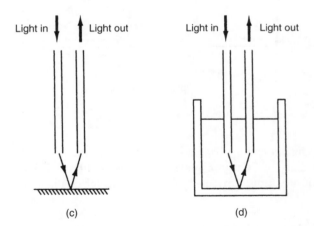

Fig. 13.8 Intensity modulating sensor: (a) simple pressure sensor; (b) roller-chain pressure sensor (microbend sensor); (c) proximity sensor; (d) pH sensor.

Cryogenic leakage is thus indicated by monitoring the light transmission characteristics of the fibre.

A further use of refractive index variation is found in devices that detect oil in water. These use a special form of cable where the cladding used is sensitive to oil. Any oil present diffuses into the cladding and changes the refractive index, thus increasing

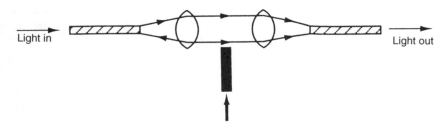

Fig. 13.9 Variable-shutter sensor.

light losses from the core. Unclad fibres are used in a similar way. In these, any oil present settles on the core and allows light to escape.

The *cross-talk sensor* measures several different variables by modulating the intensity of light transmitted. It consists of two parallel fibres that are close together and where one or more short lengths of adjacent cladding are removed from the fibres. When immersed in a transparent liquid, there are three different effects that each cause a variation in the intensity of light transmitted. Thus, the sensor can perform three separate functions. Firstly, it can measure temperature according to the temperature-induced variation in the refractive index of the liquid. Secondly, it can act as a level detector, as the transmission characteristics between the fibres change according to the depth of the liquid. Thirdly, it can measure the refractive index of the liquid itself when used under controlled temperature conditions.

The refractive index of a liquid can be measured in an alternative way by using an arrangement where light travels across the liquid between two cable ends that are fairly close together. The angle of the cone of light emitted from the source cable, and hence the amount of light transmitted into the detector, is dependent on the refractive index of the liquid.

The use of materials where the fluorescence varies according to the value of the measurand can also be used as part of intensity modulating intrinsic sensors. Fluorescence-modulating sensors can give very high sensitivity and are potentially very attractive in biomedical applications where requirements exist to measure very small quantities such as low oxygen and carbon monoxide concentrations, low blood pressure levels etc. Similarly, low concentrations of hormones, steroids etc. may be measured (Grattan, 1989).

Further examples of intrinsic fibre-optic sensors that modulate light intensity are described later in Chapter 17 (level measurement) and Chapter 19 (measuring small displacements).

As mentioned previously, light phase, polarization, wavelength and transit time can be modulated as well as intensity in intrinsic sensors. Monomode cables are used almost exclusively in these types of intrinsic sensor.

Phase modulation normally requires a coherent (laser) light source. It can provide very high sensitivity in displacement measurement but cross-sensitivity to temperature and strain degrades its performance. Additional problems are maintaining frequency stability of the light source and manufacturing difficulties in coupling the light source to the fibre. Various versions of this class of instrument exist to measure temperature, pressure, strain, magnetic fields and electric fields. Field-generated quantities such as electric current and voltage can also be measured. In each case, the measurand causes a phase change between a measuring and a reference light beam that is detected by an interferometer. Fuller details can be found in Harmer (1982) and Medlock (1986).

The principle of phase modulation has also been used in the fibre-optic accelerometer (where a mass subject to acceleration rests on a fibre), and in fibre strain gauges (where two fibres are fixed on the upper and lower surfaces of a bar under strain). These are discussed in more detail in Harmer (1982). The fibre-optic gyroscope described in Chapter 20 is a further example of a phase-modulating device.

Devices using polarization modulation require special forms of fibre that maintain polarization. Polarization changes can be effected by electrical fields, magnetic fields, temperature changes and mechanical strain. Each of these parameters can therefore be measured by polarization modulation.

Various devices that modulate the wavelength of light are used for special purposes, as described in Medlock (1986). However, the only common wavelength-modulating fibre-optic device is the form of laser Doppler flowmeter that uses fibre-optic cables, as described in Chapter 16.

Fibre-optic devices using modulation of the transit time of light are uncommon because of the speed of light. Measurement of the transit time for light to travel from a source, be reflected off an object, and travel back to a detector, is only viable for extremely large distances. However, a few special arrangements have evolved which use transit-time modulation, as described in Medlock (1986). These include instruments such as the optical resonator, which can measure both mechanical strain and temperature. Temperature-dependent wavelength variation also occurs in semiconductor crystal beads (e.g. aluminium gallium arsenide). This is bonded to the end of a fibre-optic cable and excited from an LED at the other end of the cable. Light from the LED is reflected back along the cable by the bead at a different wavelength. Measurement of the wavelength change allows temperatures in the range up to 200°C to be measured accurately. A particular advantage of this sensor is its small size, typically 0.5 mm diameter at the sensing tip. Finally, to complete the catalogue of transit-time devices, the frequency modulation in a piezoelectric quartz crystal used for gas sensing can also be regarded as a form of time domain modulation.

13.8.2 Extrinsic sensors

Extrinsic fibre-optic sensors use a fibre-optic cable, normally a multimode one, to transmit modulated light from a conventional sensor such as a resistance thermometer. A major feature of extrinsic sensors, which makes them so useful in such a large number of applications, is their ability to reach places that are otherwise inaccessible. One example of this is the insertion of fibre-optic cables into the jet engines of aircraft to measure temperature by transmitting radiation into a radiation pyrometer located remotely from the engine. Fibre-optic cable can be used in the same way to measure the internal temperature of electrical transformers, where the extreme electromagnetic fields present make other measurement techniques impossible.

An important advantage of extrinsic fibre-optic sensors is the excellent protection against noise corruption that they give to measurement signals. Unfortunately, the output of many sensors is not in a form that can be transmitted by a fibre-optic cable, and conversion into a suitable form must therefore take place prior to transmission. For example, a platinum resistance thermometer (PRT) translates temperature changes into resistance changes. The PRT therefore needs electronic circuitry to convert the resistance changes into voltage signals and thence into a modulated light form, and this in turn means that the device needs a power supply. This complicates the measurement process and means that low-voltage power cables must be routed with the fibre-optic cable to the transducer. One particular adverse effect of this is that the advantage of intrinsic safety is lost. One solution to this problem (Grattan, 1989) is to use a power source in the form of electronically generated pulses driven by a lithium battery. Alternatively (Johnson, 1994), power can be generated by transmitting light down the fibre-optic cable to a photocell. Both of these solutions provide intrinsically safe operation.

Piezoelectric sensors lend themselves particularly to use in extrinsic sensors because the modulated frequency of a quartz crystal can be readily transmitted into a fibre-optic cable by fitting electrodes to the crystal that are connected to a low power LED. Resonance of the crystal can be created either by electrical means or by optical means using the photothermal effect. The photothermal effect describes the principle where, if light is pulsed at the required oscillation frequency and directed at a quartz crystal, the localized heating and thermal stress caused in the crystal results in it oscillating at the pulse frequency. Piezoelectric extrinsic sensors can be used as part of various pressure, force and displacement sensors. At the other end of the cable, a phase-locked loop is typically used to measure the transmitted frequency.

Fibre-optic cables are also now commonly included in digital encoders, where the use of fibres to transmit light to and from the discs allows the light source and detectors to be located remotely. This allows the devices to be smaller, which is a great advantage in many applications where space is at a premium.

13.8.3 Distributed sensors

Current research is looking at ways of distributing a number of discrete sensors measuring different variables along a fibre-optic cable. Alternatively, sensors of the same type, which are located at various points along a cable, are being investigated as a means of providing distributed sensing of a single measured variable. For example, the use of a 2 km long cable to measure the temperature distribution along its entire length has been demonstrated, measuring temperature at 400 separate points to a resolution of 1°C.

13.9 Ultrasonic transducers

Ultrasonic devices are used in many fields of measurement, particularly for measuring fluid flow rates, liquid levels and translational displacements. Details of such applications can be found in later chapters. Uses of ultrasound in imaging systems will also be briefly described at the end of this section, although the coverage in this case will be brief since such applications are rather outside the scope of this text.

Ultrasound is a band of frequencies in the range above 20 kHz, that is, above the sonic range that humans can usually hear. Measurement devices that use ultrasound consist of one device that transmits an ultrasound wave and another device that receives the wave. Changes in the measured variable are determined either by measuring the change in time taken for the ultrasound wave to travel between the transmitter and receiver, or, alternatively, by measuring the change in phase or frequency of the transmitted wave.

The most common form of ultrasonic element is a piezoelectric crystal contained in a casing, as illustrated in Figure 13.10. Such elements can operate interchangeably as either a transmitter or receiver. These are available with operating frequencies that vary between 20 kHz and 15 MHz. The principles of operation, by which an alternating voltage generates an ultrasonic wave and vice versa, have already been covered in the section above on piezoelectric transducers.

For completeness, mention should also be made of capacitive ultrasonic elements. These consist of a thin, dielectric membrane between two conducting layers. The membrane is stretched across a backplate and a bias voltage is applied. When a varying voltage is applied to the element, it behaves as an ultrasonic transmitter and an ultrasound wave is produced. The system also works in the reverse direction as an ultrasonic receiver. Elements with resonant frequencies in the range between 30 kHz and 3 MHz can be obtained (Rafiq, 1991).

13.9.1 Transmission speed

The transmission speed of ultrasound varies according to the medium through which it travels. Transmission speeds for some common media are given in Table 13.1.

When transmitted through air, the speed of ultrasound is affected by environmental factors such as temperature, humidity and air turbulence. Of these, temperature has the largest effect. The velocity of sound through air varies with temperature according to:

$$V = 331.6 + 0.6T \text{ m/s} \qquad (13.2)$$

where T is the temperature in °C. Thus, even for a relatively small temperature change of 20 degrees from 0°C to 20°C, the velocity changes from 331.6 m/s to 343.6 m/s.

Fig. 13.10 Ultrasonic sensor.

Table 13.1 Transmission speed of ultrasound through different media

Medium	Velocity (m/s)
Air	331.6
Water	1440
Wood (pine)	3320
Iron	5130
Rock (granite)	6000

Humidity changes have a much smaller effect. If the relative humidity increases by 20%, the corresponding increase in the transmission velocity of ultrasound is 0.07% (corresponding to an increase from 331.6 m/s to 331.8 m/s at 0°C).

Changes in air pressure itself have negligible effect on the velocity of ultrasound. Similarly, air turbulence normally has no effect (though note that air turbulence may deflect ultrasound waves away from their original direction of travel). However, if turbulence involves currents of air at different temperatures, then random changes in ultrasound velocity occur according to equation (13.2).

13.9.2 Direction of travel of ultrasound waves

Air currents can alter the direction of travel of ultrasound waves. An air current moving with a velocity of 10 km/h has been shown experimentally to deflect an ultrasound wave by 8 mm over a distance of 1 m.

13.9.3 Directionality of ultrasound waves

Although it has perhaps been implied above that ultrasound waves travel in a narrow line away from the transmitter, this is not in fact what happens in practice. The ultrasound element actually emits a spherical wave of energy whose magnitude in any direction is a function of the angle made with respect to the direction that is normal to the face of the ultrasonic element. The peak emission always occurs along a line that is normal to the transmitting face of the ultrasonic element, and this is loosely referred to as the 'direction of travel' in the earlier paragraphs. At any angle other than the 'normal' one, the magnitude of transmitted energy is less than the peak value. Figure 13.11 shows the characteristics of the emission for a range of ultrasonic elements. This is shown in terms of the attenuation of the transmission magnitude (measured in dB) as the angle with respect to the 'normal' direction increases. For many purposes, it

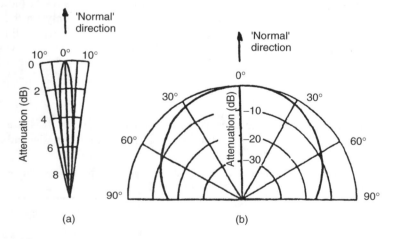

Fig. 13.11 Ultrasonic emission characteristics.

is useful to treat the transmission as a conical volume of energy, with the edges of the cone defined as the transmission angle where the amplitude of the energy in the transmission is −6 dB compared with the peak value (i.e. where the amplitude of the energy is half that in the normal direction). Using this definition, a 40 kHz ultrasonic element has a transmission cone of ±50° and a 400 kHz element has a transmission cone of ±3°.

13.9.4 Relationship between wavelength, frequency and directionality of ultrasound waves

The frequency and wavelength of ultrasound waves are related according to:

$$\lambda = v/f \qquad (13.3)$$

where λ is the wavelength, v is the velocity and f is the frequency of the ultrasound waves.

This shows that the relationship between λ and f depends on the velocity of the ultrasound and hence varies according to the nature and temperature of the medium through which it travels. Table 13.2 compares the nominal frequencies, wavelengths and transmission cones (−6 dB limits) for three different types of ultrasonic element.

It is clear from Table 13.2 that the directionality (cone angle of transmission) reduces as the nominal frequency of the ultrasound transmitter increases. However, the cone angle also depends on factors other than the nominal frequency, particularly on the shape of the transmitting horn in the element, and different models of ultrasonic element with the same nominal frequency can have substantially different cone angles.

13.9.5 Attenuation of ultrasound waves

Ultrasound waves suffer attenuation in the amplitude of the transmitted energy according to the distance travelled. The amount of attenuation also depends on the nominal frequency of the ultrasound and the adsorption characteristics of the medium through which it travels. The amount of adsorption depends not only on the type of transmission medium but also on the level of humidity and dust in the medium.

The amplitude X_d of the ultrasound wave at a distance d from the emission point can be expressed as:

$$\frac{X_d}{X_0} = \frac{\sqrt{e^{-\alpha d}}}{fd} \qquad (13.4)$$

where X_0 is the magnitude of the energy at the point of emission, f is the nominal frequency of the ultrasound and α is the attenuation constant that depends on the

Table 13.2 Comparison of frequency, wavelength and cone angle for various ultrasonic transmitters

Nominal frequency (kHz)	23	40	400
Wavelength (in air at 0°C)	14.4	8.3	0.83
Cone angle of transmission (−6 dB limits)	±80°	±50°	±3°

ultrasound frequency, the medium that the ultrasound travels through and any pollutants in the medium such as dust or water particles.

13.9.6 Ultrasound as a range sensor

The basic principles of an ultrasonic range sensor are to measure the time between transmission of a burst of ultrasonic energy from an ultrasonic transmitter and receipt of that energy by an ultrasonic receiver. Then, the distance d can be calculated from:

$$d = vt \qquad (13.5)$$

where v is the ultrasound velocity and t is the measured energy transit time. An obvious difficulty in applying this equation is the variability of v with temperature according to equation (13.2). One solution to this problem is to include an extra ultrasonic transmitter/receiver pair in the measurement system in which the two elements are positioned a known distance apart. Measurement of the transmission time of energy between this fixed pair provides the necessary measurement of velocity and hence compensation for any environmental temperature changes.

The degree of directionality in the ultrasonic elements used for range measurement is unimportant as long as the receiver and transmitter are positioned carefully so as to face each other exactly (i.e. such that the 'normal' lines to their faces are coincident). Thus, directionality imposes no restriction on the type of element suitable for range measurement. However, element choice is restricted by the attenuation characteristics of different types of element, and relatively low-frequency elements have to be used for the measurement of large ranges.

Measurement resolution and accuracy

The best measurement resolution that can be obtained with an ultrasonic ranging system is equal to the wavelength of the transmitted wave. As wavelength is inversely proportional to frequency, high-frequency ultrasonic elements would seem to be preferable. For example, whilst the wavelength and hence resolution for a 40 kHz element is 8.6 mm at room temperature (20°C), it is only 0.86 mm for a 400 kHz element. However, choice of element also depends on the required range of measurement. The range of higher-frequency elements is much reduced compared with low-frequency ones due to the greater attenuation of the ultrasound wave as it travels away from the transmitter. Hence, choice of element frequency has to be a compromise between measurement resolution and range.

The best measurement accuracy obtainable is equal to the measurement resolution value, but this is only achieved if the electronic counter used to measure the transmission time starts and stops at exactly the same point in the ultrasound cycle (usually the point in the cycle corresponding to peak amplitude is used). However, the sensitivity of the ultrasonic receiver also affects measurement accuracy. The amplitude of the ultrasound wave that is generated in the transmitter ramps up to full amplitude in the manner shown in Figure 13.12. The receiver has to be sensitive enough to detect the peak of the first cycle, which can usually be arranged. However, if the range of measurement is large, attenuation of the ultrasound wave may cause the amplitude of the first cycle to become less than the threshold level that the receiver is set to detect.

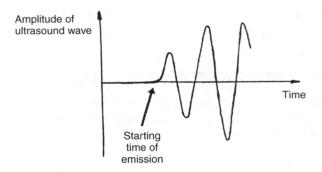

Fig. 13.12 Ramp-up of ultrasonic wave after emission.

In this case, only the second cycle will be detected and there will be an additional measurement error equal to one wavelength.

13.9.7 Use of ultrasound in tracking 3D object motion

An arrangement of the form shown in Figure 13.13 can be used to provide measurements of the position of an object moving in 3D space. In this, an ultrasonic transmitter mounted on the moving object (T) transmits bursts of energy to three receivers A, B, C located at the origin (A) and at distances q (to B) and p (to C) along the axes of an xyz co-ordinate system. If the transit times from T to A, B and C are measured, the distances a, b and c from T to the receivers can be calculated from equation (13.5). The position of T in spatial (xyz) co-ordinates can then be calculated by triangulation by solving the following set of equations:

$$x = \frac{a^2 + q^2 - b^2}{2q}; \quad y = \frac{a^2 + p^2 - c^2}{2p}; \quad z = \sqrt{a^2 - x^2 - y^2}$$

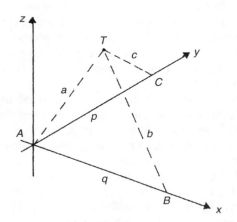

Fig. 13.13 Three-dimensional position measurement system.

13.9.8 Effect of noise in ultrasonic measurement systems

Signal levels at the output of ultrasonic measurement systems are usually of low amplitude and are therefore prone to contamination by electromagnetic noise. Because of this, it is necessary to use special precautions such as making ground (earth) lines thick, using shielded cables for transmission of the signal from the ultrasonic receiver and locating the signal amplifier as close to the receiver as possible.

Another potentially serious form of noise is background ultrasound produced by manufacturing operations in the typical industrial environment that many ultrasonic range measurement systems operate. Analysis of industrial environments has shown that ultrasound at frequencies up to 100 kHz is generated by many operations and some operations generate ultrasound at higher frequencies up to 200 kHz. There is not usually any problem if ultrasonic measurement systems operate at frequencies above 200 kHz, but these often have insufficient range for the needs of the measurement situation. In these circumstances, any objects that are likely to generate energy at ultrasonic frequencies should be covered in sound-absorbing material such that interference with ultrasonic measurement systems is minimized. The placement of sound-absorbing material around the path that the measurement ultrasound wave travels along contributes further towards reducing the effect of background noise. A natural solution to the problem is also partially provided by the fact that the same processes of distance travelled and adsorption that attenuate the amplitude of ultrasound waves travelling between the transmitter and receiver in the measurement system also attenuate ultrasound noise that is generated by manufacturing operations.

Because ultrasonic energy is emitted at angles other than the direction that is normal to the face of the transmitting element, a problem arises in respect of energy that is reflected off some object in the environment around the measurement system and back into the ultrasonic receiver. This has a longer path than the direct one between the transmitter and receiver and can cause erroneous measurements in some circumstances. One solution to this is to arrange for the transmission-time counter to stop as soon as the receiver first detects the ultrasound wave. This will usually be the wave that has travelled along the direct path, and so no measurement error is caused as long as the rate at which ultrasound pulses are emitted is such that the next burst isn't emitted until all reflections from the previous pulse have died down. However, in circumstances where the direct path becomes obstructed by some obstacle, the counter will only be stopped when the reflected signal is detected by the receiver, giving a potentially large measurement error.

13.9.9 Exploiting Doppler shift in ultrasound transmission

The Doppler effect is evident in all types of wave motion and describes the apparent change in frequency of the wave when there is relative motion between the transmitter and receiver. If a continuous ultrasound wave with velocity is v and frequency f takes t seconds to travel from a source S to a receiver R, then R will receive ft cycles of sound during time t (see Figure 13.14). Suppose now that R moves towards S at velocity r (with S stationary). R will receive rt/λ extra cycles of sound during time t, increasing the total number of sound cycles received to $(ft + rt/\lambda)$. With $(ft + rt/\lambda)$

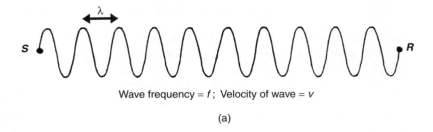

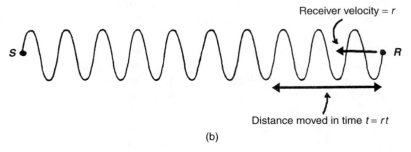

Fig. 13.14 Illustration of Doppler effect.

cycles received in t seconds, the apparent frequency f' is given by:

$$f' = \frac{ft + rt/\lambda}{t} = f + r/\lambda = f + \frac{rf}{v} = \frac{f(r+v)}{v}$$

$$\left(\text{using the relation } \frac{1}{\lambda} = \frac{f}{v} \text{ from equation 13.3}\right)$$

The frequency difference Δf can be expressed as:

$$\Delta f = f' - f = \frac{f(v+r)}{v} - f = \frac{fr}{v}$$

from which the velocity of the receiver r can be expressed as

$$r = v\Delta f/f.$$

Similarly, it can be shown that, if R moves away from S with velocity r, f' is given by:

$$f' = \frac{f(v-r)}{v}$$

and

$$\Delta f = -\frac{fr}{v}$$

If the ultrasound source moves towards the stationary receiver at velocity s, it will move a distance st in time t and the ft cycles that are emitted during time t will be compressed into a distance $(vt - st)$.

Hence, the apparent wavelength λ' will be given by:

$$\lambda' = \frac{vt - st}{ft} = \frac{v - s}{f}$$

Using equation (13.3), this can be expressed alternatively as:

$$f' = \frac{v}{\lambda'} = \frac{vf}{v - s}$$

Similarly, with S moving away from R, it can be shown that:

$$f' = \frac{vf}{v + s}$$

Thus, the velocity of an ultrasound receiver moving with respect to an ultrasound source can be calculated from the measured ratio between the real and apparent frequencies of the wave. This is used in devices like the Doppler shift flowmeter.

13.9.10 Ultrasonic imaging

The main applications of ultrasound in imaging are found in medical diagnosis and in industrial testing procedures. In both of these applications, a short burst of ultrasonic energy is transmitted from the ultrasonic element into the medium being investigated and the energy that is reflected back into the element is analysed. Ultrasonic elements in the frequency range 1 MHz to 15 MHz are used.

Ultrasound is reflected back at all interfaces between different materials, with the proportion of energy reflected being a function of the materials either side of the interface. The principal components inside a human body are water, fat, muscle and bone, and the interfaces between each of these have different reflectance characteristics. Measurement of the time between energy transmission and receipt of the reflected signal gives the depth of the interface according to equation (13.5). Therefore, in medical diagnosis procedures, the reflected energy appears as a series of peaks, with the magnitude of each peak corresponding to the type of interface that it is reflected from and the time of each peak corresponding to the depth of the interface in the body. Thus, a 'map' of fat, muscle and bone in the body is obtained. A fuller account can be found elsewhere (Webster, 1998).

Applications in industrial test procedures usually involve detecting internal flaws within components. Such flaws cause an interface between air and the material that the component is made of. By timing the reflections of ultrasound from the flaw, the depth of each flaw is determined.

13.10 Nuclear sensors

Nuclear sensors are uncommon measurement devices, partly because of the strict safety regulations that govern their use, and partly because they are usually expensive. Some very low-level radiation sources are now available that largely overcome the safety problems, but measurements are then prone to contamination by background radiation. The principle of operation of nuclear sensors is very similar to optical sensors in that radiation is transmitted between a source and a detector through some medium in which the magnitude of transmission is attenuated according to the value of the measured variable. Caesium-137 is commonly used as a gamma-ray source and a sodium iodide device is commonly used as a gamma-ray detector. The latter gives a voltage output that is proportional to the radiation incident upon it. One current use of nuclear sensors is in a non-invasive technique for measuring the level of liquid in storage tanks (see Chapter 17). They are also used in mass flow rate measurement (see Chapter 16) and in medical scanning applications (see Webster, 1998).

13.11 Microsensors

Microsensors are millimetre-sized two- and three-dimensional micromachined structures that have smaller size, improved performance, better reliability and lower production costs than many alternative forms of sensor. Currently, devices to measure temperature, pressure, force, acceleration, humidity, magnetic fields, radiation and chemical parameters are either in production or at advanced stages of research.

Microsensors are usually constructed from a silicon semiconductor material, but are sometimes fabricated from other materials such as metals, plastics, polymers, glasses and ceramics that are deposited on a silicon base. Silicon is an ideal material for sensor construction because of its excellent mechanical properties. Its tensile strength and Young's modulus is comparable to that of steel, whilst its density is less than that of aluminium. Sensors made from a single crystal of silicon remain elastic almost to the breaking point, and mechanical hysteresis is very small. In addition, silicon has a very low coefficient of thermal expansion and can be exposed to extremes of temperature and most gases, solvents and acids without deterioration.

Microengineering techniques are an essential enabling technology for microsensors, which are designed so that their electromechanical properties change in response to a change in the measured parameter. Many of the techniques used for integrated circuit (IC) manufacture are also used in sensor fabrication, common techniques being crystal growing and polishing, thin film deposition, ion implantation, wet and dry chemical and laser etching, and photolithography. However, apart from standard IC production techniques, some special techniques are also needed in addition to produce the 3D structures that are unique to some types of microsensor. The various manufacturing techniques are used to form sensors directly in silicon crystals and films. Typical structures have forms such as thin diaphragms, cantilever beams and bridges.

Whilst the small size of a microsensor is of particular benefit in many applications, it also leads to some problems that require special attention. For example, microsensors typically have very low capacitance. This makes the output signals very prone to noise contamination. Hence, it is usually necessary to integrate microelectronic circuits that perform signal processing in the device, which therefore becomes a *smart microsensor*. Another problem is that microsensors generally produce output signals of very low magnitude. This requires the use of special types of analogue-to-digital converter that can cope with such low-amplitude input signals. One suitable technique is sigma–delta conversion. This is based on charge balancing techniques and gives better than 16-bit accuracy in less than 20 ms (Riedijk, 1997). Special designs can reduce conversion time to less than 0.1 ms if necessary.

At present, almost all smart microsensors have an analogue output. However, a resonant-technology pressure-measuring device is now available with a digital output. This consists of a silicon crystal on which two H-shaped resonators are formed, one at the centre and one at the edge. If the pressure to be measured is applied to the crystal, the central resonator is compressed, changing the spring constant of the material and thus reducing its resonant frequency. At the same time, the outer resonator is stretched, increasing its resonant frequency. The resulting frequency difference produces a digital output signal that is proportional to the applied pressure. The device can also give a signal proportional to differential pressure if this is applied between the centre and periphery of the crystal.

Microsensors are used most commonly for measuring pressure, acceleration, force and chemical parameters. They are used in particularly large numbers in the automotive industry, where unit prices can be as low as £5–£10. Microsensors are also widely used in medical applications, particularly for blood pressure measurement, with unit prices down to £10.

Mechanical microsensors transform measured variables such as force, pressure and acceleration into a displacement. The displacement is usually measured by capacitive or piezoresistive techniques, although some devices use other technologies such as resonant frequency variation, resistance change, inductance change, the piezoelectric effect and changes in magnetic or optical coupling. The design of a cantilever silicon microaccelerometer is shown in Figure 13.15. The proof mass within this is about 100 µm across and the typical deflection measured is of the order of 1 micron (10^{-3} mm).

An alternative capacitive microaccelerometer provides a calibrated, compensated and amplified output. It has a capacitive silicon microsensor to measure displacement of the proof mass. This is integrated with a signal processing chip and protected by a plastic enclosure. The capacitive element has a 3D structure, which gives a higher measurement sensitivity than surface-machined elements.

Microsensors to measure many other physical variables are either in production or at advanced stages of research. Microsensors measuring magnetic field are based on a number of alternative technologies such as Hall-effect, magnetoresistors, magnetodiodes and magnetotransistors. Radiation microsensors are made from silicon p-n diodes or avalanche photodiodes and can detect radiation over wavelengths from the visible spectrum to infrared. Microsensors in the form of a micro thermistor, a p-n thermodiode

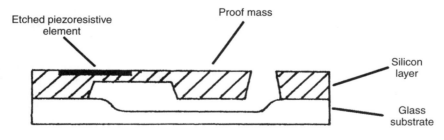

Fig. 13.15 Silicon microaccelerometer.

or a thermotransistor are used as digital thermometers. Microsensors have also enabled measurement techniques that were previously laboratory-based ones to be extended into field instruments. Examples are spectroscopic instruments and devices to measure viscosity.

References and further reading

Grattan, K.T.V. (1989) New developments in sensor technology – fibre-optics and electro-optics, *Measurement and Control*, **22**(6), pp. 165–175.

Harmer, A.L. (1982) Principles of optical fibre sensors and instrumentation, *Measurement and Control*, **15**(4), pp. 143–151.

Johnson, J.S. (1994) Optical sensors: the OSCA experience, *Measurement and Control*, 27, pp. 180–184.

Medlock, R.S. (1986) Review of modulating techniques for fibre optic sensors, *Measurement and Control*, **19**(1), pp. 4–17.

Rafiq, M. and Wykes, C. (1991) The performance of capacitive ultrasonic transducers, *J. Meas. Sci. Technology*, pp. 168–174.

Riedijk, F.R. and Huijsing, J.H. (1997) Sensor interface environment based on a serial bus interface, *Measurement and Control*, **30**, pp. 297–299.

Webster, J.G. (1998) *Medical Instrumentation*, John Wiley.

14

Temperature measurement

14.1 Principles of temperature measurement

Temperature measurement is very important in all spheres of life and especially so in the process industries. However, it poses particular problems, since temperature measurement cannot be related to a fundamental standard of temperature in the same way that the measurement of other quantities can be related to the primary standards of mass, length and time. If two bodies of lengths l_1 and l_2 are connected together end to end, the result is a body of length $l_1 + l_2$. A similar relationship exists between separate masses and separate times. However, if two bodies at the same temperature are connected together, the joined body has the same temperature as each of the original bodies.

This is a root cause of the fundamental difficulties that exist in establishing an absolute standard for temperature in the form of a relationship between it and other measurable quantities for which a primary standard unit exists. In the absence of such a relationship, it is necessary to establish fixed, reproducible reference points for temperature in the form of freezing and boiling points of substances where the transition between solid, liquid and gaseous states is sharply defined. The *International Practical Temperature Scale* (IPTS)* uses this philosophy and defines six *primary fixed points* for reference temperatures in terms of:

- the triple point of equilibrium hydrogen $-259.34°C$
- the boiling point of oxygen $-182.962°C$
- the boiling point of water $100.0°C$
- the freezing point of zinc $419.58°C$
- the freezing point of silver $961.93°C$
- the freezing point of gold $1064.43°C$
  (all at standard atmospheric pressure)

The freezing points of certain other metals are also used as *secondary fixed points* to provide additional reference points during calibration procedures.

---

* The IPTS is subject to periodic review and improvement as research produces more precise fixed reference points. The latest version was published in 1990.

272 Temperature measurement

Instruments to measure temperature can be divided into separate classes according to the physical principle on which they operate. The main principles used are:

- The thermoelectric effect
- Resistance change
- Sensitivity of semiconductor device
- Radiative heat emission
- Thermography
- Thermal expansion
- Resonant frequency change
- Sensitivity of fibre optic devices
- Acoustic thermometry
- Colour change
- Change of state of material.

14.2 Thermoelectric effect sensors (thermocouples)

Thermoelectric effect sensors rely on the physical principle that, when any two different metals are connected together, an e.m.f., which is a function of the temperature, is generated at the junction between the metals. The general form of this relationship is:

$$e = a_1T + a_2T^2 + a_3T^3 + \cdots + a_nT^n \quad (14.1)$$

where e is the e.m.f. generated and T is the absolute temperature.

This is clearly non-linear, which is inconvenient for measurement applications. Fortunately, for certain pairs of materials, the terms involving squared and higher powers of T (a_2T^2, a_3T^3 etc.) are approximately zero and the e.m.f.–temperature relationship is approximately linear according to:

$$e \approx a_1T \quad (14.2)$$

Wires of such pairs of materials are connected together at one end, and in this form are known as *thermocouples*. Thermocouples are a very important class of device as they provide the most commonly used method of measuring temperatures in industry.

Thermocouples are manufactured from various combinations of the base metals copper and iron, the base-metal alloys of alumel (Ni/Mn/Al/Si), chromel (Ni/Cr), constantan (Cu/Ni), nicrosil (Ni/Cr/Si) and nisil (Ni/Si/Mn), the noble metals platinum and tungsten, and the noble-metal alloys of platinum/rhodium and tungsten/rhenium. Only certain combinations of these are used as thermocouples and each standard combination is known by an internationally recognized type letter, for instance type K is chromel–alumel. The e.m.f.–temperature characteristics for some of these standard thermocouples are shown in Figure 14.1: these show reasonable linearity over at least part of their temperature-measuring ranges.

A typical thermocouple, made from one chromel wire and one constantan wire, is shown in Figure 14.2(a). For analysis purposes, it is useful to represent the thermocouple by its equivalent electrical circuit, shown in Figure 14.2(b). The e.m.f. generated at the point where the different wires are connected together is represented by a voltage

Measurement and Instrumentation Principles 273

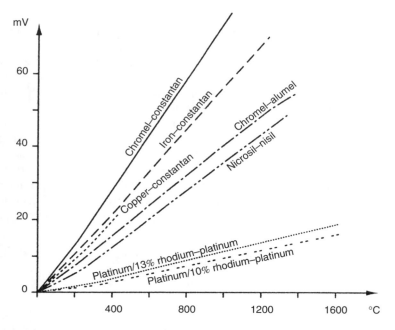

Fig. 14.1 E.m.f. temperature characteristics for some standard thermocouple materials.

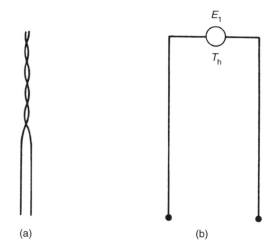

Fig. 14.2 (a) Thermocouple; (b) equivalent circuit.

source, E_1, and the point is known as the *hot junction*. The temperature of the hot junction is customarily shown as T_h on the diagram. The e.m.f. generated at the hot junction is measured at the open ends of the thermocouple, which is known as the *reference junction*.

In order to make a thermocouple conform to some precisely defined e.m.f.–temperature characteristic, it is necessary that all metals used are refined to a high degree of

274 Temperature measurement

pureness and all alloys are manufactured to an exact specification. This makes the materials used expensive, and consequently thermocouples are typically only a few centimetres long. It is clearly impractical to connect a voltage-measuring instrument at the open end of the thermocouple to measure its output in such close proximity to the environment whose temperature is being measured, and therefore *extension leads* up to several metres long are normally connected between the thermocouple and the measuring instrument. This modifies the equivalent circuit to that shown in Figure 14.3(a). There are now three junctions in the system and consequently three voltage sources, E_1, E_2 and E_3, with the point of measurement of the e.m.f. (still called the reference junction) being moved to the open ends of the extension leads.

The measuring system is completed by connecting the extension leads to the voltage-measuring instrument. As the connection leads will normally be of different materials to those of the thermocouple extension leads, this introduces two further e.m.f.-generating junctions E_4 and E_5 into the system as shown in Figure 14.3(b). The net output e.m.f. measured (E_m) is then given by:

$$E_m = E_1 + E_2 + E_3 + E_4 + E_5 \tag{14.3}$$

and this can be re-expressed in terms of E_1 as:

$$E_1 = E_m - E_2 - E_3 - E_4 - E_5 \tag{14.4}$$

In order to apply equation (14.1) to calculate the measured temperature at the hot junction, E_1 has to be calculated from equation (14.4). To do this, it is necessary to calculate the values of E_2, E_3, E_4 and E_5.

It is usual to choose materials for the extension lead wires such that the magnitudes of E_2 and E_3 are approximately zero, irrespective of the junction temperature. This avoids the difficulty that would otherwise arise in measuring the temperature of the junction between the thermocouple wires and the extension leads, and also in determining the e.m.f./temperature relationship for the thermocouple–extension lead combination.

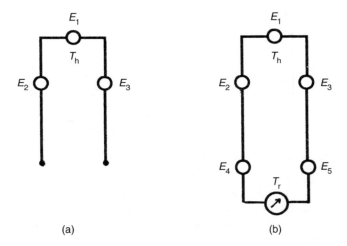

Fig. 14.3 (a) Equivalent circuit for thermocouple with extension leads; (b) equivalent circuit for thermocouple and extension leads connected to a meter.

A zero junction e.m.f. is most easily achieved by choosing the extension leads to be of the same basic materials as the thermocouple, but where their cost per unit length is greatly reduced by manufacturing them to a lower specification. However, such a solution is still prohibitively expensive in the case of noble metal thermocouples, and it is necessary in this case to search for base-metal extension leads that have a similar thermoelectric behaviour to the noble-metal thermocouple. In this form, the extension leads are usually known as *compensating leads*. A typical example of this is the use of nickel/copper–copper extension leads connected to a platinum/rhodium–platinum thermocouple. Copper compensating leads are also sometimes used with some types of base metal thermocouples and, in such cases, the law of intermediate metals can be applied to compensate for the e.m.f. at the junction between the thermocouple and compensating leads.

To analyse the effect of connecting the extension leads to the voltage-measuring instrument, a thermoelectric law known as the *law of intermediate metals* can be used. This states that the e.m.f. generated at the junction between two metals or alloys A and C is equal to the sum of the e.m.f. generated at the junction between metals or alloys A and B and the e.m.f. generated at the junction between metals or alloys B and C, where all junctions are at the same temperature. This can be expressed more simply as:

$$e_{AC} = e_{AB} + e_{BC} \tag{14.5}$$

Suppose we have an iron–constantan thermocouple connected by copper leads to a meter. We can express E_4 and E_5 in Figure 14.4 as:

$$E_4 = e_{\text{iron–copper}}; \quad E_5 = e_{\text{copper–constantan}}$$

The sum of E_4 and E_5 can be expressed as:

$$E_4 + E_5 = e_{\text{iron–copper}} + e_{\text{copper–constantan}}$$

Applying equation (14.5):

$$e_{\text{iron–copper}} + e_{\text{copper–constantan}} = e_{\text{iron–constantan}}$$

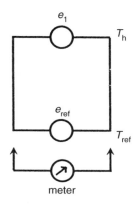

Fig. 14.4 Effective e.m.f. sources in a thermocouple measurement system.

276 Temperature measurement

Thus, the effect of connecting the thermocouple extension wires to the copper leads to the meter is cancelled out, and the actual e.m.f. at the reference junction is equivalent to that arising from an iron–constantan connection at the reference junction temperature, which can be calculated according to equation (14.1). Hence, the equivalent circuit in Figure 14.3(b) becomes simplified to that shown in Figure 14.4. The e.m.f. E_m measured by the voltage-measuring instrument is the sum of only two e.m.f.s, consisting of the e.m.f. generated at the hot junction temperature E_1 and the e.m.f. generated at the reference junction temperature E_{ref}. The e.m.f. generated at the hot junction can then be calculated as:

$$E_1 = E_m + E_{ref}$$

E_{ref} can be calculated from equation (14.1) if the temperature of the reference junction is known. In practice, this is often achieved by immersing the reference junction in an ice bath to maintain it at a reference temperature of 0°C. However, as discussed in the following section on thermocouple tables, it is very important that the ice bath remains exactly at 0°C if this is to be the reference temperature assumed, otherwise significant measurement errors can arise. For this reason, refrigeration of the reference junction at a temperature of 0°C is often preferred.

14.2.1 Thermocouple tables

Although the preceding discussion has suggested that the unknown temperature T can be evaluated from the calculated value of the e.m.f. E_1 at the hot junction using equation (14.1), this is very difficult to do in practice because equation (14.1) is a high order polynomial expression. An approximate translation between the value of E_1 and temperature can be achieved by expressing equation (14.1) in graphical form as in Figure 14.1. However, this is not usually of sufficient accuracy, and it is normal practice to use tables of e.m.f. and temperature values known as *thermocouple tables*. These include compensation for the effect of the e.m.f. generated at the reference junction (E_{ref}), which is assumed to be at 0°C. Thus, the tables are only valid when the reference junction is exactly at this temperature. Compensation for the case where the reference junction temperature is not at zero is considered later in this section.

Tables for a range of standard thermocouples are given in Appendix 4. In these tables, a range of temperatures is given in the left-hand column and the e.m.f. output for each standard type of thermocouple is given in the columns to the right. In practice, any general e.m.f. output measurement taken at random will not be found exactly in the tables, and interpolation will be necessary between the values shown in the table.

Example 14.1
If the e.m.f. output measured from a chromel–constantan thermocouple is 13.419 mV with the reference junction at 0°C, the appropriate column in the tables shows that this corresponds to a hot junction temperature of 200°C.

Example 14.2
If the measured output e.m.f. for a chromel–constantan thermocouple (reference junction at 0°C) was 10.65 mV, it is necessary to carry out linear interpolation between the

temperature of 160°C corresponding to an e.m.f. of 10.501 mV shown in the tables and the temperature of 170°C corresponding to an e.m.f. of 11.222 mV. This interpolation procedure gives an indicated hot junction temperature of 162°C.

14.2.2 Non-zero reference junction temperature

If the reference junction is immersed in an ice bath to maintain it at a temperature of 0°C so that thermocouple tables can be applied directly, the ice in the bath must be in a state of just melting. This is the only state in which ice is exactly at 0°C, and otherwise it will be either colder or hotter than this temperature. Thus, maintaining the reference junction at 0°C is not a straightforward matter, particularly if the environmental temperature around the measurement system is relatively hot. In consequence, it is common practice in many practical applications of thermocouples to maintain the reference junction at a non-zero temperature by putting it into a controlled environment maintained by an electrical heating element. In order to still be able to apply thermocouple tables, correction then has to be made for this non-zero reference junction temperature using a second thermoelectric law known as the *law of intermediate temperatures*. This states that:

$$E_{(T_h,T_0)} = E_{(T_h,T_r)} + E_{(T_r,T_0)} \tag{14.6}$$

where: $E_{(T_h,T_0)}$ is the e.m.f. with the junctions at temperatures T_h and T_0, $E_{(T_h,T_r)}$ is the e.m.f. with the junctions at temperatures T_h and T_r, and $E_{(T_r,T_0)}$ is the e.m.f. with the junctions at temperatures T_r and T_0, T_h is the hot junction measured temperature, T_0 is 0°C and T_r is the non-zero reference junction temperature that is somewhere between T_0 and T_h.

Example 14.3
Suppose that the reference junction of a chromel–constantan thermocouple is maintained at a temperature of 80°C and the output e.m.f. measured is 40.102 mV when the hot junction is immersed in a fluid.
The quantities given are $T_r = 80°C$ and $E_{(T_h,T_r)} = 40.102$ mV
From the tables, $E_{(T_r,T_0)} = 4.983$ mV
Now applying equation (14.6), $E_{(T_h,T_0)} = 40.102 + 4.983 = 45.085$ mV
Again referring to the tables, this indicates a fluid temperature of 600°C.

In using thermocouples, it is essential that they are connected correctly. Large errors can result if they are connected incorrectly, for example by interchanging the extension leads or by using incorrect extension leads. Such mistakes are particularly serious because they do not prevent some sort of output being obtained, which may look sensible even though it is incorrect, and so the mistake may go unnoticed for a long period of time. The following examples illustrate the sort of errors that may arise:

Example 14.4
This example is an exercise in the use of thermocouple tables, but it also serves to illustrate the large errors that can arise if thermocouples are used incorrectly. In a particular industrial situation, a chromel–alumel thermocouple with chromel–alumel

extension wires is used to measure the temperature of a fluid. In connecting up this measurement system, the instrumentation engineer responsible has inadvertently interchanged the extension wires from the thermocouple. The ends of the extension wires are held at a reference temperature of 0°C and the output e.m.f. measured is 14.1 mV. If the junction between the thermocouple and extension wires is at a temperature of 40°C, what temperature of fluid is indicated and what is the true fluid temperature?

Solution

The initial step necessary in solving a problem of this type is to draw a diagrammatical representation of the system and to mark on this the e.m.f. sources, temperatures etc., as shown in Figure 14.5. The first part of the problem is solved very simply by looking up in thermocouple tables what temperature the e.m.f. output of 12.1 mV indicates for a chromel–alumel thermocouple. This is 297.4°C. Then, summing e.m.f.s around the loop:

$$V = 12.1 = E_1 + E_2 + E_3 \quad \text{or} \quad E_1 = 12.1 - E_2 - E_3$$

$$E_2 = E_3 = \text{e.m.f.}_{(\text{alumel–chromel})_{40}} = -\text{e.m.f.}_{(\text{chromel–alumel})_{40}}{}^* = -1.611 \, \text{mV}$$

Hence:

$$E_1 = 12.1 + 1.611 + 1.611 = 15.322 \, \text{mV}$$

Interpolating from the thermocouple tables, this indicates that the true fluid temperature is 374.5°C.

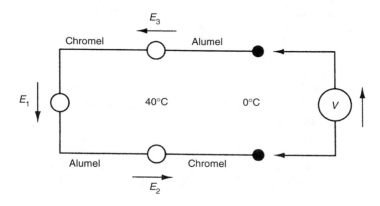

Fig. 14.5 Diagram for solution of example 14.4.

Example 14.5
This example also illustrates the large errors that can arise if thermocouples are used incorrectly. An iron–constantan thermocouple measuring the temperature of a fluid is connected by mistake with copper–constantan extension leads (such that the two constantan wires are connected together and the copper extension wire is connected to the iron thermocouple wire). If the fluid temperature was actually 200°C, and the

---

* The thermocouple tables quote e.m.f. using the convention that going from chromel to alumel is positive. Hence, the e.m.f. going from alumel to chromel is minus the e.m.f. going from chromel to alumel.

junction between the thermocouple and extension wires was at 50°C, what e.m.f. would be measured at the open ends of the extension wires if the reference junction is maintained at 0°C? What fluid temperature would be deduced from this (assuming that the connection mistake was not known about)?

Solution
Again, the initial step necessary is to draw a diagram showing the junctions, temperatures and e.m.f.s, as shown in Figure 14.6. The various quantities can then be calculated:

$$E_2 = \text{e.m.f.}_{(\text{iron}-\text{copper})_{50}}$$

By the law of intermediate metals:

$$\text{e.m.f.}_{(\text{iron}-\text{copper})_{50}} = \text{e.m.f.}_{(\text{iron}-\text{constantan})_{50}} - \text{e.m.f.}_{(\text{copper}-\text{constantan})_{50}}$$
$$= 2.585 - 2.035 \text{ (from thermocouple tables)} = 0.55\,\text{mV}$$
$$E_1 = \text{e.m.f.}_{(\text{iron}-\text{constantan})_{200}} = 10.777 \text{ (from thermocouple tables)}$$
$$V = E_1 - E_2 = 10.777 - 0.55 = 10.227$$

Using tables and interpolating, 10.227 mV indicates a temperature of:

$$\left(\frac{10.227 - 10.222}{10.777 - 10.222}\right) 10 + 190 = 190.1°\text{C}$$

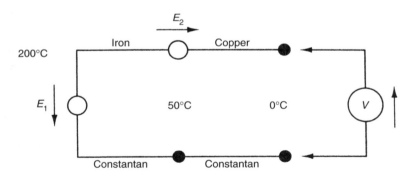

Fig. 14.6 Diagram for solution of example 14.5.

14.2.3 Thermocouple types

The five standard base-metal thermocouples are chromel–constantan (type E), iron–constantan (type J), chromel–alumel (type K), nicrosil–nisil (type N) and copper–constantan (type T). These are all relatively cheap to manufacture but they become inaccurate with age and have a short life. In many applications, performance is also affected through contamination by the working environment. To overcome this, the thermocouple can be enclosed in a *protective sheath*, but this has the adverse effect of introducing a significant time constant, making the thermocouple slow to respond to temperature changes. Therefore, as far as possible, thermocouples are used without protection.

Chromel–constantan devices give the highest measurement sensitivity of 80 µV/°C, with an inaccuracy of ±0.5% and a useful measuring range of −200°C up to 900°C. Unfortunately, whilst they can operate satisfactorily in oxidizing environments when unprotected, their performance and life are seriously affected by reducing atmospheres. Iron–constantan thermocouples have a sensitivity of 60 µV/°C and are the preferred type for general-purpose measurements in the temperature range −150°C to +1000°C, where the typical measurement inaccuracy is ±0.75%. Their performance is little affected by either oxidizing or reducing atmospheres. Copper–constantan devices have a similar measurement sensitivity of 60 µV/°C and find their main application in measuring subzero temperatures down to −200°C, with an inaccuracy of ±0.75%. They can also be used in both oxidising and reducing atmospheres to measure temperatures up to 350°C. Chromel–alumel thermocouples have a measurement sensitivity of only 45 µV/°C, although their characteristic is particularly linear over the temperature range between 700°C and 1200°C and this is therefore their main application. Like chromel–constantan devices, they are suitable for oxidizing atmospheres but not for reducing ones unless protected by a sheath. Their measurement inaccuracy is ±0.75%. Nicrosil–nisil thermocouples are a recent development that resulted from attempts to improve the performance and stability of chromel–alumel thermocouples. Their thermoelectric characteristic has a very similar shape to type K devices, with equally good linearity over a large temperature measurement range, measurement sensitivity of 40 µV/°C and measurement uncertainty of ±0.75%. The operating environment limitations are the same as for chromel–alumel devices but their long-term stability and life are at least three times better. A detailed comparison between type K and N devices can be found in Brooks, (1985).

Noble-metal thermocouples are always expensive but enjoy high stability and long life even when used at high temperatures, though they cannot be used in reducing atmospheres. Thermocouples made from platinum and a platinum–rhodium alloy (type R and type S) have a low inaccuracy of only ±0.5% and can measure temperatures up to 1500°C, but their measurement sensitivity is only 10 µV/°C. Alternative devices made from tungsten and a tungsten/rhenium alloy have a better sensitivity of 20 µV/°C and can measure temperatures up to 2300°C, though they cannot be used in either oxidizing or reducing atmospheres.

14.2.4 Thermocouple protection

Thermocouples are delicate devices that must be treated carefully if their specified operating characteristics are to be maintained. One major source of error is induced strain in the hot junction. This reduces the e.m.f. output, and precautions are normally taken to minimize induced strain by mounting the thermocouple horizontally rather than vertically. It is usual to cover most of the thermocouple wire with thermal insulation, which also provides mechanical protection, although the tip is left exposed if possible to maximize the speed of response to changes in the measured temperature. However, thermocouples are prone to contamination in some operating environments. This means

Table 14.1 Common sheath materials for thermocouples

Material	Maximum operating temperature (°C)*
Mild steel	900
Nickel–chromium	900
Fused silica	1000
Special steel	1100
Mullite	1700
Recrystallized alumina	1850
Beryllia	2300
Magnesia	2400
Zirconia	2400
Thoria	2600

*The maximum operating temperatures quoted assume oxidizing or neutral atmospheres. For operation in reducing atmospheres, the maximum allowable temperature is usually reduced.

that their e.m.f.–temperature characteristic varies from that published in standard tables. Contamination also makes them brittle and shortens their life.

Where they are prone to contamination, thermocouples have to be protected by enclosing them entirely in an insulated sheath. Some common sheath materials and their maximum operating temperatures are shown in Table 14.1. Whilst the thermocouple is a device that has a naturally first order type of step response characteristic, the time constant is usually so small as to be negligible when the thermocouple is used unprotected. However, when enclosed in a sheath, the time constant of the combination of thermocouple and sheath is significant. The size of the thermocouple and hence the diameter required for the sheath has a large effect on the importance of this. The time constant of a thermocouple in a 1 mm diameter sheath is only 0.15 s and this has little practical effect in most measurement situations, whereas a larger sheath of 6 mm diameter gives a time constant of 3.9 s that cannot be ignored so easily.

14.2.5 Thermocouple manufacture

Thermocouples are manufactured by connecting together two wires of different materials, where each material is produced so as to conform precisely with some defined composition specification. This ensures that its thermoelectric behaviour accurately follows that for which standard thermocouple tables apply. The connection between the two wires is effected by welding, soldering or in some cases just by twisting the wire ends together. Welding is the most common technique used generally, with silver soldering being reserved for copper–constantan devices.

The diameter of wire used to construct thermocouples is usually in the range between 0.4 mm and 2 mm. The larger diameters are used where ruggedness and long life are required, although these advantages are gained at the expense of increasing the measurement time constant. In the case of noble-metal thermocouples, the use of large diameter wire incurs a substantial cost penalty. Some special applications have a requirement for a very fast response time in the measurement of temperature, and in such cases wire diameters as small as 0.1 µm (0.1 microns) can be used.

14.2.6 The thermopile

The thermopile is the name given to a temperature-measuring device that consists of several thermocouples connected together in series, such that all the reference junctions are at the same cold temperature and all the hot junctions are exposed to the temperature being measured, as shown in Figure 14.7. The effect of connecting n thermocouples together in series is to increase the measurement sensitivity by a factor of n. A typical thermopile manufactured by connecting together 25 chromel–constantan thermocouples gives a measurement resolution of 0.001°C.

14.2.7 Digital thermometer

Thermocouples are also used in digital thermometers, of which both simple and intelligent versions exist (see section 14.13 for a description of the latter). A simple digital thermometer is the combination of a thermocouple, a battery-powered, dual slope digital voltmeter to measure the thermocouple output, and an electronic display. This provides a low noise, digital output that can resolve temperature differences as small as 0.1°C. The accuracy achieved is dependent on the accuracy of the thermocouple element, but reduction of measurement inaccuracy to ±0.5% is achievable.

14.2.8 The continuous thermocouple

The continuous thermocouple is one of a class of devices that detect and respond to heat. Other devices in this class include the *line-type heat detector* and *heat-sensitive cable*. The basic construction of all these devices consists of two or more strands of wire separated by insulation within a long thin cable. Whilst they sense temperature, they do not in fact provide an output measurement of temperature. Their function is to respond to abnormal temperature rises and thus prevent fires, equipment damage etc.

The advantages of continuous thermocouples become more apparent if the problems with other types of heat detector are considered. The insulation in the line-type heat

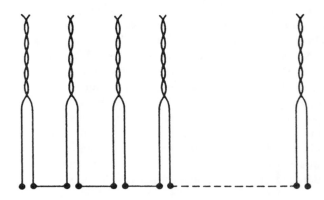

Fig. 14.7 Thermopile.

detector and heat-sensitive cable consists of plastic or ceramic material with a negative temperature coefficient (i.e. the resistance falls as the temperature rises). An alarm signal can be generated when the measured resistance falls below a certain level. Alternatively, in some versions, the insulation is allowed to break down completely, in which case the device acts as a switch. The major limitation of these devices is that the temperature change has to be relatively large, typically 50–200°C above ambient temperature, before the device responds. Also, it is not generally possible for such devices to give an output that indicates that an alarm condition is developing before it actually happens, and thus allow preventative action. Furthermore, after the device has generated an alarm it usually has to be replaced. This is particularly irksome because there is a large variation in the characteristics of detectors coming from different batches and so replacement of the device requires extensive on-site recalibration of the system.

In contrast, the continuous thermocouple suffers from very few of these problems. It differs from other types of heat detector in that the two strands of wire inside it are a pair of thermocouple materials* separated by a special, patented, mineral insulation and contained within a stainless steel protective sheath. If any part of the cable is subjected to heat, the resistance of the insulation at that point is reduced and a 'hot junction' is created between the two wires of dissimilar metals. An e.m.f. is generated at this hot junction according to normal thermoelectric principles.

The continuous thermocouple can detect temperature rises as small as 1°C above normal. Unlike other types of heat detector, it can also monitor abnormal rates of temperature rise and provide a warning of alarm conditions developing before they actually happen. Replacement is only necessary if a great degree of insulation breakdown has been caused by a substantial hot spot at some point along the detector's length. Even then, the use of thermocouple materials of standard characteristics in the detector means that recalibration is not needed if it is replaced. Calibration is not affected either by cable length, and so a replacement cable may be of a different length to the one it is replacing. One further advantage of continuous thermocouples over earlier forms of heat detector is that no power supply is needed, thus significantly reducing installation costs.

14.3 Varying resistance devices

Varying resistance devices rely on the physical principle of the variation of resistance with temperature. The devices are known as either resistance thermometers or thermistors according to whether the material used for their construction is a metal or a semiconductor, and both are common measuring devices. The normal method of measuring resistance is to use a d.c. bridge. The excitation voltage of the bridge has to be chosen very carefully because, although a high value is desirable for achieving high measurement sensitivity, the self-heating effect of high currents flowing in the temperature transducer creates an error by increasing the temperature of the device and so changing the resistance value.

---

* Normally type E, chromel–constantan, or type K, chromel–alumel.

14.3.1 Resistance thermometers (resistance temperature devices)

Resistance thermometers, which are alternatively known as *resistance temperature devices* (or RTDs), rely on the principle that the resistance of a metal varies with temperature according to the relationship:

$$R = R_0 \left(1 + a_1 T + a_2 T^2 + a_3 T^3 + \cdots + a_n T^n\right) \tag{14.7}$$

This equation is non-linear and so is inconvenient for measurement purposes. The equation becomes linear if all the terms in $a_2 T^2$ and higher powers of T are negligible such that the resistance and temperature are related according to:

$$R \approx R_0 (1 + a_1 T)$$

This equation is approximately true over a limited temperature range for some metals, notably platinum, copper and nickel, whose characteristics are summarized in Figure 14.8. Platinum has the most linear resistance–temperature characteristic, and it also has good chemical inertness, making it the preferred type of resistance thermometer in most applications. Its resistance–temperature relationship is linear within ±0.4% over the temperature range between −200°C and +40°C. Even at +1000°C, the quoted inaccuracy figure is only ±1.2%. Platinum thermometers are made in two forms, as a coil wound on a mandrel and as a film deposited on a ceramic substrate. The nominal resistance at 0°C is typically 100 Ω or 1000 Ω, though 200 Ω and 500 Ω versions also exist. Sensitivity is 0.385 Ω/°C (100 Ω type) or 3.85 Ω/°C (1000 Ω type). A high nominal resistance is advantageous in terms of higher measurement sensitivity, and the resistance of connecting leads has less effect on measurement accuracy. However, cost goes up as the nominal resistance increases.

Besides having a less linear characteristic, both nickel and copper are inferior to platinum in terms of their greater susceptibility to oxidation and corrosion. This seriously limits their accuracy and longevity. However, because platinum is very expensive compared with nickel and copper, the latter are used in resistance thermometers when cost is important. Another metal, tungsten, is also used in resistance thermometers in some circumstances, particularly for high temperature measurements. The working range of each of these four types of resistance thermometer is as shown below:

Platinum: −270°C to +1000°C (though use above 650°C is uncommon)
Copper: −200°C to +260°C
Nickel: −200°C to +430°C
Tungsten: −270°C to +1100°C

In the case of non-corrosive and non-conducting environments, resistance thermometers are used without protection. In all other applications, they are protected inside a sheath. As in the case of thermocouples, such protection reduces the speed of response of the system to rapid changes in temperature. A typical time constant for a sheathed platinum resistance thermometer is 0.4 seconds. Moisture build-up within the sheath can also impair measurement accuracy.

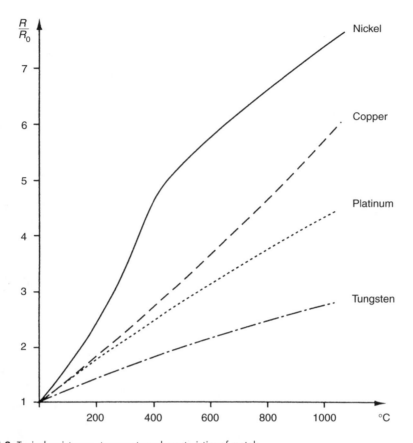

Fig. 14.8 Typical resistance–temperature characteristics of metals.

14.3.2 Thermistors

Thermistors are manufactured from beads of semiconductor material prepared from oxides of the iron group of metals such as chromium, cobalt, iron, manganese and nickel. Normally, thermistors have a negative temperature coefficient, i.e. the resistance decreases as the temperature increases, according to:

$$R = R_0 e^{[\beta(1/T - 1/T_0)]} \tag{14.8}$$

This relationship is illustrated in Figure 14.9. However, alternative forms of heavily doped thermistors are now available (at greater cost) that have a positive temperature coefficient. The form of equation (14.8) is such that it is not possible to make a linear approximation to the curve over even a small temperature range, and hence the thermistor is very definitely a non-linear sensor. However, the major advantages of thermistors are their relatively low cost and their small size. This size advantage means that the time constant of thermistors operated in sheaths is small, although the size reduction also decreases its heat dissipation capability and so makes the self-heating effect greater. In consequence, thermistors have to be operated at generally

286 Temperature measurement

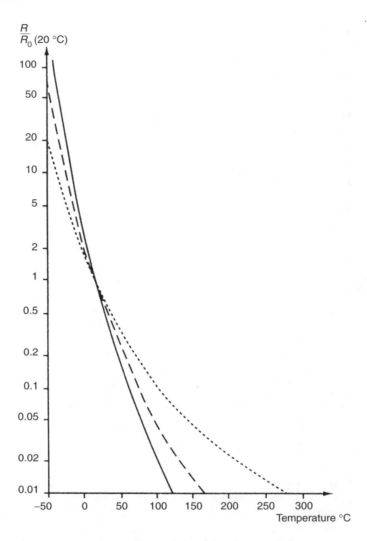

Fig. 14.9 Typical resistance–temperature characteristics of thermistor materials.

lower current levels than resistance thermometers and so the measurement sensitivity is less.

14.4 Semiconductor devices

Semiconductor devices, consisting of either diodes or integrated circuit transistors, have only been commonly used in industrial applications for a few years, but they were first invented several decades ago. They have the advantage of being relatively inexpensive, but one difficulty that affects their use is the need to provide an external power supply to the sensor.

Integrated circuit transistors produce an output proportional to the absolute temperature. Different types are configured to give an output in the form of either a varying current (typically 1 µA/K) or varying voltage (typically 10 mV/K). Current forms are normally used with a digital voltmeter that detects the current output in terms of the voltage drop across a 10 kΩ resistor. Although the devices have a very low cost (typically a few pounds) and a better linearity than either thermocouples or resistance thermometers, they only have a limited measurement range from −50°C to +150°C. Their inaccuracy is typically ±3%, which limits their range of application. However, they are widely used to monitor pipes and cables, where their low cost means that it is feasible to mount multiple sensors along the length of the pipe/cable to detect hot spots.

In diodes, the forward voltage across the device varies with temperature. Output from a typical diode package is in the microamp range. Diodes have a small size, with good output linearity and typical inaccuracy of only ±0.5%. Silicon diodes cover the temperature range from −50 to +200°C and germanium ones from −270 to +40°C.

14.5 Radiation thermometers

All objects emit electromagnetic radiation as a function of their temperature above absolute zero, and radiation thermometers (also known as radiation pyrometers) measure this radiation in order to calculate the temperature of the object. The total rate of radiation emission per second is given by:

$$E = KT^4 \tag{14.9}$$

The power spectral density of this emission varies with temperature in the manner shown in Figure 14.10. The major part of the frequency spectrum lies within the band of wavelengths between 0.3 µm and 40 µm, which corresponds to the visible (0.3–0.72 µm) and infrared (0.72–1000 µm) ranges. As the magnitude of the radiation varies with temperature, measurement of the emission from a body allows the temperature of the body to be calculated. Choice of the best method of measuring the emitted radiation depends on the temperature of the body. At low temperatures, the peak of the power spectral density function (Figure 14.10) lies in the infrared region, whereas at higher temperatures it moves towards the visible part of the spectrum. This phenomenon is observed as the red glow that a body begins to emit as its temperature is increased beyond 600°C.

Different versions of radiation thermometers are capable of measuring temperatures between −100°C and +10 000°C with measurement inaccuracy as low as ±0.05% (though this level of accuracy is not obtained when measuring very high temperatures). Portable, battery-powered, hand-held versions are also available, and these are particularly easy to use. The important advantage that radiation thermometers have over other types of temperature-measuring instrument is that there is no contact with the hot body while its temperature is being measured. Thus, the measured system is not disturbed in any way. Furthermore, there is no possibility of contamination, which is particularly important in food and many other process industries. They are especially suitable for measuring high temperatures that are beyond the capabilities of contact

288 Temperature measurement

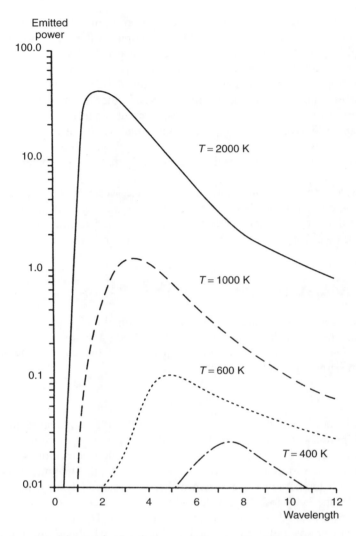

Fig. 14.10 Power spectral density of radiated energy emission at various temperatures.

instruments such as thermocouples, resistance thermometers and thermistors. They are also capable of measuring moving bodies, for instance the temperature of steel bars in a rolling mill. Their use is not as straightforward as the discussion so far might have suggested, however, because the radiation from a body varies with the composition and surface condition of the body as well as with temperature. This dependence on surface condition is quantified by the *emissivity* of the body. The use of radiation thermometers is further complicated by absorption and scattering of the energy between the emitting body and the radiation detector. Energy is scattered by atmospheric dust and water droplets and absorbed by carbon dioxide, ozone and water vapour molecules. Therefore, all radiation thermometers have to be carefully calibrated for each particular body whose temperature they are required to monitor.

Various types of radiation thermometer exist, as described below. The optical pyrometer can only be used to measure high temperatures, but various types of radiation pyrometers are available that between them cover the whole temperature spectrum. Intelligent versions (see section 14.13) also now provide full or partial solution to many of the problems described below for non-intelligent pyrometers.

14.5.1 Optical pyrometers

The optical pyrometer, illustrated in Figure 14.11, is designed to measure temperatures where the peak radiation emission is in the red part of the visible spectrum, i.e. where the measured body glows a certain shade of red according to the temperature. This limits the instrument to measuring temperatures above 600°C. The instrument contains a heated tungsten filament within its optical system. The current in the filament is increased until its colour is the same as the hot body: under these conditions the filament apparently disappears when viewed against the background of the hot body. Temperature measurement is therefore obtained in terms of the current flowing in the filament. As the brightness of different materials at any particular temperature varies according to the emissivity of the material, the calibration of the optical pyrometer must be adjusted according to the emissivity of the target. Manufacturers provide tables of standard material emissivities to assist with this.

The inherent measurement inaccuracy of an optical pyrometer is ±5°C. However, in addition to this error, there can be a further operator-induced error of ±10°C arising out of the difficulty in judging the moment when the filament 'just' disappears. Measurement accuracy can be improved somewhat by employing an optical filter within the instrument that passes a narrow band of frequencies of wavelength around 0.65 µm corresponding to the red part of the visible spectrum. This also extends the upper temperature measurable from 5000°C in unfiltered instruments up to 10 000°C.

The instrument cannot be used in automatic temperature control schemes because the eye of the human operator is an essential part of the measurement system. The

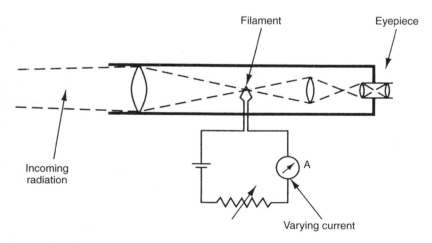

Fig. 14.11 Optical pyrometer.

reading is also affected by fumes in the sight path. Because of these difficulties and its low accuracy, hand-held radiation pyrometers are rapidly overtaking the optical pyrometer in popularity, although the instrument is still widely used in industry for measuring temperatures in furnaces and similar applications at present.

14.5.2 Radiation pyrometers

All the alternative forms of radiation pyrometer described below have an optical system that is similar to that in the optical pyrometer and focuses the energy emitted from the measured body. However, they differ by omitting the filament and eyepiece and having instead an energy detector in the same focal plane as the eyepiece was, as shown in Figure 14.12. This principle can be used to measure temperature over a range from $-100°C$ to $+3600°C$. The radiation detector is either a thermal detector, which measures the temperature rise in a black body at the focal point of the optical system, or a photon detector.

Thermal detectors respond equally to all wavelengths in the frequency spectrum, and consist of either thermopiles, resistance thermometers or thermistors. All of these typically have time constants of several milliseconds, because of the time taken for the black body to heat up and the temperature sensor to respond to the temperature change.

Photon detectors respond selectively to a particular band within the full spectrum, and are usually of the photoconductive or photovoltaic type. They respond to temperature changes very much faster than thermal detectors because they involve atomic processes, and typical measurement time constants are a few microseconds.

Fibre-optic technology is frequently used in high-temperature measurement applications to collect the incoming radiation and transmit it to a detector and processing electronics that are located remotely. This prevents exposure of the processing electronics to potentially damaging, high temperature. Fibre-optic cables are also used to apply radiation pyrometer principles in very difficult applications, such as measuring the temperature inside jet engines by collecting the radiation from inside the engine and transmitting it outside (see section 14.9).

The size of objects measured by a radiation pyrometer is limited by the optical resolution, which is defined as the ratio of target size to distance. A good ratio is 1:300, and this would allow temperature measurement of a 1 mm sized object at a range of 300 mm. With large distance/target size ratios, accurate aiming and focusing of the pyrometer at the target is essential. It is now common to find 'through the lens' viewing provided in pyrometers, using a principle similar to SLR camera technology,

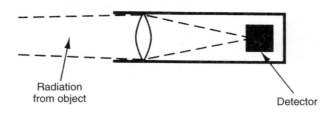

Fig. 14.12 Structure of the radiation thermometer.

as focusing and orientating the instrument for visible light automatically focuses it for infrared light. Alternatively, dual laser beams are sometimes used to ensure that the instrument is aimed correctly towards the target.

Various forms of electrical output are available from the radiation detector: these are functions of the incident energy on the detector and are therefore functions of the temperature of the measured body. Whilst this therefore makes such instruments of use in automatic control systems, their accuracy is often inferior to optical pyrometers. This reduced accuracy arises firstly because a radiation pyrometer is sensitive to a wider band of frequencies than the optical instrument and the relationship between emitted energy and temperature is less well defined. Secondly, the magnitude of energy emission at low temperatures gets very small, according to equation (14.9), increasing the difficulty of accurate measurement.

The forms of radiation pyrometer described below differ mainly in the technique used to measure the emitted radiation. They also differ in the range of energy wavelengths, and hence the temperature range, which each is designed to measure. One further difference is the material used to construct the energy-focusing lens. Outside the visible part of the spectrum, glass becomes almost opaque to infrared wavelengths, and other lens materials such as arsenic trisulphide are used.

Broad-band (unchopped) radiation pyrometers

The broadband radiation pyrometer finds wide application in industry and has a measurement inaccuracy that varies from ±0.05% of full scale in the best instruments to ±0.5% in the cheapest. However, their accuracy deteriorates significantly over a period of time, and an error of 10°C is common after 1–2 years' operation at high temperatures. As its name implies, the instrument measures radiation across the whole frequency spectrum and so uses a thermal detector. This consists of a blackened platinum disc to which a thermopile* is bonded. The temperature of the detector increases until the heat gain from the incident radiation is balanced by the heat loss due to convection and radiation. For high-temperature measurement, a two-couple thermopile gives acceptable measurement sensitivity and has a fast time constant of about 0.1 s. At lower measured temperatures, where the level of incident radiation is much less, thermopiles constructed from a greater number of thermocouples must be used to get sufficient measurement sensitivity. This increases the measurement time constant to as much as 2 s. Standard instruments of this type are available to measure temperatures between −20°C and +1800°C, although in theory much higher temperatures could be measured by this method.

Chopped broad-band radiation pyrometers

The construction of this form of pyrometer is broadly similar to that shown in Figure 14.12 except that a rotary mechanical device is included that periodically interrupts the radiation reaching the detector. The voltage output from the thermal detector thus becomes an alternating quantity that switches between two levels. This form of a.c. output can be amplified much more readily than the d.c. output coming from an unchopped instrument. This is particularly important when amplification is necessary to achieve an acceptable measurement resolution in situations where the

---

* Typically manganin–constantan.

level of incident radiation from the measured body is low. For this reason, this form of instrument is the more common when measuring body temperatures associated with peak emission in the infrared part of the frequency spectrum. For such chopped systems, the time constant of thermopiles is too long. Instead, thermistors are generally used, giving a time constant of 0.01 s. Standard instruments of this type are available to measure temperatures between $+20°C$ and $+1300°C$. This form of pyrometer suffers similar accuracy drift to unchopped forms. Its life is also limited to about two years because of motor failures.

Narrow-band radiation pyrometers

Narrow-band radiation pyrometers are highly stable instruments that suffer a drift in accuracy that is typically only 1°C in 10 years. They are also less sensitive to emissivity changes than other forms of radiation pyrometer. They use photodetectors of either the photoconductive or photovoltaic form whose performance is unaffected by either carbon dioxide or water vapour in the path between the target object and the instrument. A photoconductive detector exhibits a change in resistance as the incident radiation level changes whereas a photovoltaic cell exhibits an induced voltage across its terminals that is also a function of the incident radiation level. All photodetectors are preferentially sensitive to a particular narrow band of wavelengths in the range $0.5\,\mu m - 1.2\,\mu m$ and all have a form of output that varies in a highly non-linear fashion with temperature, and thus a microcomputer inside the instrument is highly desirable. Four commonly used materials for photodetectors are cadmium sulphide, lead sulphide, indium antimonide and lead–tin telluride. Each of these is sensitive to a different band of wavelengths and therefore all find application in measuring the particular temperature ranges corresponding to each of these bands.

The output from the narrow-band radiation pyrometer is normally chopped into an a.c. signal in the same manner as used in the chopped broad-band pyrometer. This simplifies the amplification of the output signal, which is necessary to achieve an acceptable measurement resolution. The typical time constant of a photon detector is only $5\,\mu s$, which allows high chopping frequencies up to 20 kHz. This gives such instruments an additional advantage in being able to measure fast transients in temperature as short as $10\,\mu s$.

Two-colour pyrometer (ratio pyrometer)

As stated earlier, the emitted radiation–temperature relationship for a body depends on its emissivity. This is very difficult to calculate, and therefore in practice all pyrometers have to be calibrated to the particular body they are measuring. The two-colour pyrometer (alternatively known as a ratio pyrometer) is a system that largely overcomes this problem by using the arrangement shown in Figure 14.13. Radiation from the body is split equally into two parts, which are applied to separate narrow-band filters. The outputs from the filters consist of radiation within two narrow bands of wavelength λ_1 and λ_2. Detectors sensitive to these frequencies produce output voltages V_1 and V_2 respectively. The ratio of these outputs, (V_1/V_2), can be shown (see Dixon, 1987) to be a function of temperature and to be independent of the emissivity provided that the two wavelengths λ_1 and λ_2 are close together.

The theoretical basis of the two-colour pyrometer is that the output is independent of emissivity because the emissivities at the two wavelengths λ_1 and λ_2 are equal.

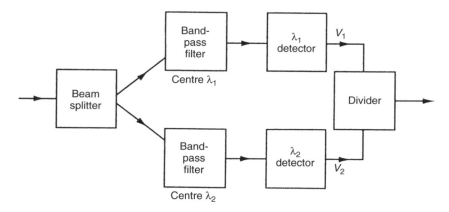

Fig. 14.13 Two-colour pyrometer system.

This is based on the assumption that λ_1 and λ_2 are very close together. In practice, this assumption does not hold and therefore the accuracy of the two-colour pyrometer tends to be relatively poor. However, the instrument is still of great use in conditions where the target is obscured by fumes or dust, which is a common problem in the cement and mineral processing industries. Two-colour pyrometers typically cost 50%–100% more than other types of pyrometer.

Selected waveband pyrometer

The selected waveband pyrometer is sensitive to one waveband only, e.g. 5 µm, and is dedicated to particular, special situations where other forms of pyrometer are inaccurate. One example of such a situation is measuring the temperature of steel billets that are being heated in a furnace. If an ordinary radiation pyrometer is aimed through the furnace door at a hot billet, it receives radiation from the furnace walls (by reflection off the billet) as well as radiation from the billet itself. If the temperature of the furnace walls is measured by a thermocouple, a correction can be made for the reflected radiation, but variations in transmission losses inside the furnace through fumes etc. make this correction inaccurate. However, if a carefully chosen selected-waveband pyrometer is used, this transmission loss can be minimized and the measurement accuracy is thereby greatly improved.

14.6 Thermography (thermal imaging)

Thermography, or thermal imaging, involves scanning an infrared radiation detector across an object. The information gathered is then processed and an output in the form of the temperature distribution across the object is produced. Temperature measurement over the range from $-20°C$ up to $+1500°C$ is possible. Elements of the system are shown in Figure 14.14.

The radiation detector uses the same principles of operation as a radiation pyrometer in inferring the temperature of the point that the instrument is focused on from a measurement of the incoming infrared radiation. However, instead of providing a

294 Temperature measurement

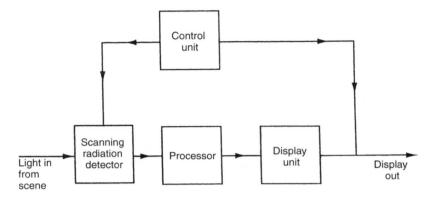

Fig. 14.14 Thermography (thermal imaging) system.

measurement of the temperature of a single point at the focal point of the instrument, the detector is scanned across a body or scene, and thus provides information about temperature distributions. Because of the scanning mode of operation of the instrument, radiation detectors with a very fast response are required, and only photoconductive or photovoltaic sensors are suitable. These are sensitive to the portion of the infrared spectrum between wavelengths of 2 µm and 14 µm.

Simpler versions of thermal imaging instruments consist of hand-held viewers that are pointed at the object of interest. The output from an array of infrared detectors is directed onto a matrix of red light-emitting diodes assembled behind a glass screen, and the output display thus consists of different intensities of red on a black background, with the different intensities corresponding to different temperatures. Measurement resolution is high, with temperature differences as small as 0.1°C being detectable. Such instruments are used in a wide variety of applications such as monitoring product flows through pipework, detecting insulation faults, and detecting hot spots in furnace linings, electrical transformers, machines, bearings etc. The number of applications is extended still further if the instrument is carried in a helicopter, where uses include scanning electrical transmission lines for faults, searching for lost or injured people and detecting the source and spread pattern of forest fires.

More complex thermal imaging systems comprise a tripod-mounted detector connected to a desktop computer and display system. Multi-colour displays are commonly used in such systems, where up to 16 different colours represent different bands of temperature across the measured range. The heat distribution across the measured body or scene is thus displayed graphically as a contoured set of coloured bands representing the different temperature levels. Such colour-thermography systems find many applications such as inspecting electronic circuit boards and monitoring production processes. There are also medical applications in body scanning.

14.7 Thermal expansion methods

Thermal expansion methods make use of the fact that the dimensions of all substances, whether solids, liquids or gases, change with temperature. Instruments operating on this physical principle include the liquid-in-glass thermometer, the bimetallic thermometer and the pressure thermometer.

14.7.1 Liquid-in-glass thermometers

The liquid-in-glass thermometer is a well-known temperature-measuring instrument that is used in a wide range of applications. The fluid used is usually either mercury or coloured alcohol, and this is contained within a bulb and capillary tube, as shown in Figure 14.15(a). As the temperature rises, the fluid expands along the capillary tube and the meniscus level is read against a calibrated scale etched on the tube. The process of estimating the position of the curved meniscus of the fluid against the scale introduces some error into the measurement process and a measurement inaccuracy less than ±1% of full-scale reading is hard to achieve.

However, an inaccuracy of only ±0.15% can be obtained in the best industrial instruments. Industrial versions of the liquid-in-glass thermometer are normally used to measure temperature in the range between −200°C and +1000°C, although instruments are available to special order that can measure temperatures up to 1500°C.

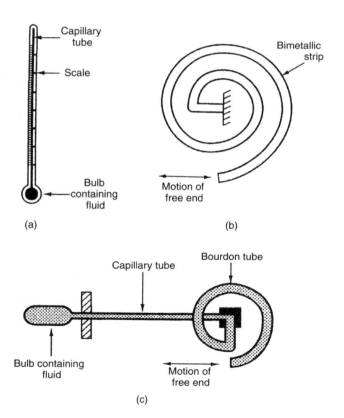

Fig. 14.15 Thermal expansion devices: (a) liquid-in-glass thermometer; (b) bimetallic thermometer; (c) Pressure thermometer.

14.7.2 Bimetallic thermometer

The bimetallic principle is probably more commonly known in connection with its use in thermostats. It is based on the fact that if two strips of different metals are bonded together, any temperature change will cause the strip to bend, as this is the only way in which the differing rates of change of length of each metal in the bonded strip can be accommodated. In the bimetallic thermostat, this is used as a switch in control applications. If the magnitude of bending is measured, the bimetallic device becomes a thermometer. For such purposes, the strip is often arranged in a spiral or helical configuration, as shown in Figure 14.15(b), as this gives a relatively large displacement of the free end for any given temperature change. The measurement sensitivity is increased further by choosing the pair of materials carefully such that the degree of bending is maximized, with Invar (a nickel–steel alloy) or brass being commonly used.

The system used to measure the displacement of the strip must be carefully designed. Very little resistance must be offered to the end of the strip, otherwise the spiral or helix will distort and cause a false reading in the measurement of the displacement. The device is normally just used as a temperature indicator, where the end of the strip is made to turn a pointer that moves against a calibrated scale. However, some versions produce an electrical output, using either a linear variable differential transformer (LVDT) or a fibre-optic shutter sensor to transduce the output displacement.

Bimetallic thermometers are used to measure temperatures between $-75°C$ and $+1500°C$. The inaccuracy of the best instruments can be as low as $\pm 0.5\%$ but such devices are quite expensive. Many instrument applications do not require this degree of accuracy in temperature measurements, and in such cases much cheaper bimetallic thermometers with substantially inferior accuracy specifications are used.

14.7.3 Pressure thermometers

Pressure thermometers have now been superseded by other alternatives in most applications, but they still remain useful in a few applications such as furnace temperature measurement when the level of fumes prevents the use of optical or radiation pyrometers. Examples can also still be found of their use as temperature sensors in pneumatic control systems. The sensing element in a pressure thermometer consists of a stainless-steel bulb containing a liquid or gas. If the fluid were not constrained, temperature rises would cause its volume to increase. However, because it is constrained in a bulb and cannot expand, its pressure rises instead. As such, the pressure thermometer does not strictly belong to the thermal expansion class of instruments but is included because of the relationship between volume and pressure according to Boyle's law: $PV = KT$.

The change in pressure of the fluid is measured by a suitable pressure transducer such as the Bourdon tube (see Chapter 15). This transducer is located remotely from the bulb and connected to it by a capillary tube as shown in Figure 14.15(c). The need to protect the pressure-measuring instrument from the environment where the temperature is being measured can require the use of capillary tubes up to 5 m long, and the temperature gradient, and hence pressure gradient, along the tube acts as a modifying input that can introduce a significant measurement error. Pressure thermometers can be used to measure temperatures in the range between −250°C and +2000°C and their typical inaccuracy is ±0.5% of full-scale reading. However, the instrument response has a particularly long time constant.

14.8 Quartz thermometers

The quartz thermometer makes use of the principle that the resonant frequency of a material such as quartz is a function of temperature, and thus enables temperature changes to be translated into frequency changes. The temperature-sensing element consists of a quartz crystal enclosed within a probe (sheath). The probe commonly consists of a stainless steel cylinder, which makes the device physically larger than devices like thermocouples and resistance thermometers. The crystal is connected electrically so as to form the resonant element within an electronic oscillator. Measurement of the oscillator frequency therefore allows the measured temperature to be calculated.

The instrument has a very linear output characteristic over the temperature range between −40°C and +230°C, with a typical inaccuracy of ±0.1%. Measurement resolution is typically 0.1°C but versions can be obtained with resolutions as small as 0.0003°C. The characteristics of the instrument are generally very stable over long periods of time and therefore only infrequent calibration is necessary. The frequency-change form of output means that the device is insensitive to noise. However, it is very expensive, with a typical cost of £3000 ($5000).

14.9 Fibre-optic temperature sensors

Fibre-optic cables can be used as either intrinsic or extrinsic temperature sensors, as discussed in Chapter 13, though special attention has to be paid to providing a suitable protective coating when high temperatures are measured. Cost varies from £1000 to £4000, according to type, and the normal temperature range covered is 250°C to 3000°C, though special devices can detect down to 100°C and others can detect up to 3600°C. Their main application is measuring temperatures in hard-to-reach locations, though they are also used when very high measurement accuracy is required. Some laboratory versions have an inaccuracy as low as ±0.01%, which is better than a type S thermocouple, although versions used in industry have a more typical inaccuracy of ±1.0%. Whilst it is often assumed that fibre-optic sensors are intrinsically safe, it has been shown (Johnson, 1994) that flammable gas might be ignited by the optical power levels available from some laser diodes. Thus, the power level used with optical fibres must be carefully chosen, and certification of intrinsic safety is necessary if such sensors are to be used in hazardous environments.

One type of intrinsic sensor uses cable where the core and cladding have similar refractive indices but different temperature coefficients. Temperature rises cause the refractive indices to become even closer together and losses from the core to increase, thus reducing the quantity of light transmitted. Other types of intrinsic temperature sensor include the cross-talk sensor, phase modulating sensor and optical resonator, as described in Chapter 13. Research into the use of distributed temperature sensing using fibre-optic cable has also been reported. This can be used to measure things like the temperature distribution along an electricity supply cable. It works by measuring the reflection characteristics of light transmitted down a fibre-optic cable that is bonded to the electrical cable. By analysing the back-scattered radiation, a table of temperature versus distance along the cable can be produced, with a measurement inaccuracy of only ±0.5°C.

A common form of extrinsic sensor uses fibre-optic cables to transmit light from a remote targeting lens into a standard radiation pyrometer. This technique can be used with all types of radiation pyrometer, including the two-colour version, and a particular advantage is that this method of measurement is intrinsically safe. However, it is not possible to measure very low temperatures, because the very small radiation levels that exist at low temperatures are badly attenuated during transmission along the fibre-optic cable. Therefore, the minimum temperature that can be measured is about 50°C, and the light guide for this must not exceed 600 mm in length. At temperatures exceeding 1000°C, lengths of fibre up to 20 m long can be successfully used as a light guide.

One extremely accurate device that uses this technique is known as the Accufibre sensor. This is a form of radiation pyrometer that has a black box cavity at the focal point of the lens system. A fibre-optic cable is used to transmit radiation from the black box cavity to a spectrometric device that computes the temperature. This has a measurement range 500°C to 2000°C, a resolution of 10^{-5}°C and an inaccuracy of only ±0.0025% of full scale.

Several other types of device that are marketed as extrinsic fibre-optic temperature sensors consist of a conventional temperature sensor (e.g. a resistance thermometer) connected to a fibre-optic cable so that the transmission of the signal from the measurement point is free of noise. Such devices must include an electricity supply for the electronic circuit that is needed to convert the sensor output into light variations in the cable. Thus, low-voltage power cables must be routed with the fibre-optic cable, and the device is therefore not intrinsically safe.

14.10 Acoustic thermometers

The principle of acoustic thermometry was discovered as long ago as 1873 and uses the fact that the velocity of sound through a gas varies with temperature according to the equation:

$$v = \sqrt{\alpha RT/M} \qquad (14.10)$$

where v is the sound velocity, T is the gas temperature, M is the molecular weight of the gas and both R and α are constants. Until very recently, it had only been used for measuring cryogenic (very low) temperatures, but it is now also used for measuring higher temperatures and can potentially measure right up to 20 000°C. However, typical inaccuracy is ±5%, and the devices are expensive (typically £6000 or $10 000). The various versions of acoustic thermometer that are available differ according to the technique used for generating sound and measuring its velocity in the gas. If ultrasonic generation is used, the instrument is often known as an *ultrasonic thermometer*. Further information can be found in Michalski, (1991).

14.11 Colour indicators

The colour of various substances and objects changes as a function of temperature. One use of this is in the optical pyrometer as discussed earlier. The other main use of colour change is in special colour indicators that are widely used in industry to determine whether objects placed in furnaces have reached the required temperature. Such colour indicators consist of special paints or crayons that are applied to an object before it is placed in a furnace. The colour-sensitive component within these is some form of metal salt (usually of chromium, cobalt or nickel). At a certain temperature, a chemical reaction takes place and a permanent colour change occurs in the paint or crayon, although this change does not occur instantaneously but only happens over a period of time.

Hence, the colour change mechanism is complicated by the fact that the time of exposure as well as the temperature is important. Such crayons or paints usually have a dual rating that specifies the temperature and length of exposure time required for the colour change to occur. If the temperature rises above the rated temperature, then the colour change will occur in less than the rated exposure time. This causes little problem if the rate of temperature rise is slow with respect to the specified exposure time required for colour change to occur. However, if the rate of rise of temperature is high, the object will be significantly above the rated change temperature of the paint/crayon by the time that the colour change happens. Besides wasting energy by leaving the object in the furnace longer than necessary, this can also cause difficulty if excess temperature can affect the required metallurgical properties of the heated object.

Paints and crayons are available to indicate temperatures between 50°C and 1250°C. A typical exposure time rating is 30 minutes, i.e. the colour change will occur if the paint/crayon is exposed to the rated temperature for this length of time. They have the advantage of low cost, typically a few pounds per application. However, they adhere strongly to the heated object, which can cause difficulty if they have to be cleaned off the object later.

Some liquid crystals also change colour at a certain temperature. According to the design of sensors using such liquid crystals, the colour change can either occur gradually during a temperature rise of perhaps 50°C or else change abruptly at some specified temperature. The latter kind of sensors are able to resolve temperature changes as small as 0.1°C and, according to type, are used over the temperature range from −20°C to +100°C.

14.12 Change of state of materials

Temperature-indicating devices known as Seger cones or pyrometric cones are commonly used in the ceramics industry. They consist of a fused oxide and glass material that is formed into a cone shape. The tip of the cone softens and bends over when a particular temperature is reached. Cones are available that indicate temperatures over the range from 600°C to +2000°C.

14.13 Intelligent temperature-measuring instruments

Intelligent temperature transmitters have now been introduced into the catalogues of most instrument manufacturers, and they bring about the usual benefits associated with intelligent instruments. Such transmitters are separate boxes designed for use with transducers that have either a d.c. voltage output in the mV range or an output in the form of a resistance change. They are therefore suitable for use in conjunction with thermocouples, thermopiles, resistance thermometers, thermistors and broad-band radiation pyrometers. All of the transmitters presently available have non-volatile memories where all constants used in correcting output values for modifying inputs etc. are stored, thus enabling the instrument to survive power failures without losing such information. Facilities in transmitters now available include adjustable damping, noise rejection, self-adjustment for zero and sensitivity drifts and expanded measurement range. These features allow an inaccuracy level of ±0.05% of full scale to be specified.

Mention must be made particularly of intelligent pyrometers, as some versions of these are able to measure the emissivity of the target body and automatically provide an emissivity-corrected output. This particular development provides an alternative to the two-colour pyrometer when emissivity measurement and calibration for other types of pyrometer pose difficulty.

Digital thermometers (see section 14.2) also exist in intelligent versions, where the inclusion of a microprocessor allows a number of alternative thermocouples and resistance thermometers to be offered as options for the primary sensor.

The cost of intelligent temperature transducers is significantly more than their non-intelligent counterparts, and justification purely on the grounds of their superior accuracy is hard to make. However, their expanded measurement range means immediate savings are made in terms of the reduction in the number of spare instruments needed to cover a number of measurement ranges. Their capability for self-diagnosis and self-adjustment means that they require attention much less frequently, giving additional savings in maintenance costs.

14.14 Choice between temperature transducers

The suitability of different instruments in any particular measurement situation depends substantially on whether the medium to be measured is a solid or a fluid. For measuring the temperature of solids, it is essential that good contact is made between the body and the transducer unless a radiation thermometer is used. This restricts the range of suitable transducers to thermocouples, thermopiles, resistance thermometers, thermistors, semiconductor devices and colour indicators. On the other hand, fluid temperatures can be measured by any of the instruments described in this chapter, with the exception of radiation thermometers.

The most commonly used device in industry for temperature measurement is the base-metal thermocouple. This is relatively cheap, with prices varying widely from a few pounds upwards according to the thermocouple type and sheath material used. Typical inaccuracy is ±0.5% of full scale over the temperature range −250°C to +1200°C. Noble metal thermocouples are much more expensive, but are chemically inert and can measure temperatures up to 2300°C with an inaccuracy of ±0.2% of full scale. However, all types of thermocouple have a low-level output voltage, making them prone to noise and therefore unsuitable for measuring small temperature differences.

Resistance thermometers are also in common use within the temperature range −270°C to +650°C, with a measurement inaccuracy of ±0.5%. Whilst they have a smaller temperature range than thermocouples, they are more stable and can measure small temperature differences. The platinum resistance thermometer is generally regarded as offering the best ratio of price to performance for measurement in the temperature range −200°C to +500°C, with prices starting from £15.

Thermistors are another relatively common class of devices. They are small and cheap, with a typical cost of around £5. They give a fast output response to temperature changes, with good measurement sensitivity, but their measurement range is quite limited.

Dual diverse sensors are a new development that include a thermocouple and a resistance thermometer inside the same sheath. Both of these devices are affected by various factors in the operating environment, but each tends to be sensitive to different things in different ways. Thus, comparison of the two outputs means that any change in characteristics is readily detected, and appropriate measures to replace or recalibrate the sensors can be taken.

Pulsed sensors are a further recent development. They consist of a water-cooled thermocouple or resistance thermometer, and enable temperature measurement to be made well above the normal upper temperature limit for these devices. At the measuring instant, the water-cooling is temporarily stopped, causing the temperature in the sensor to rise towards the process temperature. Cooling is restarted before the sensor temperature rises to the level where the sensor would be damaged, and the process temperature is then calculated by extrapolating from the measured temperature according to the exposure time.

Semiconductor devices have a better linearity than thermocouples and resistance thermometers and a similar level of accuracy. Thus they are a viable alternative to these in many applications. Integrated circuit transistor sensors are particularly cheap (from £10 each), although their accuracy is relatively poor and they have a very limited measurement range (−50°C to +150°C). Diode sensors are much more accurate and have a wider temperature range (−270°C to +200°C), though they are also more expensive (typical costs are anywhere from £50 to £500).

A major virtue of radiation thermometers is their non-contact, non-invasive mode of measurement. Costs vary from £250 up to £3000 according to type. Although calibration for the emissivity of the measured object often poses difficulties, some instruments now provide automatic calibration. Optical pyrometers are used to monitor temperatures above 600°C in industrial furnaces etc., but their inaccuracy is typically ±5%. Various forms of radiation pyrometer are used over the temperature range between −20°C and +1800°C and can give measurement inaccuracies as low as ±0.05%. One particular merit of narrow-band radiation pyrometers is their ability to measure fast temperature transients of duration as small as 10 µs. No other instrument can measure transients anywhere near as fast as this.

The range of instruments working on the thermal expansion principle are mainly used as temperature indicating devices rather than as components within automatic control schemes. Temperature ranges and costs are: mercury-in-glass thermometers up to +1000°C (cost from a few pounds), bi-metallic thermometers up to +1500°C (cost £50 to £100) and pressure thermometers up to +2000°C (cost £100 to £500). The usual measurement inaccuracy is in the range ±0.5% to ±1.0%. The bimetallic thermometer is more rugged than liquid-in-glass types but less accurate (however, the greater inherent accuracy of liquid-in-glass types can only be realized if the liquid meniscus level is read carefully).

Fibre optic devices are more expensive than most other forms of temperature sensor (costing up to £4000) but provide a means of measuring temperature in very inaccessible locations. Inacccuracy varies from ±1% down to ±0.01% in some laboratory versions. Measurement range also varies with type, but up to +3600°C is possible.

The quartz thermometer provides very high resolution (0.0003°C is possible with special versions) but is expensive because of the complex electronics required to analyse the frequency-change form of output. A typical price is £3000 ($5000). It only operates over the limited temperature range of −40°C to +230°C, but gives a low measurement inaccuracy of ±0.1% within this range.

Acoustic thermometers provide temperature measurement over a very wide range (−150°C to +20 000°C). However, their inaccuracy is relatively high (typically ±5%) and they are very expensive (typically £6000 or $10 000).

Colour indicators are widely used to determine when objects in furnaces have reached the required temperature. These indicators work well if the rate of rise of temperature of the object in the furnace is relatively slow but, because temperature indicators only change colour over a period of time, the object will be above the required temperature by the time that the indicator responds if the rate of rise of temperature is large. Cost is low, for example a crayon typically costs £3.

14.15 Self-test questions

14.1 The output e.m.f. from a chromel–alumel thermocouple (type K), with its reference junction maintained at 0°C, is 12.207 mV. What is the measured temperature?

14.2 The output e.m.f. from a nicrosil–nisil thermocouple (type N), with its reference junction maintained at 0°C, is 4.21 mV. What is the measured temperature?

14.3 A platinum/10% rhodium–platinum (type S) thermocouple is used to measure the temperature of a furnace. The output e.m.f., with the reference junction maintained at 50°C, is 5.975 mV. What is the temperature of the furnace?

14.4 In a particular industrial situation, a nicrosil–nisil thermocouple with nicrosil–nisil extension wires is used to measure the temperature of a fluid. In connecting up this measurement system, the instrumentation engineer responsible has inadvertently interchanged the extension wires from the thermocouple. The ends of the extension wires are held at a reference temperature of 0°C and the output e.m.f. measured is 21.0 mV. If the junction between the thermocouple and extension wires is at a temperature of 50°C, what temperature of fluid is indicated and what is the true fluid temperature?

14.5 A chromel–constantan thermocouple measuring the temperature of a fluid is connected by mistake with copper–constantan extension leads (such that the two constantan wires are connected together and the copper extension wire is connected to the chromel thermocouple wire). If the fluid temperature was actually 250°C, and the junction between the thermocouple and extension wires was at 80°C, what e.m.f. would be measured at the open ends of the extension wires if the reference junction is maintained at 0°C? What fluid temperature would be deduced from this (assuming that the connection mistake was not known about)? (Hint: apply the law of intermediate metals for the thermocouple-extension lead junction.)

References and further reading

Brookes, C. (1985) Nicrosil–nisil thermocouples, *Journal of Measurement and Control*, **18**(7), pp. 245–248.

Dixon, J. (1987) Industrial radiation thermometry, *Journal of Measurement and Control*, **20**(6), pp. 11–16.

Editorial (1996) Control Engineering, September, p. 93.

Johnson, J.S. (1994) Optical sensors: the OCSA experience, *Measurement and Control*, **27**(7), pp. 180–184.

Michalski, L., Eckersdorf, K. and McGhee, J. (1991) *Temperature Measurement*, John Wiley.

15

Pressure measurement

Pressure measurement is a very common requirement for most industrial process control systems and many different types of pressure-sensing and pressure-measurement systems are available. However, before considering these in detail, it is important to explain some terms used in pressure measurement and to define the difference between absolute pressure, gauge pressure and differential pressure.

Absolute pressure: This is the difference between the pressure of the fluid and the absolute zero of pressure.

Gauge pressure: This describes the difference between the pressure of a fluid and atmospheric pressure. Absolute and gauge pressure are therefore related by the expression:

$$\text{Absolute pressure} = \text{Gauge pressure} + \text{Atmospheric pressure}$$

Thus, gauge pressure varies as the atmospheric pressure changes and is therefore not a fixed quantity.

Differential pressure: This term is used to describe the difference between two absolute pressure values, such as the pressures at two different points within the same fluid (often between the two sides of a flow restrictor in a system measuring volume flow rate).

In most applications, the typical values of pressure measured range from 1.013 bar (the mean atmospheric pressure) up to 7000 bar. This is considered to be the 'normal' pressure range, and a large number of pressure sensors are available that can measure pressures in this range. Measurement requirements outside this range are much less common. Whilst some of the pressure sensors developed for the 'normal' range can also measure pressures that are either lower or higher than this, it is preferable to use special instruments that have been specially designed to satisfy such low- and high-pressure measurement requirements.

The discussion below summarizes the main types of pressure sensor that are in use. This discussion is primarily concerned only with the measurement of static pressure, because the measurement of dynamic pressure is a very specialized area that is not of general interest. In general, dynamic pressure measurement requires special instruments, although modified versions of diaphragm-type sensors can also be used if they contain a suitable displacement sensor (usually either a piezoelectric crystal or a capacitive element).

15.1 Diaphragms

The diaphragm, shown schematically in Figure 15.1, is one of three types of elastic-element pressure transducer. Applied pressure causes displacement of the diaphragm and this movement is measured by a displacement transducer. Different versions of diaphragm sensors can measure both absolute pressure (up to 50 bar) and gauge pressure (up to 2000 bar) according to whether the space on one side of the diaphragm is respectively evacuated or is open to the atmosphere. A diaphragm can also be used to measure differential pressure (up to 2.5 bar) by applying the two pressures to the two sides of the diaphragm. The diaphragm can be either plastic, metal alloy, stainless steel or ceramic. Plastic diaphragms are cheapest, but metal diaphragms give better accuracy. Stainless steel is normally used in high temperature or corrosive environments. Ceramic diaphragms are resistant even to strong acids and alkalis, and are used when the operating environment is particularly harsh.

The typical magnitude of diaphragm displacement is 0.1 mm, which is well suited to a strain-gauge type of displacement-measuring transducer, although other forms of displacement measurement are also used in some kinds of diaphragm-based sensors. If the displacement is measured with strain gauges, it is normal to use four strain gauges arranged in a bridge circuit configuration. The output voltage from the bridge is a function of the resistance change due to the strain in the diaphragm. This arrangement automatically provides compensation for environmental temperature changes. Older pressure transducers of this type used metallic strain gauges bonded to a diaphragm typically made of stainless steel. However, apart from manufacturing difficulties arising from the problem of bonding the gauges, metallic strain gauges have a low gauge factor, which means that the low output from the strain gauge bridge has to be amplified by an expensive d.c. amplifier. The development of semiconductor (piezoresistive) strain gauges provided a solution to the low-output problem, as they have gauge factors up

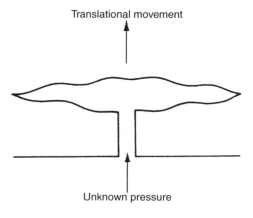

Fig. 15.1 Schematic representation of diaphragm pressure sensor.

to one hundred times greater than metallic gauges. However, the difficulty of bonding gauges to the diaphragm remained and a new problem emerged regarding the highly non-linear characteristic of the strain–output relationship.

The problem of strain-gauge bonding was solved with the emergence of monolithic piezoresistive pressure transducers. These have a typical measurement uncertainty of $\pm 0.5\%$ and are now the most commonly used type of diaphragm pressure transducer. The monolithic cell consists of a diaphragm made of a silicon sheet into which resistors are diffused during the manufacturing process. Such pressure transducers can be made to be very small and are often known as *micro-sensors*. Also, besides avoiding the difficulty with bonding, such monolithic silicon measuring cells have the advantage of being very cheap to manufacture in large quantities. Although the inconvenience of a non-linear characteristic remains, this is normally overcome by processing the output signal with an active linearization circuit or incorporating the cell into a microprocessor-based intelligent measuring transducer. The latter usually provides analogue-to-digital conversion and interrupt facilities within a single chip and gives a digital output that is readily integrated into computer control schemes. Such instruments can also offer automatic temperature compensation, built-in diagnostics and simple calibration procedures. These features allow measurement inaccuracy to be reduced to a figure as low as $\pm 0.1\%$ of full-scale reading.

15.2 Capacitive pressure sensor

A capacitive pressure sensor is simply a diaphragm-type device in which the diaphragm displacement is determined by measuring the capacitance change between the diaphragm and a metal plate that is close to it. Such devices are in common use. It is also possible to fabricate capacitive elements in a silicon chip and thus form very small *micro-sensors*. These have a typical measurement uncertainty of $\pm 0.2\%$.

15.3 Fibre-optic pressure sensors

Fibre-optic sensors provide an alternative method of measuring displacements in diaphragm and Bourdon tube pressure sensors by optoelectronic means, and enable the resulting sensors to have lower mass and size compared with sensors in which the displacement is measured by other methods. The shutter sensor described earlier in Chapter 13 is one form of fibre-optic displacement sensor. Another form is the Fotonic sensor shown in Figure 15.2 in which light travels from a light source, down an optical fibre, is reflected back from a diaphragm, and then travels back along a second fibre to a photodetector. There is a characteristic relationship between the light reflected and the distance from the fibre ends to the diaphragm, thus making the amount of reflected light dependent upon the diaphragm displacement and hence the measured pressure.

Apart from the mass and size advantages of fibre-optic displacement sensors, the output signal is immune to electromagnetic noise. However, the measurement accuracy is usually inferior to that provided by alternative displacement sensors, and choice of such sensors also incurs a cost penalty. Thus, sensors using fibre optics to measure diaphragm or Bourdon tube displacement tend to be limited to applications where

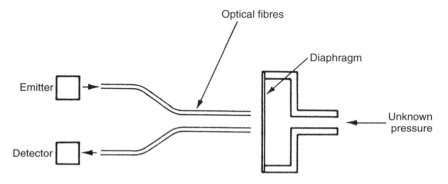

Fig. 15.2 Fotonic sensor.

their small size, low mass and immunity to electromagnetic noise are particularly advantageous.

Apart from the limited use above within diaphragm and Bourdon tube sensors, fibre-optic cables are also used in several other ways to measure pressure. A form of fibre-optic pressure sensor known as a *microbend sensor* is sketched in Figure 13.7(a). In this, the refractive index of the fibre (and hence of the intensity of light transmitted) varies according to the mechanical deformation of the fibre caused by pressure. The sensitivity of pressure measurement can be optimized by applying the pressure via a roller chain such that the bending is applied periodically (see Figure 13.7(b)). The optimal pitch for the chain varies according to the radius, refractive index and type of cable involved. Microbend sensors are typically used to measure the small pressure changes generated in vortex shedding flowmeters. When fibre-optic sensors are used in this flow-measurement role, the alternative arrangement shown in Figure 15.3 can be used, where a fibre-optic cable is merely stretched across the pipe. This often simplifies the detection of vortices.

Phase-modulating fibre-optic pressure sensors also exist. The mode of operation of these was discussed in Chapter 13.

15.4 Bellows

The bellows, schematically illustrated in Figure 15.4, is another elastic-element type of pressure sensor that operates on very similar principles to the diaphragm pressure sensor. Pressure changes within the bellows, which is typically fabricated as a seamless tube of either metal or metal alloy, produce translational motion of the end of the bellows that can be measured by capacitive, inductive (LVDT) or potentiometric transducers. Different versions can measure either absolute pressure (up to 2.5 bar) or gauge pressure (up to 150 bar). Double-bellows versions also exist that are designed to measure differential pressures of up to 30 bar.

Bellows have a typical measurement uncertainty of only ±0.5%, but they have a relatively high manufacturing cost and are prone to failure. Their principal attribute in the past has been their greater measurement sensitivity compared with diaphragm sensors. However, advances in electronics mean that the high-sensitivity requirement

Pressure measurement

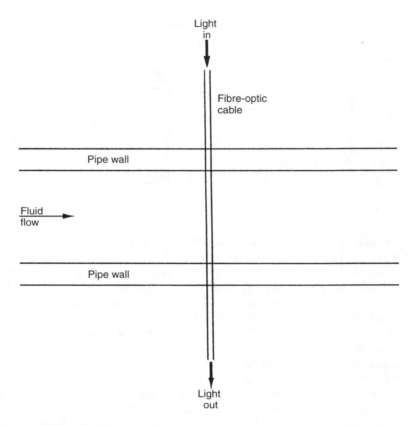

Fig. 15.3 Simple fibre-optic vortex detector.

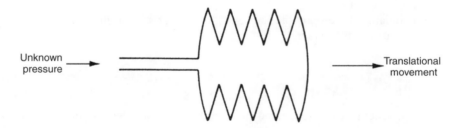

Fig. 15.4 Bellows.

can usually be satisfied now by diaphragm-type devices, and usage of bellows is therefore falling.

15.5 Bourdon tube

The Bourdon tube is also an elastic element type of pressure transducer. It is relatively cheap and is commonly used for measuring the gauge pressure of both gaseous and liquid fluids. It consists of a specially shaped piece of oval-section, flexible, metal tube that is fixed at one end and free to move at the other end. When pressure is applied at the open, fixed end of the tube, the oval cross-section becomes more circular. In consequence, there is a displacement of the free end of the tube. This displacement is measured by some form of displacement transducer, which is commonly a potentiometer or LVDT. Capacitive and optical sensors are also sometimes used to measure the displacement.

The three common shapes of Bourdon tube are shown in Figure 15.5. The maximum possible deflection of the free end of the tube is proportional to the angle subtended by the arc through which the tube is bent. For a C-type tube, the maximum value for this arc is somewhat less than 360°. Where greater measurement sensitivity and resolution are required, spiral and helical tubes are used. These both give a much greater deflection at the free end for a given applied pressure. However, this increased measurement performance is only gained at the expense of a substantial increase in manufacturing difficulty and cost compared with C-type tubes, and is also associated with a large decrease in the maximum pressure that can be measured. Spiral and helical types are sometimes provided with a rotating pointer that moves against a scale to give a visual indication of the measured pressure.

C-type tubes are available for measuring pressures up to 6000 bar. A typical C-type tube of 25 mm radius has a maximum displacement travel of 4 mm, giving a moderate level of measurement resolution. Measurement inaccuracy is typically quoted at ±1% of full-scale deflection. Similar accuracy is available from helical and spiral types, but whilst the measurement resolution is higher, the maximum pressure measurable is only 700 bar.

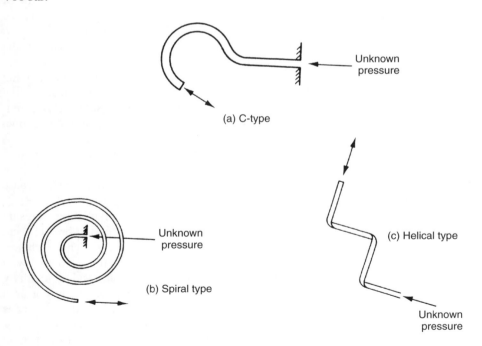

Fig. 15.5 Bourdon tubes.

The existence of one potentially major source of error in Bourdon tube pressure measurement has not been widely documented, and few manufacturers of Bourdon tubes make any attempt to warn users of their products appropriately. The problem is concerned with the relationship between the fluid being measured and the fluid used for calibration. The pointer of Bourdon tubes is normally set at zero during manufacture, using air as the calibration medium. However, if a different fluid, especially a liquid, is subsequently used with a Bourdon tube, the fluid in the tube will cause a non-zero deflection according to its weight compared with air, resulting in a reading error of up to 6%. This can be avoided by calibrating the Bourdon tube with the fluid to be measured instead of with air, assuming of course that the user is aware of the problem. Alternatively, correction can be made according to the calculated weight of the fluid in the tube. Unfortunately, difficulties arise with both of these solutions if air is trapped in the tube, since this will prevent the tube being filled completely by the fluid. Then, the amount of fluid actually in the tube, and its weight, will be unknown.

In conclusion, therefore, Bourdon tubes only have guaranteed accuracy limits when measuring gaseous pressures. Their use for accurate measurement of liquid pressures poses great difficulty unless the gauge can be totally filled with liquid during both calibration and measurement, a condition that is very difficult to fulfil practically.

15.6 Manometers

Manometers are passive instruments that give a visual indication of pressure values. Various types exist.

The *U-tube manometer*, shown in Figure 15.6(a), is the most common form of manometer. Applied pressure causes a displacement of liquid inside the U-shaped glass tube, and the output pressure reading P is made by observing the difference h between the level of liquid in the two halves of the tube A and B, according to the equation $P = h\rho g$, where ρ is the specific gravity of the fluid. If an unknown pressure is applied to side A, and side B is open to the atmosphere, the output reading is gauge pressure. Alternatively, if side B of the tube is sealed and evacuated, the output reading is absolute pressure. The U-tube manometer also measures the differential pressure $(p_1 - p_2)$, according to the expression $(p_1 - p_2) = h\rho g$, if two unknown pressures p_1 and p_2 are applied respectively to sides *A* and *B* of the tube.

Output readings from U-tube manometers are subject to error, principally because it is very difficult to judge exactly where the meniscus levels of the liquid are in the two halves of the tube. In absolute pressure measurement, an addition error occurs because it is impossible to totally evacuate the closed end of the tube.

U-tube manometers are typically used to measure gauge and differential pressures up to about 2 bar. The type of liquid used in the instrument depends on the pressure and characteristics of the fluid being measured. Water is a cheap and convenient choice, but it evaporates easily and is difficult to see. Nevertheless, it is used extensively, with the major obstacles to its use being overcome by using coloured water and by regularly topping up the tube to counteract evaporation. However, water is definitely not used when measuring the pressure of fluids that react with or dissolve in water. Water is also unsuitable when high-pressure measurements are required. In such circumstances, liquids such as aniline, carbon tetrachloride, bromoform, mercury or transformer oil are used instead.

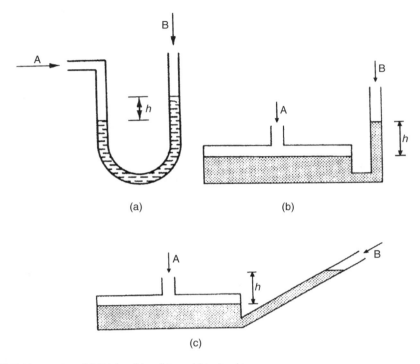

Fig. 15.6 Manometers: (a) U-tube; (b) well type; (c) inclined type.

The *well-type or cistern manometer*, shown in Figure 15.6(b), is similar to a U-tube manometer but one half of the tube is made very large so that it forms a well. The change in the level of the well as the measured pressure varies is negligible. Therefore, the liquid level in only one tube has to be measured, which makes the instrument much easier to use than the U-tube manometer. If an unknown pressure p_1 is applied to port A, and port B is open to the atmosphere, the gauge pressure is given by $p_1 = h\rho$. It might appear that the instrument would give a better measurement accuracy than the U-tube manometer because the need to subtract two liquid level measurements in order to arrive at the pressure value is avoided. However, this benefit is swamped by errors that arise due to the typical cross-sectional area variations in the glass used to make the tube. Such variations do not affect the accuracy of the U-tube manometer to the same extent.

The *inclined manometer* or *draft gauge*, shown in Figure 15.6(c), is a variation on the well-type manometer in which one leg of the tube is inclined to increase measurement sensitivity. However, similar comments to those above apply about accuracy.

15.7 Resonant-wire devices

A typical resonant-wire device is shown schematically in Figure 15.7. Wire is stretched across a chamber containing fluid at unknown pressure subjected to a magnetic field.

312 Pressure measurement

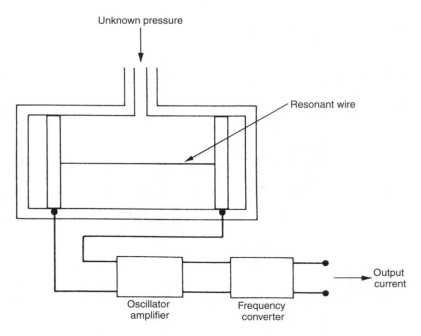

Fig. 15.7 Resonant-wire device.

The wire resonates at its natural frequency according to its tension, which varies with pressure. Thus pressure is calculated by measuring the frequency of vibration of the wire. Such frequency measurement is normally carried out by electronics integrated into the cell. These devices are highly accurate, with a typical inaccuracy figure being ±0.2% full-scale reading. They are also particularly insensitive to ambient condition changes and can measure pressures between 5 mbar and 2 bar.

15.8 Dead-weight gauge

The dead-weight gauge, as shown in Figure 2.3, is a null-reading type of measuring instrument in which weights are added to the piston platform until the piston is adjacent to a fixed reference mark, at which time the downward force of the weights on top of the piston is balanced by the pressure exerted by the fluid beneath the piston. The fluid pressure is therefore calculated in terms of the weight added to the platform and the known area of the piston. The instrument offers the ability to measure pressures to a high degree of accuracy but is inconvenient to use. Its major application is as a reference instrument against which other pressure-measuring devices are calibrated. Various versions are available that allow measurement of gauge pressures up to 7000 bar.

15.9 Special measurement devices for low pressures

A number of special devices have been developed for measurement of pressures in the vacuum range below atmospheric pressure (<1.013 bar). These special devices include the thermocouple gauge, the Pirani gauge, the thermistor gauge, the McLeod gauge and the ionization gauge, and they are covered in more detail below. Unfortunately, all of these specialized instruments are quite expensive.

The *thermocouple gauge* is one of a group of gauges working on the thermal conductivity principal. The paranoia and thermistor gauges also belong to this group. At low pressure, the kinematic theory of gases predicts a linear relationship between pressure and thermal conductivity. Thus measurement of thermal conductivity gives an indication of pressure. Figure 15.8 shows a sketch of a thermocouple gauge. Operation of the gauge depends on the thermal conduction of heat between a thin hot metal strip in the centre and the cold outer surface of a glass tube (that is normally at room temperature). The metal strip is heated by passing a current through it and its temperature is measured by a thermocouple. The temperature measured depends on the thermal conductivity of the gas in the tube and hence on its pressure. A source of error in this instrument is the fact that heat is also transferred by radiation as well as conduction. This error is of a constant magnitude, independent of pressure. Hence, it can be measured, and thus correction can be made for it. However, it is usually more convenient to design for low radiation loss by choosing a heated element with low emissivity. Thermocouple gauges are typically used to measure pressures in the range 10^{-4} mbar up to 1 mbar.

A typical form of *Pirani gauge* is shown in Figure 15.9(a). This is similar to a thermocouple gauge but has a heated element that consists of four coiled tungsten wires connected in parallel. Two identical tubes are normally used, connected in a bridge circuit as shown in Figure 15.9(b), with one containing the gas at unknown pressure and the other evacuated to a very low pressure. Current is passed through the tungsten element, which attains a certain temperature according to the thermal conductivity of the gas. The resistance of the element changes with temperature and causes an imbalance of the measurement bridge. Thus, the Pirani gauge avoids the use

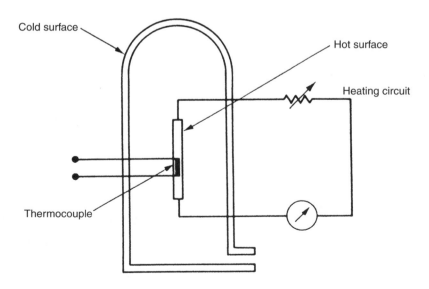

Fig. 15.8 Thermocouple gauge.

314 Pressure measurement

of a thermocouple to measure temperature (as in the thermocouple gauge) by effectively using a resistance thermometer as the heated element. Such gauges cover the pressure range 10^{-5} mbar to 1 mbar.

The *thermistor gauge* operates on identical principles to the Pirani gauge but uses semiconductor materials for the heated elements instead of metals. The normal pressure range covered is 10^{-4} mbar to 1 mbar.

Figure 15.10(a) shows the general form of a *McLeod gauge*, in which low-pressure fluid is compressed to a higher pressure that is then read by manometer techniques. In

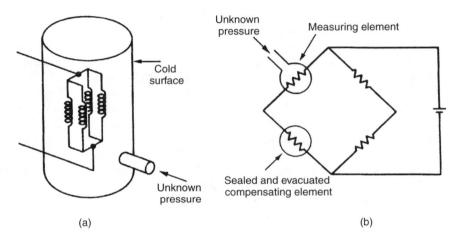

Fig. 15.9 (a) Pirani gauge; (b) Wheatstone bridge circuit to measure output.

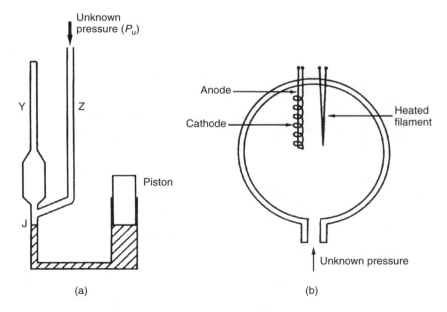

Fig. 15.10 Other low-pressure gauges: (a) McLeod gauge; (b) ionization gauge.

essence, the gauge can be visualized as a U-tube manometer that is sealed at one end, and where the bottom of the U can be blocked at will. To operate the gauge, the piston is first withdrawn. This causes the level of mercury in the lower part of the gauge to fall below the level of the junction J between the two tubes marked Y and Z in the gauge. Fluid at unknown pressure P_u is then introduced via the tube marked Z, from where it also flows into the tube of cross-sectional area A marked Y. Next, the piston is pushed in, moving the mercury level up to block the junction J. At the stage where J is just blocked, the fluid in tube Y is at pressure P_u and is contained in a known volume V_u. Further movement of the piston compresses the fluid in tube Y and this process continues until the mercury level in tube Z reaches a zero mark. Measurement of the height (h) above the mercury column in tube Y then allows calculation of the compressed volume of the fluid V_c as $V_c = hA$.
Then, by Boyle's law:

$$P_u V_u = P_c V_c$$

Also, applying the normal manometer equation:

$$P_c = P_u + h\rho g$$

where ρ is the mass density of mercury, the pressure P_u can be calculated as:

$$P_u = \frac{Ah^2 \rho g}{V_u - Ah} \qquad (15.1)$$

The compressed volume V_c is often very much smaller than the original volume, in which case equation (15.1) approximates to:

$$P_u = \frac{Ah^2 \rho g}{V_u} \quad \text{for} \quad Ah \ll V_u \qquad (15.2)$$

Although the smallest inaccuracy achievable with McLeod gauges is $\pm 1\%$, this is still better than that which is achievable with most other gauges that are available for measuring pressures in this range. Therefore, the McLeod gauge is often used as a standard against which other gauges are calibrated. The minimum pressure normally measurable is 10^{-4} bar, although lower pressures can be measured if pressure-dividing techniques are applied.

The *ionization gauge* is a special type of instrument used for measuring very low pressures in the range 10^{-13} to 10^{-3} bar. Gas of unknown pressure is introduced into a glass vessel containing free electrons discharged from a heated filament, as shown in Figure 15.10(b). Gas pressure is determined by measuring the current flowing between an anode and cathode within the vessel. This current is proportional to the number of ions per unit volume, which in turn is proportional to the gas pressure. Ionization gauges are normally only used in laboratory conditions.

15.10 High-pressure measurement (greater than 7000 bar)

Measurement of pressures above 7000 bar is normally carried out electrically by monitoring the change of resistance of wires of special materials. Materials having

316 Pressure measurement

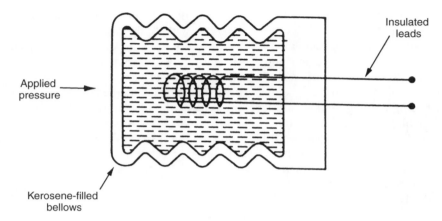

Fig. 15.11 High-pressure measurement—wire coil in bellows.

resistance-pressure characteristics that are suitably linear and sensitive include manganin and gold–chromium alloys. A coil of such wire is enclosed in a sealed, kerosene filled, flexible bellows, as shown in Figure 15.11. The unknown pressure is applied to one end of the bellows, which transmits the pressure to the coil. The magnitude of the applied pressure is then determined by measuring the coil resistance. Pressures up to 30 000 bar can be measured by devices like the manganin-wire pressure sensor, with a typical inaccuracy of ±0.5%.

15.11 Intelligent pressure transducers

Adding microprocessor power to pressure transducers brings about substantial improvements in their characteristics. Measurement sensitivity improvement, extended measurement range, compensation for hysteresis and other non-linearities, and correction for ambient temperature and pressure changes are just some of the facilities offered by intelligent pressure transducers. For example, inaccuracy figures as low as ±0.1% can be achieved with silicon piezoresistive-bridge devices.

Inclusion of microprocessors has also enabled the use of novel techniques of displacement measurement, for example the optical method of displacement measurement shown in Figure 15.12. In this, the motion is transmitted to a vane that progressively shades one of two monolithic photodiodes that are exposed to infrared radiation. The second photodiode acts as a reference, enabling the microprocessor to compute a ratio signal that is linearized and is available as either an analogue or digital measurement of pressure. The typical measurement inaccuracy is ±0.1%. Versions of both diaphragms and Bourdon tubes that use this technique are available.

15.12 Selection of pressure sensors

Choice between the various types of instrument available for measuring mid-range pressures (1.013–7000 bar) is usually strongly influenced by the intended application.

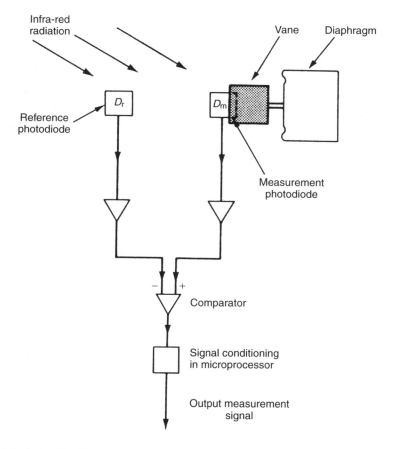

Fig. 15.12 Example of intelligent pressure-measuring instrument.

Manometers are commonly used when just a visual indication of pressure level is required, and deadweight gauges, because of their superior accuracy, are used in calibration procedures of other pressure-measuring devices. When an electrical form of output is required, the choice is usually either one of the several types of diaphragm sensor (strain gauge, capacitive or fibre optic) or, less commonly, a Bourdon tube. Bellows-type instruments are also sometimes used for this purpose, but much less frequently. If very high measurement accuracy is required, the resonant-wire device is a popular choice.

In the case of pressure measurement in the vacuum range (less than atmospheric pressure, i.e. below 1.013 bar), adaptations of most of the types of pressure transducer described earlier can be used. Special forms of Bourdon tubes measure pressures down to 10 mbar, manometers and bellows-type instruments measure pressures down to 0.1 mbar, and diaphragms can be designed to measure pressures down to 0.001 mbar. However, a number of more specialized instruments have also been developed to measure vacuum pressures, as discussed in section 15.9. These generally give better measurement accuracy and sensitivity compared with instruments that are primarily designed for measuring mid-range pressures. This improved accuracy is particularly evident at low pressures. Therefore, only the special instruments described in section 15.9 are used to measure pressures below 10^{-4} mbar.

At high pressures (>7000 bar), the only devices in common use are the manganin-wire sensor and similar devices based on alternative alloys to manganin.

For differential pressure measurement, diaphragm-type sensors are the preferred option, with double-bellows sensors being used occasionally. Manometers are also sometimes used to give visual indication of differential pressure values (especially in liquid flow-rate indicators). These are passive instruments that have the advantage of not needing a power supply.

16

Flow measurement

The rate at which fluid flows through a closed pipe can be quantified by either measuring the mass flow rate or measuring the volume flow rate. Of these alternatives, mass flow measurement is more accurate, since mass, unlike volume, is invariant. In the case of the flow of solids, the choice is simpler, since only mass flow measurement is appropriate.

16.1 Mass flow rate

The method used to measure mass flow rate is largely determined by whether the measured quantity is in a solid, liquid or gaseous state. The main techniques available are summarized below. A more comprehensive discussion can be found in Medlock (1990).

16.1.1 Conveyor-based methods

These methods are concerned with measurement of the flow of solids that are in the form of small particles. Such particles are usually produced by crushing or grinding procedures in process industries, and the particles are usually transported by some form of conveyor. This mode of transport allows the mass flow rate to be calculated in terms of the mass of material on a given length of conveyor multiplied by the speed of the conveyor. Figure 16.1 shows a typical measurement system. A load cell measures the mass M of material distributed over a length L of the conveyor. If the conveyor velocity is v, the mass flow rate, Q, is given by:

$$Q = Mv/L$$

As an alternative to weighing the flowing material, a *nuclear mass-flow sensor* can be used, in which a gamma-ray source is directed at the material being transported along the conveyor. The material absorbs some radiation, and the amount of radiation received by a detector on the other side of the material indicates the amount of material on the conveyor. This technique has obvious safety concerns, and is therefore subject to licensing and strict regulation.

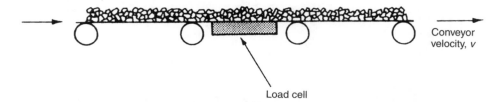

Fig. 16.1 Conveyor-based mass flow rate measurement.

16.1.2 Coriolis flowmeter

The Coriolis flowmeter is primarily used to measure the mass flow rate of liquids, although it has also been successfully used in some gas-flow measurement applications. The flowmeter consists of either a pair of parallel vibrating tubes or else a single vibrating tube that is formed into a configuration that has two parallel sections.

The two vibrating tubes (or the two parallel sections of a single tube) deflect according to the mass flow rate of the measured fluid that is flowing inside. Tubes are made of various materials, of which stainless steel is the most common. They are also manufactured in different shapes such as B-shaped, D-shaped, U-shaped, triangular-shaped, helix-shaped and straight. These alternative shapes are sketched in Figure 16.2(a) and a U-shaped tube is shown in more detail in Figure 16.2(b). The tubes are anchored at two points. An electromechanical drive unit, positioned midway between the two anchors, excites vibrations in each tube at the tube resonant frequency. The vibrations in the two tubes, or the two parallel sections of a single tube, are 180 degrees out of phase. The vibratory motion of each tube causes forces on the particles in the flowing fluid. These forces induce motion of the fluid particles in a direction that is orthogonal to the direction of flow, and this produces a Coriolis force. This Coriolis force causes a deflection of the tubes that is superimposed on top of the vibratory motion. The net deflection of one tube relative to the other is given by $d = kfR$, where k is a constant, f is the frequency of the tube vibration and R is the mass flow rate of the fluid inside the tube. This deflection is measured by a suitable sensor. A full account of the theory of operation can be found in Figliola (1995).

Coriolis meters give excellent accuracy, with measurement uncertainties of ±0.2% being typical. They also have low maintenance requirements. However, apart from being expensive (typical cost is £4000), they suffer from a number of operational problems. Failure may occur after a period of use because of mechanical fatigue in the tubes. Tubes are also subject to both corrosion caused by chemical interaction with the measured fluid and abrasion caused by particles within the fluid. Diversion of the flowing fluid around the flowmeter causes it to suffer a significant pressure drop, though this is much less evident in straight tube designs.

16.1.3 Thermal mass flow measurement

Thermal mass flowmeters are primarily used to measure the flow rate of gases. The principle of operation is to direct the flowing material past a heated element. The mass flow rate is inferred in one of two ways, (a) by measuring the temperature rise in the

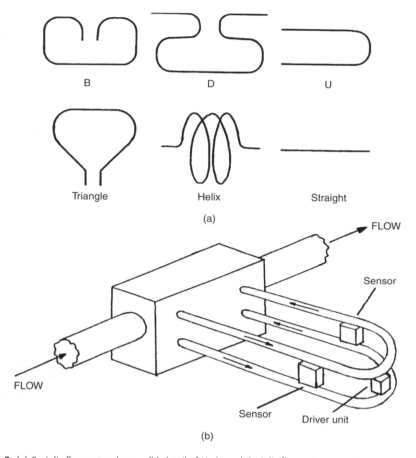

Fig. 16.2 (a) Coriolis flowmeter shapes; (b) detail of U-shaped Coriolis flowmeter.

flowing material or (b) by measuring the heater power required to achieve a constant set temperature in the flowing material. Typical measurement uncertainty is ±2%.

16.1.4 Joint measurement of volume flow rate and fluid density

Before the advent of the Coriolis meter, the usual way of measuring mass flow rate was to compute this from separate, simultaneous measurements of the volume flow rate and the fluid density. In many circumstances, this is still the cheapest option, although measurement accuracy is substantially inferior to that provided by a Coriolis meter.

16.2 Volume flow rate

Volume flow rate is an appropriate way of quantifying the flow of all materials that are in a gaseous, liquid or semi-liquid slurry form (where solid particles are suspended in a liquid host), although measurement accuracy is inferior to mass flow measurement as noted earlier. Materials in these forms are carried in pipes, and various instruments can be used to measure the volume flow rate as described below.

16.2.1 Differential pressure (obstruction-type) meters

Differential pressure meters involve the insertion of some device into a fluid-carrying pipe that causes an obstruction and creates a pressure difference on either side of the device. Such meters are sometimes known as obstruction-type meters or flow-restriction meters. Devices used to obstruct the flow include the *orifice plate*, the *Venturi tube*, the *flow nozzle* and the *Dall flow tube*, as illustrated in Figure 16.3. When such a restriction is placed in a pipe, the velocity of the fluid through the restriction increases and the pressure decreases. The volume flow rate is then proportional to the square root of the pressure difference across the obstruction. The manner in which this pressure difference is measured is important. Measuring the two pressures with different instruments and calculating the difference between the two measurements is not satisfactory because of the large measurement error which can arise when the pressure difference is small, as explained in Chapter 3. Therefore, the normal procedure is to use a differential pressure transducer, which is commonly a diaphragm type.

The *Pitot static tube* is a further device that measures flow by creating a pressure difference within a fluid-carrying pipe. However, in this case, there is negligible obstruction of flow in the pipe. The Pitot tube is a very thin tube that obstructs only a small part of the flowing fluid and thus measures flow at a single point across the cross-section of the pipe. This measurement only equates to average flow velocity in the pipe for the case of uniform flow. The *Annubar* is a type of multi-port Pitot tube that does measure the average flow across the cross-section of the pipe by forming the mean value of several local flow measurements across the cross-section of the pipe.

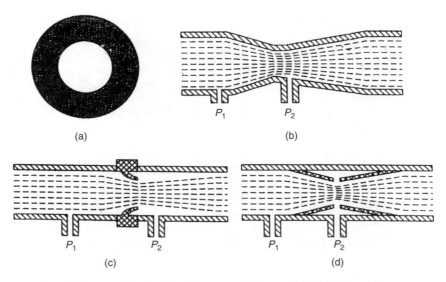

Fig. 16.3 Obstuction devices: (a) orifice plate; (b) venturi; (c) flow nozzle; (d) Dall flow tube.

Measurement and Instrumentation Principles 323

All applications of this method of flow measurement assume that flow conditions upstream of the obstruction device are in steady state, and a certain minimum length of straight run of pipe ahead of the flow measurement point is specified to ensure this. The minimum lengths required for various pipe diameters are specified in British Standards tables (and also in alternative but equivalent national standards used in other countries), but a useful rule of thumb widely used in the process industries is to specify a length of ten times the pipe diameter. If physical restrictions make this impossible to achieve, special flow smoothing vanes can be inserted immediately ahead of the measurement point.

Flow-restriction type instruments are popular because they have no moving parts and are therefore robust, reliable and easy to maintain. One disadvantage of this method is that the obstruction causes a permanent loss of pressure in the flowing fluid. The magnitude and hence importance of this loss depends on the type of obstruction element used, but where the pressure loss is large, it is sometimes necessary to recover the lost pressure by an auxiliary pump further down the flow line. This class of device is not normally suitable for measuring the flow of slurries as the tappings into the pipe to measure the differential pressure are prone to blockage, although the Venturi tube can be used to measure the flow of dilute slurries.

Figure 16.4 illustrates approximately the way in which the flow pattern is interrupted when an orifice plate is inserted into a pipe. The other obstruction devices also have a similar effect to this. Of particular interest is the fact that the minimum cross-sectional area of flow occurs not within the obstruction but at a point downstream of there. Knowledge of the pattern of pressure variation along the pipe, as shown in Figure 16.5, is also of importance in using this technique of volume flow rate measurement. This shows that the point of minimum pressure coincides with the point of minimum cross-section flow, a little way downstream of the obstruction. Figure 16.5 also shows that there is a small rise in pressure immediately before the obstruction. It is therefore important not only to position the instrument measuring P_2 exactly at the point of minimum pressure, but also to measure the pressure P_1 at a point upstream of the point where the pressure starts to rise before the obstruction.

In the absence of any heat transfer mechanisms, and assuming frictionless flow of an incompressible fluid through the pipe, the theoretical volume flow rate of the fluid,

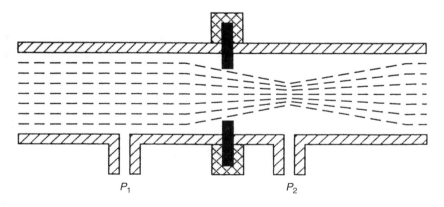

Fig. 16.4 Profile of flow across orifice plate.

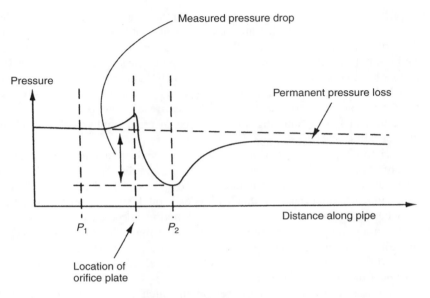

Fig. 16.5 Pattern of pressure variation either side of orifice plate.

Q, is given by:

$$Q = \left[\frac{A_2}{\sqrt{1 - (A_2/A_1)^2}}\right] \left[\sqrt{\frac{2(P_1 - P_2)}{\rho}}\right] \quad (16.1)$$

where A_1 and P_1 are the cross-sectional area and pressure of the fluid flow before the obstruction, A_2 and P_2 are the cross-sectional area and pressure of the fluid flow at the narrowest point of the flow beyond the obstruction, and ρ is the fluid density.

Equation (16.1) is never applicable in practice for several reasons. Firstly, frictionless flow is never achieved. However, in the case of turbulent flow through smooth pipes, friction is low and it can be adequately accounted for by a variable called the Reynolds number, which is a measurable function of the flow velocity and the viscous friction. The other reasons for the nonapplicability of equation (16.1) are that the initial cross-sectional area of the fluid flow is less than the diameter of the pipe carrying it and that the minimum cross-sectional area of the fluid is less than the diameter of the obstruction. Therefore, neither A_1 nor A_2 can be measured. These problems are taken account of by modifying equation (16.1) to the following:

$$Q = \left[\frac{C_D A'_2}{\sqrt{1 - (A'_2/A'_1)^2}}\right] \left[\sqrt{\frac{2(P_1 - P_2)}{\rho}}\right] \quad (16.2)$$

where A'_1 and A'_2 are the pipe diameters before and at the obstruction and C_D is a constant, known as the discharge coefficient, which accounts for the Reynolds number and the difference between the pipe and flow diameters.

Before equation (16.2) can be evaluated, the discharge coefficient must be calculated. As this varies between each measurement situation, it would appear at first sight that the discharge coefficient must be determined by practical experimentation in each case. However, provided that certain conditions are met, standard tables can be used to obtain the value of the discharge coefficient appropriate to the pipe diameter and fluid involved.

One particular problem with all flow restriction devices is that the pressure drop $(P_1 - P_2)$ varies as the square of the flow rate Q according to equation (16.2). The difficulty of measuring small pressure differences accurately has already been noted earlier. In consequence, the technique is only suitable for measuring flow rates that are between 30% and 100% of the maximum flow rate that a given device can handle. This means that alternative flow measurement techniques have to be used in applications where the flow rate can vary over a large range that can drop to below 30% of the maximum rate.

Orifice plate

The orifice plate is a metal disc with a concentric hole in it, which is inserted into the pipe carrying the flowing fluid. Orifice plates are simple, cheap and available in a wide range of sizes. In consequence, they account for almost 50% of the instruments used in industry for measuring volume flow rate. One limitation of the orifice plate is that its inaccuracy is typically at least ±2% and may approach ±5%. Also, the permanent pressure loss caused in the measured fluid flow is between 50% and 90% of the magnitude of the pressure difference $(P_1 - P_2)$. Other problems with the orifice plate are a gradual change in the discharge coefficient over a period of time as the sharp edges of the hole wear away, and a tendency for any particles in the flowing fluid to stick behind the hole and thereby gradually reduce its diameter as the particles build up. The latter problem can be minimized by using an orifice plate with an eccentric hole. If this hole is close to the bottom of the pipe, solids in the flowing fluid tend to be swept through, and build-up of particles behind the plate is minimized. A very similar problem arises if there are any bubbles of vapour or gas in the flowing fluid when liquid flow is involved. These also tend to build up behind an orifice plate and distort the pattern of flow. This difficulty can be avoided by mounting the orifice plate in a vertical run of pipe.

Venturis and similar devices

A number of obstruction devices are available that are specially designed to minimize the pressure loss in the measured fluid. These have various names such as Venturi, flow nozzle and Dall flow tube. They are all much more expensive than an orifice plate but have better performance. The smooth internal shape means that they are not prone to solid particles or bubbles of gas sticking in the obstruction, as is likely to happen in an orifice plate. The smooth shape also means that they suffer much less wear, and consequently have a longer life than orifice plates. They also require less maintenance and give greater measurement accuracy.

The **Venturi** has a precision-engineered tube of a special shape. This offers measurement uncertainty of only ±1%. However, the complex machining required to manufacture it means that it is the most expensive of all the obstruction devices discussed. Permanent pressure loss in the measured system is 10–15% of the pressure difference $(P_1 - P_2)$ across it.

The **Dall flow tube** consists of two conical reducers inserted into the fluid-carrying pipe. It has a very similar internal shape to the Venturi, except that it lacks a throat. This construction is much easier to manufacture and this gives the Dall flow tube an advantage in cost over the Venturi, although the typical measurement inaccuracy is a little higher ($\pm 1.5\%$). Another advantage of the Dall flow tube is its shorter length, which makes the engineering task of inserting it into the flow line easier. The Dall tube has one further operational advantage, in that the permanent pressure loss imposed on the measured system is only about 5% of the measured pressure difference ($P_1 - P_2$).

The **flow nozzle** is of simpler construction still, and is therefore cheaper than either a Venturi or a Dall flow tube, but the pressure loss imposed on the flowing fluid is 30–50% of the measured pressure difference ($P_1 - P_2$).

Pitot static tube

The Pitot static tube is mainly used for making temporary measurements of flow, although it is also used in some instances for permanent flow monitoring. It measures the local velocity of flow at a particular point within a pipe rather than the average flow velocity as measured by other types of flowmeter. This may be very useful where there is a requirement to measure local flow rates across the cross-section of a pipe in the case of non-uniform flow. Multiple Pitot tubes are normally used to do this.

The instrument depends on the principle that a tube placed with its open end in a stream of fluid, as shown in Figure 16.6, will bring to rest that part of the fluid which impinges on it, and the loss of kinetic energy will be converted to a measurable increase in pressure inside the tube. This pressure (P_1), as well as the static pressure of the undisturbed free stream of flow (P_2), is measured. The flow velocity can then be calculated from the formula:

$$v = C\sqrt{2g(P_1 - P_2)}$$

The constant C, known as the Pitot tube coefficient, is a factor which corrects for the fact that not all fluid incident on the end of the tube will be brought to rest: a proportion will slip around it according to the design of the tube. Having calculated v, the volume flow rate can then be calculated by multiplying v by the cross-sectional area of the flow pipe, A.

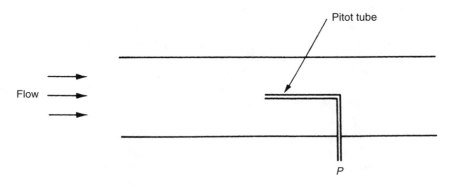

Fig. 16.6 Pitot tube.

Pitot tubes have the advantage that they cause negligible pressure loss in the flow. They are also cheap, and the installation procedure consists of the very simple process of pushing them down a small hole drilled in the flow-carrying pipe. Their main failing is that the measurement inaccuracy is typically about ±5%, although more expensive versions can reduce inaccuracy down to ±1%. The *annubar* is a development of the Pitot tube that has multiple sensing ports distributed across the cross-section of the pipe. It thus provides only an approximate measurement of the mean flow rate across the pipe.

16.2.2 Variable area flowmeters (Rotameters)

In the variable area flowmeter (which is also sometimes known as a Rotameter), the differential pressure across a variable aperture is used to adjust the area of the aperture. The aperture area is then a measure of the flow rate. The instrument is reliable and cheap and used extensively throughout industry, accounting for about 20% of all flowmeters sold. Normally, this type of instrument only gives a visual indication of flow rate, and so it is of no use in automatic control schemes. However, special versions of variable area flowmeters are now available that incorporate fibre optics. In these, a row of fibres detects the position of the float by sensing the reflection of light from it, and an electrical signal output can be derived from this.

In its simplest form, shown in Figure 16.7, the instrument consists of a tapered glass tube containing a float which takes up a stable position where its submerged weight is balanced by the upthrust due to the differential pressure across it. The position of the float is a measure of the effective annular area of the flow passage and hence of

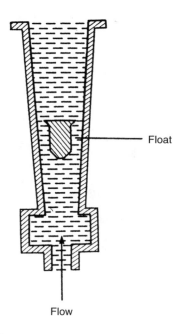

Fig. 16.7 Variable area flowmeter.

328 Flow measurement

the flow rate. The inaccuracy of the cheapest instruments is typically ±5%, but more expensive versions offer measurement inaccuracies as low as ±0.5%.

16.2.3 Positive displacement flowmeters

Positive displacement flowmeters account for nearly 10% of the total number of flowmeters used in industry and are used in large numbers for metering domestic gas and water consumption. The cheapest instruments have a typical inaccuracy of about ±2%, but the inaccuracy in more expensive ones can be as low as ±0.5%. These higher quality instruments are used extensively within the oil industry, as such applications can justify the high cost of such instruments.

All positive displacement meters operate by using mechanical divisions to displace discrete volumes of fluid successively. Whilst this principle of operation is common, many different mechanical arrangements exist for putting the principle into practice. However, all versions of positive displacement meter are low friction, low maintenance and long-life devices, although they do impose a small permanent pressure loss on the flowing fluid. Low friction is especially important when measuring gas flows, and meters with special mechanical arrangements to satisfy this requirement have been developed.

The *rotary piston meter* is a common type of positive displacement meter, and the principles of operation of this are shown in Figure 16.8. It consists of a slotted cylindrical piston moving inside a cylindrical working chamber that has an inlet port and an outlet port. The piston moves round the chamber such that its outer surface maintains contact with the inner surface of the chamber, and, as this happens, the piston slot slides up and down a fixed division plate in the chamber. At the start of each piston motion cycle, liquid is admitted to volume B from the inlet port. The fluid

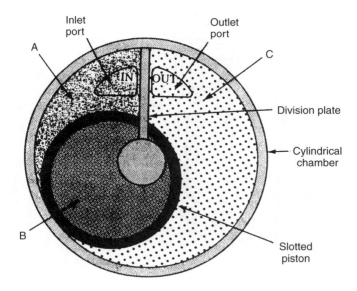

Fig. 16.8 Rotaty piston form of positive displacement flowmeter.

pressure causes the piston to start to rotate around the chamber, and, as this happens, liquid in volume C starts to flow out of the outlet port, and also liquid starts to flow from the inlet port into volume A. As the piston rotates further, volume B becomes shut off from the inlet port, whilst liquid continues to be admitted into A and pushed out of C. When the piston reaches the endpoint of its motion cycle, the outlet port is opened to volume B, and the liquid which has been transported round inside the piston is expelled. After this, the piston pivots about the contact point between the top of its slot and the division plate, and volume A effectively becomes volume C ready for the start of the next motion cycle. A peg on top of the piston causes a reciprocating motion of a lever attached to it. This is made to operate a counter, and the flow rate is therefore determined from the count in unit time multiplied by the quantity (fixed) of liquid transferred between the inlet and outlet ports for each motion cycle.

16.2.4 Turbine meters

A turbine flowmeter consists of a multi-bladed wheel mounted in a pipe along an axis parallel to the direction of fluid flow in the pipe, as shown in Figure 16.9. The flow of fluid past the wheel causes it to rotate at a rate that is proportional to the volume flow rate of the fluid. This rate of rotation has traditionally been measured by constructing the flowmeter such that it behaves as a variable reluctance tachogenerator. This is achieved by fabricating the turbine blades from a ferromagnetic material and placing a permanent magnet and coil inside the meter housing. A voltage pulse is induced in the coil as each blade on the turbine wheel moves past it, and if these pulses are measured by a pulse counter, the pulse frequency and hence flow rate can be deduced. In recent instruments, fibre optics are also now sometimes used to count the rotations by detecting reflections off the tip of the turbine blades.

Provided that the turbine wheel is mounted in low friction bearings, measurement inaccuracy can be as low as ±0.2%. However, turbine flowmeters are less rugged and

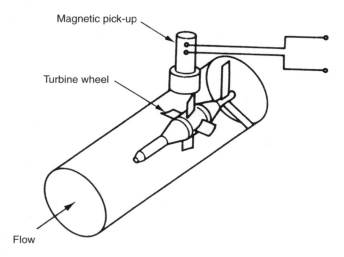

Fig. 16.9 Turbine flowmeter.

reliable than flow-restriction type instruments, and are badly affected by any particulate matter in the flowing fluid. Bearing wear is a particular problem and they also impose a permanent pressure loss on the measured system. Turbine meters are particularly prone to large errors when there is any significant second phase in the fluid measured. For instance, using a turbine meter calibrated on pure liquid to measure a liquid containing 5% air produces a 50% measurement error. As an important application of the turbine meter is in the petrochemical industries, where gas/oil mixtures are common, special procedures are being developed to avoid such large measurement errors. The most promising approach is to homogenize the two gas/oil phases prior to flow measurement (King, 1988).

Turbine meters have a similar cost and market share to positive displacement meters, and compete for many applications, particularly in the oil industry. Turbine meters are smaller and lighter than the latter and are preferred for low-viscosity, high-flow measurements. However, positive-displacement meters are superior in conditions of high viscosity and low flow rate.

16.2.5 Electromagnetic flowmeters

Electromagnetic flowmeters are limited to measuring the volume flow rate of electrically conductive fluids. The typical measurement inaccuracy of around $\pm 1\%$ is acceptable in many applications, but the instrument is expensive both in terms of the initial purchase cost and also in running costs, mainly due to its electricity consumption. A further reason for high cost is the need for careful calibration of each instrument individually during manufacture, as there is considerable variation in the properties of the magnetic materials used.

The instrument, shown in Figure 16.10, consists of a stainless steel cylindrical tube, fitted with an insulating liner, which carries the measured fluid. Typical lining materials used are Neoprene, polytetrafluoroethylene (PTFE) and polyurethane. A magnetic field is created in the tube by placing mains-energized field coils either side of it, and the voltage induced in the fluid is measured by two electrodes inserted into opposite sides of the tube. The ends of these electrodes are usually flush with the inner surface of the cylinder. The electrodes are constructed from a material which is unaffected by most types of flowing fluid, such as stainless steel, platinum–iridium alloys, Hastelloy, titanium and tantalum. In the case of the rarer metals in this list, the electrodes account for a significant part of the total instrument cost.

By Faraday's law of electromagnetic induction, the voltage, E, induced across a length, L, of the flowing fluid moving at velocity, v, in a magnetic field of flux density, B, is given by:

$$E = BLv \qquad (16.3)$$

L is the distance between the electrodes, which is the diameter of the tube, and B is a known constant. Hence, measurement of the voltage E induced across the electrodes allows the flow velocity v to be calculated from equation (16.3). Having thus calculated v, it is a simple matter to multiply v by the cross-sectional area of the tube to obtain a value for the volume flow rate. The typical voltage signal measured across the electrodes is 1 mV when the fluid flow rate is 1 m/s.

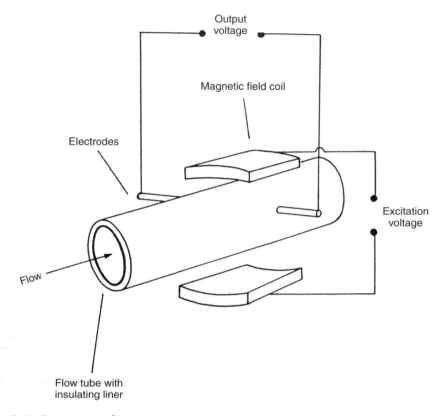

Fig. 16.10 Electromagnetic flowmeter.

The internal diameter of magnetic flowmeters is normally the same as that of the rest of the flow-carrying pipework in the system. Therefore, there is no obstruction to the fluid flow and consequently no pressure loss associated with measurement. Like other forms of flowmeter, the magnetic type requires a minimum length of straight pipework immediately prior to the point of flow measurement in order to guarantee the accuracy of measurement, although a length equal to five pipe diameters is usually sufficient.

Whilst the flowing fluid must be electrically conductive, the method is of use in many applications and is particularly useful for measuring the flow of slurries in which the liquid phase is electrically conductive. Corrosive fluids can be handled providing a suitable lining material is used. At the present time, magnetic flowmeters account for about 15% of the new flowmeters sold and this total is slowly growing. One operational problem is that the insulating lining is subject to damage when abrasive fluids are being handled, and this can give the instrument a limited life.

Current new developments in electromagnetic flowmeters are producing physically smaller instruments and employing better coil designs which reduce electricity consumption and make battery-powered versions feasible (these are now commercially available). Also, whereas conventional electromagnetic flowmeters require a minimum fluid conductivity of 10 μmho/cm³, new versions can cope with fluid conductivities as low as 1 μmho/cm³.

16.2.6 Vortex-shedding flowmeters

The vortex-shedding flowmeter is a relatively new type of instrument which is rapidly gaining in popularity and is being used as an alternative to traditional differential pressure meters in more and more applications. The operating principle of the instrument is based on the natural phenomenon of vortex shedding, created by placing an unstreamlined obstacle (known as a bluff body) in a fluid-carrying pipe, as indicated in Figure 16.11. When fluid flows past the obstacle, boundary layers of viscous, slow-moving fluid are formed along the outer surface. Because the obstacle is not streamlined, the flow cannot follow the contours of the body on the downstream side, and the separate layers become detached and roll into eddies or vortices in the low-pressure region behind the obstacle. The shedding frequency of these alternately shed vortices is proportional to the fluid velocity past the body. Various thermal, magnetic, ultrasonic and capacitive vortex detection techniques are employed in different instruments.

Such instruments have no moving parts, operate over a wide flow range, have a low power consumption, require little maintenance and have a similar cost to measurement using an orifice plate. They can measure both liquid and gas flows and a common inaccuracy figure quoted is ±1% of full-scale reading, though this can be seriously downgraded in the presence of flow disturbances upstream of the measurement point and a straight run of pipe before the measurement point of 50 pipe diameters is recommended. Another problem with the instrument is its susceptibility to pipe vibrations, although new designs are becoming available which have a better immunity to such vibrations.

16.2.7 Ultrasonic flowmeters

The ultrasonic technique of volume flow rate measurement is, like the magnetic flowmeter, a non-invasive method. It is not restricted to conductive fluids, however, and

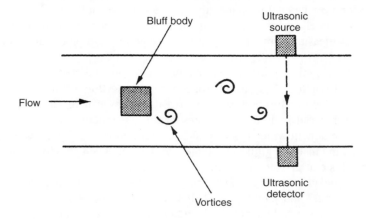

Fig. 16.11 Vortex-shedding flowmeter.

is particularly useful for measuring the flow of corrosive fluids and slurries. Besides its high reliability and low maintenance requirements, a further advantage of an ultrasonic flowmeter over a magnetic flowmeter is that the instrument can be clamped externally onto existing pipework rather than being inserted as an integral part of the flow line. As the procedure of breaking into a pipeline to insert a flowmeter can be as expensive as the cost of the flowmeter itself, the ultrasonic flowmeter has enormous cost advantages. Its clamp-on mode of operation has significant safety advantages in avoiding the possibility of personnel installing flowmeters coming into contact with hazardous fluids such as poisonous, radioactive, flammable or explosive ones. Also, any contamination of the fluid being measured (e.g. food substances and drugs) is avoided. Ultrasonic meters are still less common than differential pressure or electromagnetic flowmeters, though usage continues to expand year by year.

Two different types of ultrasonic flowmeter exist which employ distinct technologies, one based on Doppler shift and the other on transit time. In the past, the existence of these alternative technologies has not always been readily understood, and has resulted in ultrasonic technology being rejected entirely when one of these two forms has been found to be unsatisfactory in a particular application. This is unfortunate, because the two technologies have distinct characteristics and areas of application, and many situations exist where one form is very suitable and the other not suitable. To reject both, having only tried out one, is therefore a serious mistake.

Particular care has to be taken to ensure a stable flow profile in ultrasonic flowmeter applications. It is usual to increase the normal specification of the minimum length of straight pipe-run prior to the point of measurement, expressed as a number of pipe diameters, from a figure of 10 up to 20 or in some cases even 50 diameters. Analysis of the reasons for poor performance in many instances of ultrasonic flowmeter application has shown failure to meet this stable flow-profile requirement to be a significant factor.

Doppler shift ultrasonic flowmeter

The principle of operation of the Doppler shift flowmeter is shown in Figure 16.12. A fundamental requirement of these instruments is the presence of scattering elements within the flowing fluid, which deflect the ultrasonic energy output from the transmitter such that it enters the receiver. These can be provided by either solid particles, gas bubbles or eddies in the flowing fluid. The scattering elements cause a frequency shift between the transmitted and reflected ultrasonic energy, and measurement of this shift enables the fluid velocity to be inferred.

The instrument consists essentially of an ultrasonic transmitter–receiver pair clamped onto the outside wall of a fluid-carrying vessel. Ultrasonic energy consists of a train of short bursts of sinusoidal waveforms at a frequency between 0.5 MHz and 20 MHz. This frequency range is described as ultrasonic because it is outside the range of human hearing. The flow velocity, v, is given by:

$$v = \frac{c(f_t - f_r)}{2 f_t \cos(\theta)} \tag{16.4}$$

where f_t and f_r are the frequencies of the transmitted and received ultrasonic waves respectively, c is the velocity of sound in the fluid being measured, and θ is the angle that the incident and reflected energy waves make with the axis of flow in the pipe.

334 Flow measurement

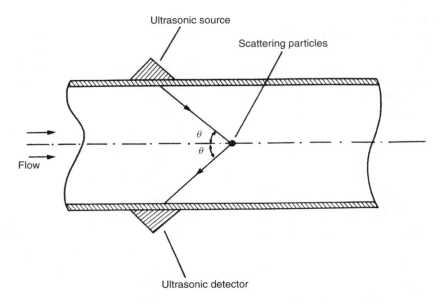

Fig. 16.12 Doppler shift ultrasonic flowmeter.

Volume flow rate is then readily calculated by multiplying the measured flow velocity by the cross-sectional area of the fluid-carrying pipe.

The electronics involved in Doppler-shift flowmeters is relatively simple and therefore cheap. Ultrasonic transmitters and receivers are also relatively inexpensive, being based on piezoelectric oscillator technology. As all of its components are cheap, the Doppler shift flowmeter itself is inexpensive. The measurement accuracy obtained depends on many factors such as the flow profile, the constancy of pipe-wall thickness, the number, size and spatial distribution of scatterers, and the accuracy with which the speed of sound in the fluid is known. Consequently, accurate measurement can only be achieved by the tedious procedure of carefully calibrating the instrument in each particular flow measurement application. Otherwise, measurement errors can approach $\pm 10\%$ of the reading, and for this reason Doppler shift flowmeters are often used merely as flow indicators, rather than for accurate quantification of the volume flow rate.

Versions are now available which avoid the problem of variable pipe thickness by being fitted inside the flow pipe, flush with its inner surface. A low inaccuracy level of $\pm 0.5\%$ is claimed for such devices. Other recent developments are the use of multiple-path ultrasonic flowmeters that use an array of ultrasonic elements to obtain an average velocity measurement that substantially reduces the error due to non-uniform flow profiles. There is a substantial cost penalty involved in this, however.

Transit-time ultrasonic flowmeter

The transit-time ultrasonic flowmeter is an instrument designed for measuring the volume flow rate in clean liquids or gases. It consists of a pair of ultrasonic transducers mounted along an axis aligned at an angle θ with respect to the fluid-flow axis, as shown in Figure 16.13. Each transducer consists of a transmitter–receiver pair, with the transmitter emitting ultrasonic energy which travels across to the receiver on the opposite

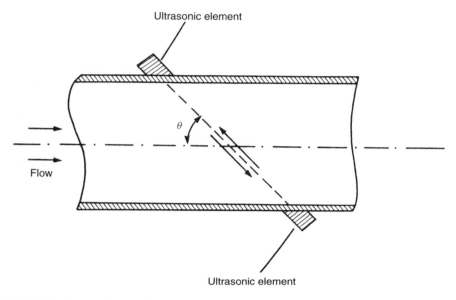

Fig. 16.13 Transit-time ultrasonic flowmeter.

side of the pipe. These ultrasonic elements are normally piezoelectric oscillators of the same type as used in Doppler shift flowmeters. Fluid flowing in the pipe causes a time difference between the transit times of the beams travelling upstream and downstream, and measurement of this difference allows the flow velocity to be calculated. The typical magnitude of this time difference is 100 ns in a total transit time of 100 µs, and high-precision electronics are therefore needed to measure it. There are three distinct ways of measuring the time shift. These are direct measurement, conversion to a phase change and conversion to a frequency change. The third of these options is particularly attractive, as it obviates the need to measure the speed of sound in the measured fluid as required by the first two methods. A scheme applying this third option is shown in Figure 16.14. This also multiplexes the transmitting and receiving functions, so that only one ultrasonic element is needed in each transducer. The forward and backward transit times across the pipe, T_f and T_b, are given by:

$$T_f = \frac{L}{c + v\cos(\theta)}; \quad T_b = \frac{L}{c - v\cos(\theta)}$$

where c is the velocity of sound in the fluid, v is the flow velocity, L is the distance between the ultrasonic transmitter and receiver, and θ is the angle of the ultrasonic beam with respect to the fluid flow axis.

The time difference δT is given by:

$$\delta T = T_b - T_f = \frac{2vL\cos(\theta)}{c^2 - v^2\cos^2(\theta)}$$

This requires knowledge of c before it can be solved. However, a solution can be found much more simply if the receipt of a pulse is used to trigger the transmission of the

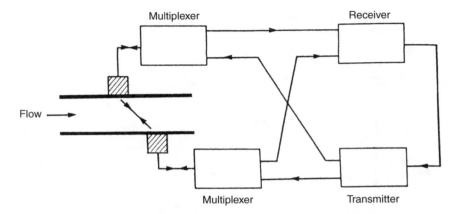

Fig. 16.14 Transit-time measurement system.

next ultrasonic energy pulse. Then, the frequencies of the forward and backward pulse trains are given by:

$$F_f = \frac{1}{T_f} = \frac{c - v\cos(\theta)}{L}; \quad F_b = \frac{1}{T_b} = \frac{c + v\cos(\theta)}{L}$$

If the two frequency signals are now multiplied together, the resulting beat frequency is given by:

$$\delta F = F_b - F_f = \frac{2v\cos(\theta)}{L}$$

c has now been eliminated and v can be calculated from a measurement of δF as:

$$v = \frac{L\delta F}{2\cos(\theta)}$$

This is often known as the *sing-around flowmeter*.

Transit-time flowmeters are of more general use than Doppler shift flowmeters, particularly where the pipe diameter involved is large and hence the transit time is consequently sufficiently large to be measured with reasonable accuracy. It is possible then to reduce the inaccuracy figure to ±0.5%. The instrument costs more than a Doppler shift flowmeter, however, because of the greater complexity of the electronics needed to make accurate transit-time measurements.

16.2.8 Other types of flowmeter for measuring volume flow rate

The **gate meter** consists of a spring-loaded, hinged flap mounted at right angles to the direction of fluid flow in the fluid-carrying pipe. The flap is connected to a pointer outside the pipe. The fluid flow deflects the flap and pointer and the flow rate is indicated by a graduated scale behind the pointer. The major difficulty with such devices is in preventing leaks at the hinge point. A variation on this principle is the *air-vane meter*, which measures deflection of the flap by a potentiometer inside the pipe. This is commonly used to measure airflow within automotive fuel-injection systems. Another similar device is the *target meter*. This consists of a circular disc-shaped flap in the pipe. Fluid flow rate is inferred from the force exerted on the disc measured by strain gauges bonded to it. This meter is very useful for measuring the flow of dilute slurries but it does not find wide application elsewhere as it has a relatively high cost. Measurement uncertainty in all of these types of meter varies between 1% and 5% according to cost and design of each instrument.

The **cross-correlation flowmeter** has not yet achieved widespread practical use in industry. Much development work is still going on, and it therefore mainly only exists in prototype form in research laboratories. However, it is included here because use is likely to become much more widespread in the future. The instrument requires some detectable random variable to be present in the flowing fluid. This can take forms such as velocity turbulence and temperature fluctuations. When such a stream of variables is detected by a sensor, the output signal generated consists of noise with a wide frequency spectrum.

Cross-correlation flowmeters use two such sensors placed a known distance apart in the fluid-carrying pipe and cross-correlation techniques are applied to the two output signals from these sensors. This procedure compares one signal with progressively time-shifted versions of the other signal until the best match is obtained between the two waveforms. If the distance between the sensors is divided by this time shift, a measurement of the flow velocity is obtained. A digital processor is an essential requirement to calculate the cross-correlation function, and therefore the instrument must be properly described as an intelligent one.

In practice, the existence of random disturbances in the flow is unreliable, and their detection is difficult. To answer this problem, ultrasonic cross-correlation flowmeters are under development. These use ultrasonic transducers to inject disturbances into the flow and also to detect the disturbances further downstream.

Further information about cross-correlation flowmeters can be found in Medlock (1985).

The **Laser Doppler flowmeter** gives direct measurements of flow velocity for liquids containing suspended particles flowing in a transparent pipe. Light from a laser is focused by an optical system to a point in the flow, with fibre-optic cables being commonly used to transmit the light. The movement of particles causes a Doppler shift of the scattered light and produces a signal in a photodetector that is related to the fluid velocity. A very wide range of flow velocities between 10 µm/s and 105 m/s can be measured by this technique.

Sufficient particles for satisfactory operation are normally present naturally in most liquid and gaseous fluids, and the introduction of artificial particles is rarely needed. The technique is advantageous in measuring flow velocity directly rather than inferring it from a pressure difference. It also causes no interruption in the flow and, as the instrument can be made very small, it can measure velocity in confined areas. One limitation is that it measures local flow velocity in the vicinity of the focal point of the light beam, which can lead to large errors in the estimation of mean volume flow rate if the flow profile is not uniform. However, this limitation is often used constructively in applications of the instrument where the flow profile across the cross-section of a pipe is determined by measuring the velocity at a succession of points.

Whilst the **Coriolis meter** is primarily intended to be a mass flow measuring instrument, it can also be used to measure volume flow rate when high measurement accuracy is required. However, its high cost means that alternative instruments are normally used for measuring volume flow rate.

16.3 Intelligent flowmeters

All the usual benefits associated with intelligent instruments are applicable to most types of flowmeter. Indeed, all types of mass flowmeter routinely have intelligence as an integral part of the instrument. For volume flow rate measurement, intelligent differential pressure measuring instruments can be used to good effect in conjunction with obstruction type flow transducers. One immediate benefit of this in the case of the commonest flow restriction device, the orifice plate, is to extend the lowest flow measurable with acceptable accuracy down to 20% of the maximum flow value. In positive displacement meters, intelligence allows compensation for thermal expansion of meter components and temperature-induced viscosity changes. Correction for variations in flow pressure is also provided for. Intelligent electromagnetic flowmeters are also available, and these have a self-diagnosis and self-adjustment capability. The usable instrument range is typically from 3% to 100% of the full-scale reading and the quoted maximum inaccuracy is $\pm 0.5\%$. It is also normal to include a non-volatile memory to protect constants used for correcting for modifying inputs, etc., against power supply failures. Intelligent turbine meters are able to detect their own bearing wear and also report deviations from initial calibration due to blade damage, etc. Some versions also have self-adjustment capability.

The trend is now moving towards total flow computers which can process inputs from almost any type of transducer. Such devices allow user input of parameters like specific gravity, fluid density, viscosity, pipe diameters, thermal expansion coefficients, discharge coefficients, etc. Auxiliary inputs from temperature transducers are also catered for. After processing the raw flow transducer output with this additional data, flow computers are able to produce measurements of flow to a very high degree of accuracy.

16.4 Choice between flowmeters for particular applications

The number of relevant factors to be considered when specifying a flowmeter for a particular application is very large. These include the temperature and pressure of the fluid, its density, viscosity, chemical properties and abrasiveness, whether it contains particles, whether it is a liquid or gas, etc. This narrows the field to a subset of instruments that are physically capable of making the measurement. Next, the required performance factors of accuracy, rangeability, acceptable pressure drop, output signal characteristics, reliability and service life must be considered. Accuracy requirements vary widely across different applications, with measurement uncertainty of $\pm 5\%$ being acceptable in some and less than $\pm 0.5\%$ being demanded in others.

Finally, the economic viability must be assessed and this must take account not only of purchase cost, but also of reliability, installation difficulties, maintenance requirements and service life.

Where only a visual indication of flow rate is needed, the variable-area meter is popular. Where a flow measurement in the form of an electrical signal is required, the choice of available instruments is very large. The orifice plate is used extremely commonly for such purposes and accounts for almost 50% of the instruments currently in use in industry. Other forms of differential pressure meter and electromagnetic flowmeters are used in significant numbers. Currently, there is a trend away from rotating devices such as turbine meters and positive displacement meters. At the same time, usage of ultrasonic and vortex meters is expanding. A survey of the current market share enjoyed by different types can be found in Control Engineering (1998).

References and further reading

Control Engineering (Editorial) (April 1998), pp. 119–128.

Figiola, R.S. and Beasley, D.E. (1995) *Theory and Design of Mechanical Measurements*, John Wiley.

Instrument Society of America (1988) *Flowmeters – a comprehensive survey and guide to selection*, Pittsburgh.

King, N.W. (1988) Multi-phase flow measurement at NEL, *Measurement and Control*, **21**, pp. 237–239.

Medlock, R.S. (1985) Cross-correlation flow measurement, *Measurement and Control*, **18**(8), pp. 293–298.

Medlock, R.S. and Furness, R.A. (1990) Mass flow measurement – a state of the art review, *Measurement and Control*, **23**(4), pp. 100–113.

17

Level measurement

A wide variety of instruments are available for measuring the level of liquids. Some of these can also be used to measure the levels of solids that are in the form of powders or small particles. In some applications, only a rough indication of level is needed, and simple devices such as dipsticks or float systems are adequate. However, in other cases where high accuracy is demanded, other types of instrument must be used. The sections below cover the various kinds of level-measuring device available.

17.1 Dipsticks

Dipsticks offer a simple means of measuring level approximately. The *ordinary dipstick* is the cheapest device available. This consists of a metal bar on which a scale is etched, as shown in Figure 17.1(a). The bar is fixed at a known position in the liquid-containing vessel. A level measurement is made by removing the instrument from the vessel and reading off how far up the scale the liquid has wetted. As a human operator is required to remove and read the dipstick, this method can only be used in relatively small and shallow vessels.

The *optical dipstick*, illustrated in Figure 17.1(b), is an alternative form that allows a reading to be obtained without removing the dipstick from the vessel, and so is applicable to larger, deeper tanks. Light from a source is reflected from a mirror, passes round the chamfered end of the dipstick, and enters a light detector after reflection by a second mirror. When the chamfered end comes into contact with liquid, its internal reflection properties are altered and light no longer enters the detector. By using a suitable mechanical drive system to move the instrument up and down and measure its position, the liquid level can be monitored.

17.2 Float systems

Float systems, whereby the position of a float on the surface of a liquid is measured by means of a suitable transducer, have a typical measurement inaccuracy of $\pm 1\%$. This method is also simple, cheap and widely used. The system using a potentiometer, shown earlier in Figure 2.2, is very common, and is well known for its application

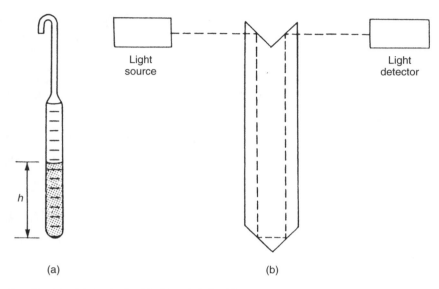

Fig. 17.1 Dipsticks: (a) simple dipstick; (b) optical dipstick.

to monitoring the level in motor vehicle fuel tanks. An alternative system, which is used in greater numbers, is called the *float and tape gauge* (or *tank gauge*). This has a tape attached to the float that passes round a pulley situated vertically above the float. The other end of the tape is attached to either a counterweight or a negative-rate counter-spring. The amount of rotation of the pulley, measured by either a synchro or a potentiometer, is then proportional to the liquid level. These two essentially mechanical systems of measurement are popular in many applications, but the maintenance requirements of them are always high.

17.3 Pressure-measuring devices (hydrostatic systems)

The hydrostatic pressure due to a liquid is directly proportional to its depth and hence to the level of its surface. Several instruments are available that use this principle, and they are widely used in many industries, particularly in harsh chemical environments. In the case of open-topped vessels (or covered ones that are vented to the atmosphere), the level can be measured by inserting a pressure sensor at the bottom of the vessel, as shown in Figure 17.2(a). The liquid level h is then related to the measured pressure P according to $h = P/\rho g$, where ρ is the liquid density and g is the acceleration due to gravity. One source of error in this method can be imprecise knowledge of the liquid density. This can be a particular problem in the case of liquid solutions and mixtures (especially hydrocarbons), and in some cases only an estimate of density is available. Even with single liquids, the density is subject to variation with temperature, and therefore temperature measurement may be required if very accurate level measurements are needed.

Where liquid-containing vessels are totally sealed, the liquid level can be calculated by measuring the differential pressure between the top and bottom of the tank, as

342 Level measurement

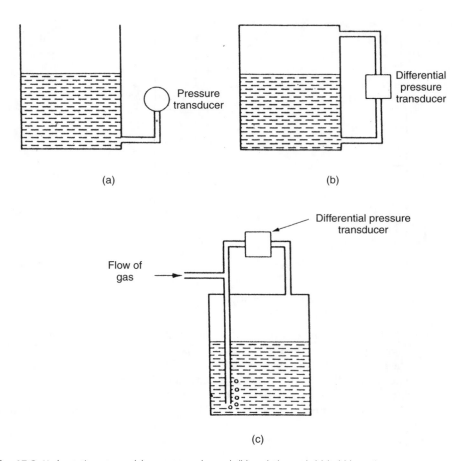

Fig. 17.2 Hydrostatic systems: (a) open-topped vessel; (b) sealed vessel; (c) bubbler unit.

shown in Figure 17.2(b). The differential pressure transducer used is normally a standard diaphragm type, although silicon-based microsensors are being used in increasing numbers. The liquid level is related to the differential pressure measured, δP, according to $h = \delta P / \rho g$. The same comments as for the case of the open vessel apply regarding uncertainty in the value of ρ. An additional problem that can occur is an accumulation of liquid on the side of the differential pressure transducer that is measuring the pressure at the top of the vessel. This can arise because of temperature fluctuations, which allow liquid to alternately vaporize from the liquid surface and then condense in the pressure tapping at the top of the vessel. The effect of this on the accuracy of the differential pressure measurement is severe, but the problem is easily avoided by placing a drain pot in the system.

A final pressure-related system of level measurement is the *bubbler unit* shown in Figure 17.2(c). This uses a dip pipe that reaches to the bottom of the tank and is purged free of liquid by a steady flow of gas through it. The rate of flow is adjusted until gas bubbles are just seen to emerge from the end of the tube. The pressure in the tube, measured by a pressure transducer, is then equal to the liquid pressure at the bottom of the tank. It is important that the gas used is inert with respect to the liquid in the vessel. Nitrogen, or sometimes just air, is suitable in most cases. Gas consumption is low, and a cylinder of nitrogen may typically last for six months. The method is suitable for measuring the liquid pressure at the bottom of both open and sealed tanks. It is particularly advantageous in avoiding the large maintenance problem associated with leaks at the bottom of tanks at the site of the pressure tappings required by alternative methods.

Measurement uncertainty varies according to the application and the condition of the measured material. A typical value would be ±0.5% of full-scale reading, although ±0.1% can be achieved in some circumstances.

17.4 Capacitive devices

Capacitive devices are widely used for measuring the level of both liquids and solids in powdered or granular form. They perform well in many applications, but become inaccurate if the measured substance is prone to contamination by agents that change the dielectric constant. Ingress of moisture into powders is one such example of this. They are also suitable for use in extreme conditions measuring liquid metals (high temperatures), liquid gases (low temperatures), corrosive liquids (acids, etc.) and high-pressure processes. Two versions are used according to whether the measured substance

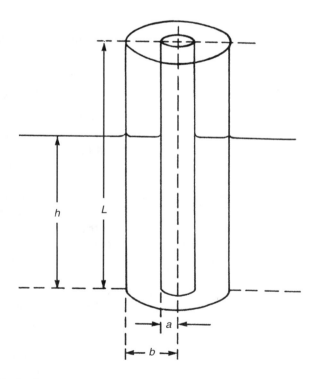

Fig. 17.3 Capacitive level sensor.

is conducting or not. For non-conducting substances (less than $0.1\,\mu\text{mho/cm}^3$), two bare-metal capacitor plates in the form of concentric cylinders are immersed in the substance, as shown in Figure 17.3. The substance behaves as a dielectric between the plates according to the depth of the substance. For concentric cylinder plates of radius a and b ($b > a$), and total height L, the depth of the substance h is related to the measured capacitance C by:

$$h = \frac{C \log_e (b/a) - 2\pi\varepsilon_0}{2\pi\varepsilon_0 (\varepsilon - 1)} \qquad (17.1)$$

where ε is the relative permittivity of the measured substance and ε_0 is the permittivity of free space. In the case of conducting substances, exactly the same measurement techniques are applied, but the capacitor plates are encapsulated in an insulating material. The relationship between C and h in equation (17.1) then has to be modified to allow for the dielectric effect of the insulator. Measurement uncertainty is typically 1–2%.

17.5 Ultrasonic level gauge

Ultrasonic level measurement is one of a number of non-contact techniques available. The principle of the ultrasonic level gauge is that energy from an ultrasonic source above the liquid is reflected back from the liquid surface into an ultrasonic energy detector, as illustrated in Figure 17.4. Measurement of the time of flight allows the liquid level to be inferred. In alternative versions, the ultrasonic source is placed at the bottom of the vessel containing the liquid, and the time of flight between emission, reflection off the liquid surface and detection back at the bottom of the vessel is measured.

Ultrasonic techniques are especially useful in measuring the position of the interface between two immiscible liquids contained in the same vessel, or measuring the sludge or precipitate level at the bottom of a liquid-filled tank. In either case, the method employed is to fix the ultrasonic transmitter–receiver transducer at a known height in the upper liquid, as shown in Figure 17.5. This establishes the level of the liquid/liquid or liquid/sludge level in absolute terms. When using ultrasonic instruments, it is essential that proper compensation is made for the working temperature if this differs from the calibration temperature, since the speed of ultrasound through air varies with temperature (see Chapter 13). Ultrasound speed also has a small sensitivity to humidity, air pressure and carbon dioxide concentration, but these factors are usually insignificant. Temperature compensation can be achieved in two ways. Firstly, the operating temperature can be measured and an appropriate correction made. Secondly, and preferably, a comparison method can be used in which the system is calibrated each time it is used by measuring the transit time of ultrasonic energy between two known reference points. This second method takes account of humidity, pressure and carbon dioxide concentration variations as well as providing temperature compensation. With appropriate care, measurement uncertainty can be reduced to about ±1%.

Measurement and Instrumentation Principles **345**

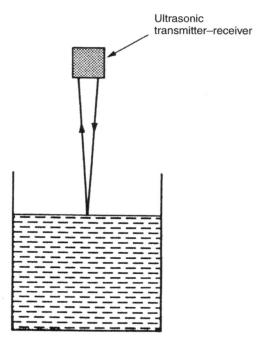

Fig. 17.4 Ultrasonic level gauge.

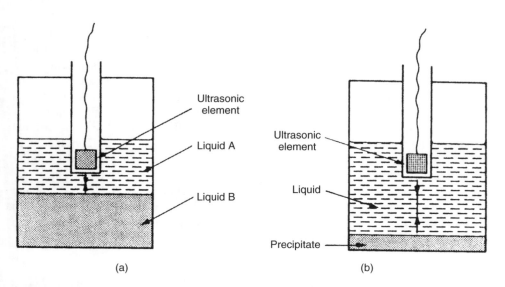

Fig. 17.5 Measuring interface positions: (a) liquid/liquid interface; (b) liquid/precipitate interface.

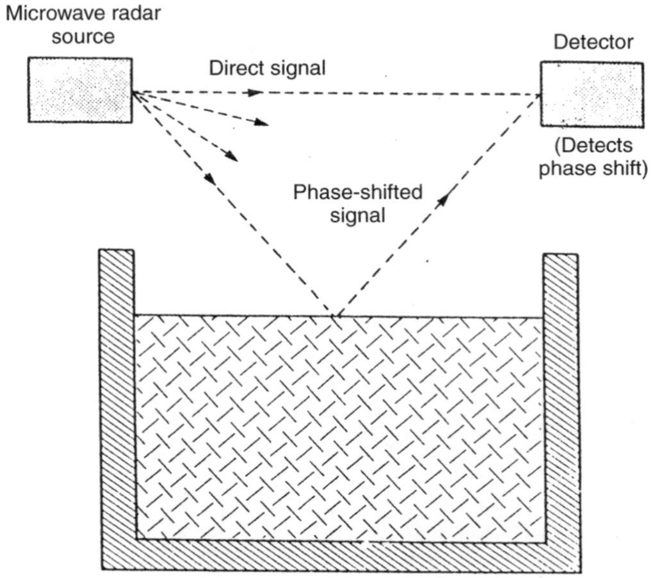

Fig. 17.6 Radar level detector.

17.6 Radar (microwave) methods

Level-measuring instruments using microwave radar are an alternative technique for non-contact measurement. Currently, they are still very expensive (∼£3000), but prices are falling and usage is expanding rapidly. They are able to provide successful level measurement in applications that are otherwise very difficult, such as measurement in closed tanks, measurement where the liquid is turbulent, and measurement in the presence of obstructions and steam condensate. The technique involves directing a constant-amplitude, frequency-modulated microwave signal at the liquid surface. A receiver measures the phase difference between the reflected signal and the original signal transmitted directly through air to it, as shown in Figure 17.6. This measured phase difference is linearly proportional to the liquid level. The system is similar in principle to ultrasonic level measurement, but has the important advantage that the transmission time of radar through air is almost totally unaffected by ambient temperature and pressure fluctuations. However, as the microwave frequency is within the band used for radio communications, strict conditions on amplitude levels have to be satisfied, and the appropriate licences have to be obtained.

17.7 Radiation methods

The radiation method is an expensive technique, which uses a radiation source and detector system located outside a liquid-filled tank in the manner shown in Figure 17.6. The non-invasive nature of this technique in using a source and detector system outside

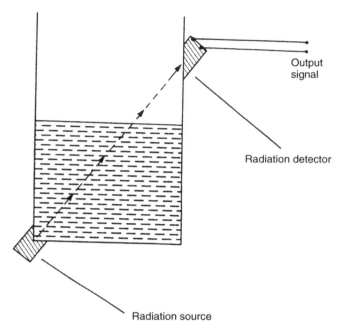

Fig. 17.7 Using a radiation source to measure level.

the tank is particularly attractive. The absorption of both beta rays and gamma rays varies with the amount of liquid between the source and detector, and hence is a function of liquid level. Caesium-137 is a commonly used gamma-ray source. The radiation level measured by the detector I is related to the length of liquid in the path x according to:

$$I = I_0 \exp(-\mu\rho x) \qquad (17.2)$$

where I_0 is the intensity of radiation that would be received by the detector in the absence of any liquid, μ is the mass absorption coefficient for the liquid and ρ is the mass density of the liquid.

In the arrangement shown in Figure 17.7, the radiation follows a diagonal path across the liquid, and therefore some trigonometrical manipulation has to be carried out to determine the liquid level h from x. In some applications, the radiation source can be located in the centre of the bottom of the tank, with the detector vertically above it. Where this is possible, the relationship between the radiation detected and liquid level is obtained by directly substituting h in place of x in equation (17.2). Apart from use with liquids at normal temperatures, this method is commonly used for measuring the level of hot, liquid metals. However, because of the obvious dangers associated with using radiation sources, very strict safety regulations have to be satisfied when applying this technique. Very low activity radiation sources are used in some systems to overcome safety problems but the system is then sensitive to background radiation and special precautions have to be taken regarding the provision of adequate shielding. Because of the many difficulties in using this technique, it is only used in special applications.

17.8 Other techniques

17.8.1 Vibrating level sensor

The principle of the vibrating level sensor is illustrated in Figure 17.8. The instrument consists of two piezoelectric oscillators fixed to the inside of a hollow tube that generate flexural vibrations in the tube at its resonant frequency. The resonant frequency of the tube varies according to the depth of its immersion in the liquid. A phase-locked loop circuit is used to track these changes in resonant frequency and adjust the excitation frequency applied to the tube by the piezoelectric oscillators. Liquid level measurement is therefore obtained in terms of the output frequency of the oscillator when the tube is resonating.

17.8.2 Hot-wire elements/carbon resistor elements

Figure 17.9 shows a level measurement system that uses a series of hot-wire elements or carbon resistors placed at regular intervals along a vertical line up the side of a tank. The heat transfer coefficient of such elements differs substantially depending upon whether the element is immersed in air or in the liquid in the tank. Consequently, elements in the liquid have a different temperature and therefore a different resistance to those in air. This method of level measurement is a simple one, but the measurement resolution is limited to the distance between sensors.

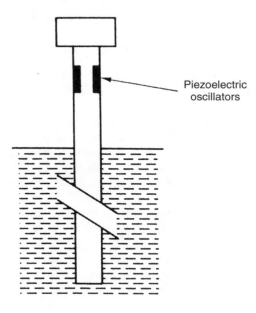

Fig. 17.8 Vibrating level sensor.

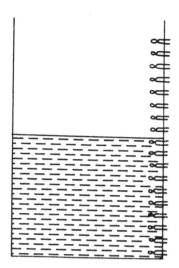

Fig. 17.9 Hot-wire-element level sensor.

17.8.3 Laser methods

One laser-based method is the *reflective level sensor*. This sensor uses light from a laser source that is reflected off the surface of the measured liquid into a line array of charge-coupled devices, as shown in Figure 17.10. Only one of these will sense light, according to the level of the liquid. An alternative, laser-based technique operates on the same general principles as the radar method described above but uses laser-generated pulses of infrared light directed at the liquid surface. This is immune to environmental conditions, and can be used with sealed vessels provided that a glass window is provided in the top of the vessel.

17.8.4 Fibre-optic level sensors

The *fibre-optic cross-talk sensor*, as described in Chapter 13, is one example of a fibre-optic sensor that can be used to measure liquid level. Another light-loss fibre-optic level sensor is the simple *loop sensor* shown in Figure 17.11. The amount of light loss depends on the proportion of cable that is submerged in the liquid. This effect is magnified if the alternative arrangement shown in Figure 17.12 is used, where light is reflected from an input fibre, round a prism, and then into an output fibre. Light is lost from this path into the liquid according to the depth of liquid surrounding the prism.

17.8.5 Thermography

Thermal imaging instruments, as discussed in Chapter 14, are a further means of detecting the level of liquids in tanks. Such instruments are capable of discriminating

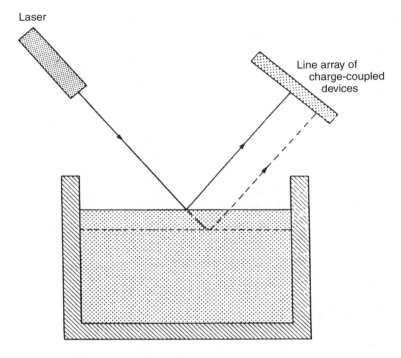

Fig. 17.10 Reflective level sensor.

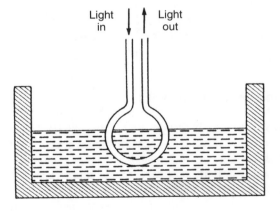

Fig. 17.11 Loop level sensor.

temperature differences as small as 0.1°C. Differences of this magnitude will normally be present at the interface between the liquid, which tends to remain at a constant temperature, and the air above, which constantly fluctuates in temperature by small amounts. The upper level of solids stored in hoppers is often detectable on the same principles.

Measurement and Instrumentation Principles

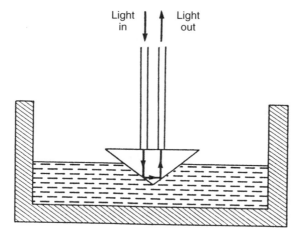

Fig. 17.12 Prism level sensor.

17.9 Intelligent level-measuring instruments

Most types of level gauge are now available in intelligent form. The pressure-measuring devices (section 17.3) are obvious candidates for inclusion within intelligent level-measuring instruments, and versions claiming ±0.05% accuracy are now on the market. Such instruments can also carry out additional functions, such as providing automatic compensation for liquid density variations. Microprocessors are also used to simplify installation and set-up procedures.

17.10 Choice between different level sensors

Two separate classes of level sensors can be distinguished according to whether they make contact or not with the material whose level is being measured. Contact devices are less reliable for a number of reasons, and therefore non-contact devices such as radar, laser, radiation or ultrasonic devices are preferred when there is a particular need for high reliability. According to the application, sensors that are relatively unaffected by changes in the temperature, composition, moisture content or density of the measured material may be preferred. In these respects, radar (microwave) and radiation sensors have the best immunity to such changes. Further guidance can be found in Liptak, (1995).

References and further reading

Liptak, B.G. (1995) *Instrument Engineers Handbook: Process Measurement and Analysis*, Chilton, Pennsylvania.

18

Mass, force and torque measurement

18.1 Mass (weight) measurement

Mass describes the quantity of matter that a body contains. Load cells are the most common instrument used to measure mass, especially in industrial applications. Most load cells are now electronic, although pneumatic and hydraulic types also exist. The alternatives to load cells are either mass-balance instruments or the spring balance.

18.1.1 Electronic load cell (electronic balance)

In an electronic load cell, the gravitational force on the body being measured is applied to an elastic element. This deflects according to the magnitude of the body mass. Mass measurement is thereby translated into a displacement measurement task. Electronic load cells have significant advantages over most other forms of mass-measuring instrument in terms of their relatively low cost, wide measurement range, tolerance of dusty and corrosive environments, remote measurement capability, tolerance of shock loading and ease of installation. The electronic load cell uses the physical principle that a force applied to an elastic element produces a measurable deflection. The elastic elements used are specially shaped and designed, some examples of which are shown in Figure 18.1. The design aims are to obtain a linear output relationship between the applied force and the measured deflection and to make the instrument insensitive to forces that are not applied directly along the sensing axis. Load cells exist in both compression and tension forms. In the compression type, the measured mass is placed on top of a platform resting on the load cell, which therefore compresses the cell. In the alternative tension type, the mass is hung from the load cell, thereby putting the cell into tension.

One problem that can affect the performance of load cells is the phenomenon of creep. Creep describes the permanent deformation that an elastic element undergoes after it has been under load for a period of time. This can lead to significant measurement errors in the form of a bias on all readings if the instrument is not recalibrated from time to time. However, careful design and choice of materials can largely eliminate the problem.

Measurement and Instrumentation Principles 353

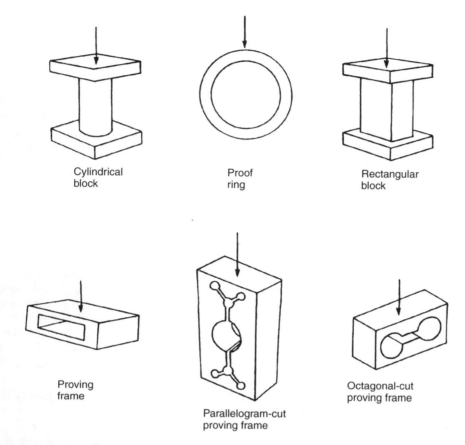

Fig. 18.1 Elastic elements used in load cells.

Various types of displacement transducer are used to measure the deflection of the elastic elements. Of these, the strain gauge is used most commonly, since this gives the best measurement accuracy, with an inaccuracy figure less than ±0.05% of full-scale reading being obtainable. Load cells including strain gauges are used to measure masses over a very wide range between 0 and 3000 tonnes. The measurement capability of an individual instrument designed to measure masses at the bottom end of this range would typically be 0.1–5 kg, whereas instruments designed for the top of the range would have a typical measurement span of 10–3000 tonnes.

Elastic force transducers based on differential transformers (LVDTs) to measure defections are used to measure masses up to 25 tonnes. Apart from having a lower maximum measuring capability, they are also inferior to strain gauge-based instruments in terms of their ±0.2% inaccuracy figure. Their major advantage is their longevity and almost total lack of maintenance requirements.

The final type of displacement transducer used in this class of instrument is the piezoelectric device. Such instruments are used to measure masses in the range 0 to 1000 tonnes. Piezoelectric crystals replace the specially designed elastic member

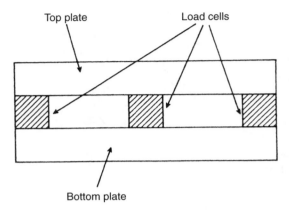

Fig. 18.2 Load-cell-based electronic balance.

normally used in this class of instrument, allowing the device to be physically small. As discussed previously, such devices can only measure dynamically changing forces because the output reading results from an induced electrical charge whose magnitude leaks away with time. The fact that the elastic element consists of the piezoelectric crystal means that it is very difficult to design such instruments to be insensitive to forces applied at an angle to the sensing axis. Therefore, special precautions have to be taken in applying these devices. Although such instruments are relatively cheap, their lowest inaccuracy is ±1% of full-scale reading, and they also have a high temperature coefficient.

The *electronic balance* is a device that contains several compression-type load cells, as illustrated in Figure 18.2. Commonly, either three or four load cells are used in the balance, with the output mass measurement being formed from the sum of the outputs of each cell. Where appropriate, the upper platform can be replaced by a tank for weighing liquids, powders etc.

18.1.2 Pneumatic/hydraulic load cells

Pneumatic and hydraulic load cells translate mass measurement into a pressure measurement task. A pneumatic load cell is shown schematically in Figure 18.3. Application of a mass to the cell causes deflection of a diaphragm acting as a variable restriction in a nozzle–flapper mechanism. The output pressure measured in the cell is approximately proportional to the magnitude of the gravitational force on the applied mass. The instrument requires a flow of air at its input of around 0.25 m^3/hour at a pressure of 4 bar. Standard cells are available to measure a wide range of masses. For measuring small masses, instruments are available with a full-scale reading of 25 kg, whilst at the top of the range, instruments with a full-scale reading of 25 tonnes are obtainable. Inaccuracy is typically ±0.5% of full scale in pneumatic load cells.

The alternative, hydraulic load cell is shown in Figure 18.4. In this, the gravitational force due to the unknown mass is applied, via a diaphragm, to oil contained within an enclosed chamber. The corresponding increase in oil pressure is measured by a suitable pressure transducer. These instruments are designed for measuring much larger masses

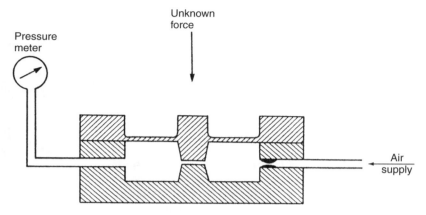

Fig. 18.3 Pneumatic load cell.

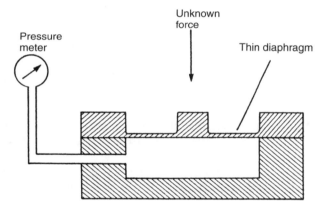

Fig. 18.4 Hydraulic load cell.

than pneumatic cells, with a load capacity of 500 tonnes being common. Special units can be obtained to measure masses as large as 50 000 tonnes. Besides their much greater measuring range, hydraulic load cells are much more accurate than pneumatic cells, with an inaccuracy figure of ±0.05% of full scale being typical. However, in order to obtain such a level of accuracy, correction for the local value of g (acceleration due to gravity) is necessary. A measurement resolution of 0.02% is attainable.

18.1.3 Intelligent load cells

Intelligent load cells are formed by adding a microprocessor to a standard cell. This brings no improvement in accuracy because the load cell is already a very accurate device. What it does produce is an intelligent weighing system that can compute total cost from the measured weight, using stored cost per unit weight information, and provide an output in the form of a digital display. Cost per weight figures can be pre-stored for a large number of substances, making such instruments very flexible in their operation.

In applications where the mass of an object is measured by several load cells used together (for example, load cells located at the corners of a platform in an electronic balance), the total mass can be computed more readily if the individual cells have a microprocessor providing digital output. In addition, it is also possible to use significant differences in the relative readings between different load cells as a fault detection mechanism in the system.

18.1.4 Mass-balance (weighing) instruments

Mass-balance instruments are based on comparing the gravitational force on the measured mass with the gravitational force on another body of known mass. This principle of mass measurement is commonly known as *weighing*, and is used in instruments like the beam balance, weigh beam, pendulum scale and electromagnetic balance.

Beam balance (equal-arm balance)

In the beam balance, shown in Figure 18.5, standard masses are added to a pan on one side of a pivoted beam until the magnitude of the gravity force on them balances the magnitude of the gravitational force on the unknown mass acting at the other end of the beam. This equilibrium position is indicated by a pointer that moves against a calibrated scale.

Instruments of this type are capable of measuring a wide span of masses. Those at the top of the range can typically measure masses up to 1000 grams whereas those at the bottom end of the range can measure masses of less than 0.01 gram. Measurement resolution can be as good as 1 part in 10^7 of the full-scale reading if the instrument is designed and manufactured very carefully. The lowest measurement inaccuracy figure attainable is $\pm 0.002\%$.

One serious disadvantage of this type of instrument is its lack of ruggedness. Continuous use and the inevitable shock loading that will occur from time to time both cause

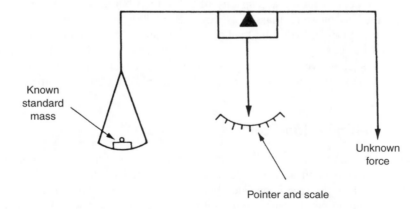

Fig. 18.5 Beam balance (equal-arm balance).

damage to the knife edges, leading to problems in measurement accuracy and resolution. A further problem in industrial use is the relatively long time needed to make each measurement. For these reasons, the beam balance is normally reserved as a calibration standard and is not used in day-to-day production environments.

Weigh beam

The weigh beam, sketched in two alternative forms in Figure 18.6, operates on similar principles to the beam balance but is much more rugged. In the first form, standard masses are added to balance the unknown mass and fine adjustment is provided by a known mass that is moved along a notched, graduated bar until the pointer is brought to the null, balance point. The alternative form has two or more graduated bars (three bars shown in Figure 18.6). Each bar carries a different standard mass and these are moved to appropriate positions on the notched bar to balance the unknown mass. Versions of these instruments are used to measure masses up to 50 tonnes.

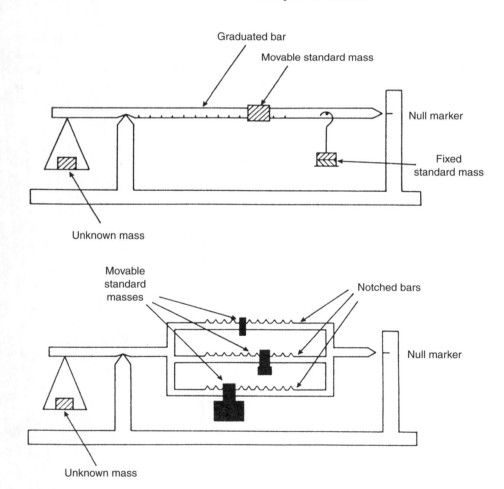

Fig. 18.6 Two alternative forms of weigh beam.

Pendulum scale

The pendulum scale, sketched in Figure 18.7, is another instrument that works on the mass-balance principle. The unknown mass is put on a platform that is attached by steel tapes to a pair of cams. Downward motion of the platform, and hence rotation of the cams, under the influence of the gravitational force on the mass, is opposed by the gravitational force acting on two pendulum type masses attached to the cams. The amount of rotation of the cams when the equilibrium position is reached is determined by the deflection of a pointer against a scale. The shape of the cams is such that this output deflection is linearly proportional to the applied mass.

This instrument is particularly useful in some applications because it is a relatively simple matter to replace the pointer and scale system by a rotational displacement transducer that gives an electrical output. Various versions of the instrument can measure masses in the range between 1 kg and 500 tonnes, with a typical measurement inaccuracy of $\pm 0.1\%$.

One potential source of difficulty with the instrument is oscillation of the weigh platform when the mass is applied. Where necessary, in instruments measuring larger masses, dashpots are incorporated into the cam system to damp out such oscillations. A further possible problem can arise, mainly when measuring large masses, if the mass is not placed centrally on the platform. This can be avoided by designing a second platform to hold the mass, which is hung from the first platform by knife edges. This lessens the criticality of mass placement.

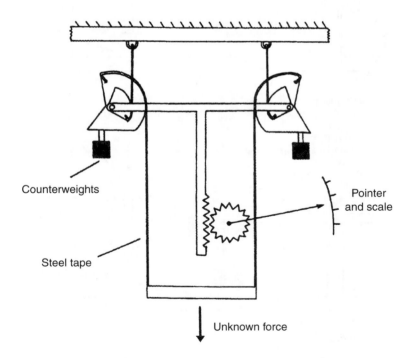

Fig. 18.7 Pendulum scale.

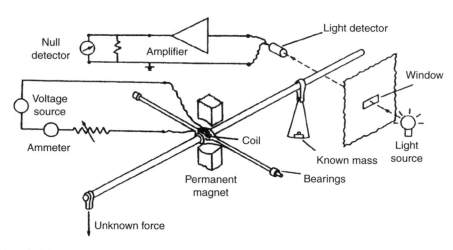

Fig. 18.8 Electromagnetic balance.

Electromagnetic balance

The electromagnetic balance uses the torque developed by a current-carrying coil suspended in a permanent magnetic field to balance the unknown mass against the known gravitational force produced on a standard mass, as shown in Figure 18.8. A light source and detector system is used to determine the null balance point. The voltage output from the light detector is amplified and applied to the coil, thus creating a servosystem where the deflection of the coil in equilibrium is proportional to the applied force. Its advantages over beam balances, weigh beams and pendulum scales include its smaller size, its insensitivity to environmental changes (modifying inputs) and its electrical form of output.

18.1.5 Spring balance

Spring balances provide a method of mass measurement that is both simple and cheap. The mass is hung on the end of a spring and the deflection of the spring due to the downwards gravitational force on the mass is measured against a scale. Because the characteristics of the spring are very susceptible to environmental changes, measurement accuracy is usually relatively poor. However, if compensation is made for the changes in spring characteristics, then a measurement inaccuracy less than ±0.2% is achievable. According to the design of the instrument, masses between 0.5 kg and 10 tonnes can be measured.

18.2 Force measurement

If a force of magnitude, F, is applied to a body of mass, M, the body will accelerate at a rate, A, according to the equation:

$$F = MA$$

The standard unit of force is the *Newton*, this being the force that will produce an acceleration of one metre per second squared in the direction of the force when it is applied to a mass of one kilogram. One way of measuring an unknown force is therefore to measure the acceleration when it is applied to a body of known mass. An alternative technique is to measure the variation in the resonant frequency of a vibrating wire as it is tensioned by an applied force.

18.2.1 Use of accelerometers

The technique of applying a force to a known mass and measuring the acceleration produced can be carried out using any type of accelerometer. Unfortunately, the method is of very limited practical value because, in most cases, forces are not free entities but are part of a system (from which they cannot be decoupled) in which they are acting on some body that is not free to accelerate. However, the technique can be of use in measuring some transient forces, and also for calibrating the forces produced by thrust motors in space vehicles.

18.2.2 Vibrating wire sensor

This instrument, illustrated in Figure 18.9, consists of a wire that is kept vibrating at its resonant frequency by a variable-frequency oscillator. The resonant frequency of a wire under tension is given by:

$$f = \frac{0.5}{L}\sqrt{\left(\frac{M}{T}\right)}$$

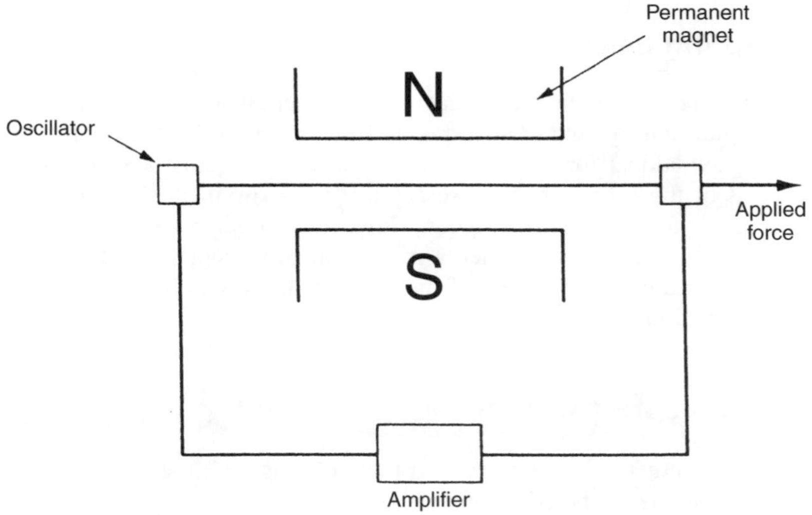

Fig. 18.9 Vibrating-wire sensor.

where M is the mass per unit length of the wire, L is the length of the wire, and T is the tension due to the applied force, F. Thus, measurement of the output frequency of the oscillator allows the force applied to the wire to be calculated.

18.3 Torque measurement

Measurement of applied torques is of fundamental importance in all rotating bodies to ensure that the design of the rotating element is adequate to prevent failure under shear stresses. Torque measurement is also a necessary part of measuring the power transmitted by rotating shafts. The three traditional methods of measuring torque consist of (i) measuring the reaction force in cradled shaft bearings, (ii) the 'Prony brake' method and (iii) measuring the strain produced in a rotating body due to an applied torque. However, recent developments in electronics and optic-fibre technology now offer an alternative method as described in paragraph 18.3.4 below.

18.3.1 Reaction forces in shaft bearings

Any system involving torque transmission through a shaft contains both a power source and a power absorber where the power is dissipated. The magnitude of the transmitted torque can be measured by cradling either the power source or the power absorber end of the shaft in bearings, and then measuring the reaction force, F, and the arm length L, as shown in Figure 18.10. The torque is then calculated as the simple product, FL. Pendulum scales are very commonly used for measuring the reaction force. Inherent errors in the method are bearing friction and windage torques.

18.3.2 Prony brake

The principle of the Prony brake is illustrated in Figure 18.11. It is used to measure the torque in a rotating shaft and consists of a rope wound round the shaft. One end of the rope is attached to a spring balance and the other end carries a load in the form of a standard mass, m. If the measured force in the spring balance is F_s, then the effective force, F_e, exerted by the rope on the shaft is given by:

$$F_e = mg - F_s$$

If the radius of the shaft is R_s and that of the rope is R_r, then the effective radius, R_e, of the rope and drum with respect to the axis of rotation of the shaft is given by:

$$R_e = R_s + R_r$$

The torque in the shaft, T, can then be calculated as:

$$T = F_e R_e$$

Whilst this is a well-known method of measuring shaft torque, a lot of heat is generated because of friction between the rope and shaft, and water cooling is usually necessary.

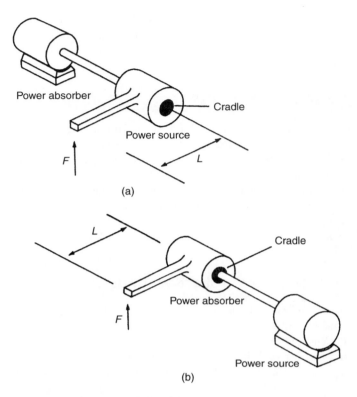

Fig. 18.10 Measuring reaction forces in cradled shaft bearings.

18.3.3 Measurement of induced strain

Measuring the strain induced in a shaft due to an applied torque has been the most common method used for torque measurement in recent years. It is a very attractive method because it does not disturb the measured system by introducing friction torques in the same way as the last two methods described do. The method involves bonding four strain gauges onto the shaft as shown in Figure 18.12, where the strain gauges are arranged in a d.c. bridge circuit. The output from the bridge circuit is a function of the strain in the shaft and hence of the torque applied. It is very important that the positioning of the strain gauges on the shaft is precise, and the difficulty in achieving this makes the instrument relatively expensive.

The technique is ideal for measuring the stalled torque in a shaft before rotation commences. However, a problem is encountered in the case of rotating shafts because a suitable method then has to be found for making the electrical connections to the strain gauges. One solution to this problem found in many commercial instruments is to use a system of slip rings and brushes for this, although this increases the cost of the instrument still further.

Measurement and Instrumentation Principles 363

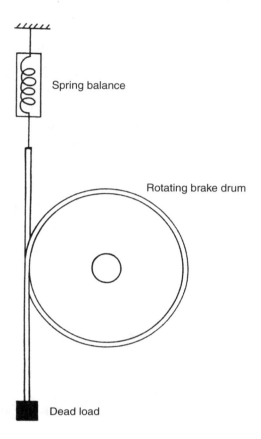

Fig. 18.11 The Prony brake.

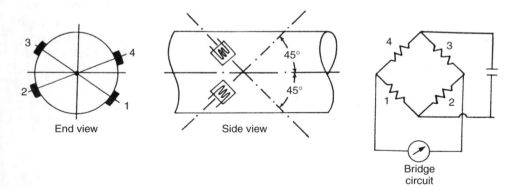

Fig. 18.12 Position of torque-measuring strain gauges on shaft.

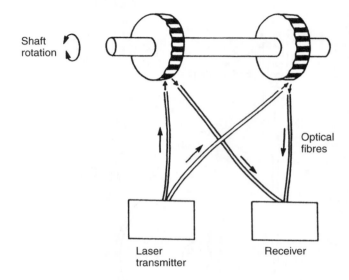

Fig. 18.13 Optical torque measurement.

18.3.4 Optical torque measurement

Optical techniques for torque measurement have become available recently with the development of laser diodes and fibre-optic light transmission systems. One such system is shown in Figure 18.13. Two black-and-white striped wheels are mounted at either end of the rotating shaft and are in alignment when no torque is applied to the shaft. Light from a laser diode light source is directed by a pair of optic-fibre cables onto the wheels. The rotation of the wheels causes pulses of reflected light and these are transmitted back to a receiver by a second pair of fibre-optic cables. Under zero torque conditions, the two pulse trains of reflected light are in phase with each other. If torque is now applied to the shaft, the reflected light is modulated. Measurement by the receiver of the phase difference between the reflected pulse trains therefore allows the magnitude of torque in the shaft to be calculated. The cost of such instruments is relatively low, and an additional advantage in many applications is their small physical size.

19

Translational motion transducers

19.1 Displacement

Translational displacement transducers are instruments that measure the motion of a body in a straight line between two points. Apart from their use as a primary transducer measuring the motion of a body, translational displacement transducers are also widely used as a secondary component in measurement systems, where some other physical quantity such as pressure, force, acceleration or temperature is translated into a translational motion by the primary measurement transducer. Many different types of translational displacement transducer exist and these, along with their relative merits and characteristics, are discussed in the following sections of this chapter. The factors governing the choice of a suitable type of instrument in any particular measurement situation are considered in the final section at the end of the chapter.

19.1.1 The resistive potentiometer

The resistive potentiometer is perhaps the best-known displacement-measuring device. It consists of a resistance element with a movable contact as shown in Figure 19.1. A voltage V_s is applied across the two ends A and B of the resistance element and an output voltage V_0 is measured between the point of contact C of the sliding element and the end of the resistance element A. A linear relationship exists between the output voltage V_0 and the distance AC, which can be expressed by:

$$\frac{V_0}{V_s} = \frac{AC}{AB} \qquad (19.1)$$

The body whose motion is being measured is connected to the sliding element of the potentiometer, so that translational motion of the body causes a motion of equal magnitude of the slider along the resistance element and a corresponding change in the output voltage V_0.

Three different types of potentiometer exist, wire-wound, carbon-film and plastic-film, so named according to the material used to construct the resistance element. Wire-wound

366 Translational motion transducers

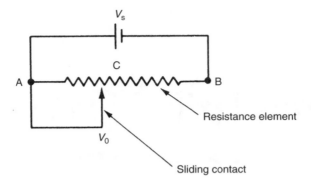

Fig. 19.1 The resistive potentiometer.

potentiometers consist of a coil of resistance wire wound on a non-conducting former. As the slider moves along the potentiometer track, it makes contact with successive turns of the wire coil. This limits the resolution of the instrument to the distance from one coil to the next. Much better measurement resolution is obtained from potentiometers using either a carbon film or a conducting plastic film for the resistance element. Theoretically, the resolution of these is limited only by the grain size of the particles in the film, suggesting that measurement resolutions up to 10^{-4} ought to be attainable. In practice, the resolution is limited by mechanical difficulties in constructing the spring system that maintains the slider in contact with the resistance track, although these types are still considerably better than wire-wound types.

Operational problems of potentiometers all occur at the point of contact between the sliding element and the resistance track. The most common problem is dirt under the slider, which increases the resistance and thereby gives a false output voltage reading, or in the worst case causes a total loss of output. High-speed motion of the slider can also cause the contact to bounce, giving an intermittent output. Friction between the slider and the track can also be a problem in some measurement systems where the body whose motion is being measured is moved by only a small force of a similar magnitude to these friction forces.

The life expectancy of potentiometers is normally quoted as a number of reversals, i.e. as the number of times the slider can be moved backwards and forwards along the track. The figures quoted for wire-wound, carbon-film and plastic-film types are respectively 1 million, 5 million and 30 million. In terms of both life expectancy and measurement resolution, therefore, the carbon and plastic film types are clearly superior, although wire-wound types do have one advantage in respect of their lower temperature coefficient. This means that wire-wound types exhibit much less variation in their characteristics in the presence of varying ambient temperature conditions.

A typical inaccuracy figure that is quoted for translational motion resistive potentiometers is $\pm 1\%$ of full-scale reading. Manufacturers produce potentiometers to cover a large span of measurement ranges. At the bottom end of this span, instruments with a range of ± 2 mm are available whilst at the top end, instruments with a range of ± 1 m are produced.

The resistance of the instrument measuring the output voltage at the potentiometer slider can affect the value of the output reading, as discussed in Chapter 3. As the slider moves along the potentiometer track, the ratio of the measured resistance to that of the measuring instrument varies, and thus the linear relationship between the measured displacement and the voltage output is distorted as well. This effect is minimized when the potentiometer resistance is small relative to that of the measuring instrument. This is achieved firstly by using a very high-impedance measuring instrument and secondly by keeping the potentiometer resistance as small as possible. Unfortunately, the latter is incompatible with achieving high measurement sensitivity since this requires a high potentiometer resistance. A compromise between these two factors is therefore necessary. The alternative strategy of obtaining high measurement sensitivity by keeping the potentiometer resistance low and increasing the excitation voltage is not possible in practice because of the power rating limitation. This restricts the allowable power loss in the potentiometer to its heat dissipation capacity.

The process of choosing the best potentiometer from a range of instruments that are available, taking into account power rating and measurement linearity considerations, is illustrated in the example below.

Example
The output voltage from a translational motion potentiometer of stroke length 0.1 metre is to be measured by an instrument whose resistance is $10\,\text{k}\Omega$. The maximum measurement error, which occurs when the slider is positioned two-thirds of the way along the element (i.e. when $AC = 2AB/3$ in Figure 19.1), must not exceed 1% of the full-scale reading. The highest possible measurement sensitivity is also required. A family of potentiometers having a power rating of 1 watt per 0.01 metre and resistances ranging from $100\,\Omega$ to $10\,\text{k}\Omega$ in $100\,\Omega$ steps is available. Choose the most suitable potentiometer from this range and calculate the sensitivity of measurement that it gives.

Solution
Referring to the labelling used in Figure 19.1, let the resistance of portion AC of the resistance element R_i and that of the whole length AB of the element be R_t. Also, let the resistance of the measuring instrument be R_m and the output voltage measured by it be V_m. When the voltage-measuring instrument is connected to the potentiometer, the net resistance across AC is the sum of two resistances in parallel (R_i and R_m) given by:

$$R_{AC} = \frac{R_i R_m}{R_i + R_m}$$

Let the excitation voltage applied across the ends AB of the potentiometer be V and the resultant current flowing between A and B be I. Then I and V are related by:

$$I = \frac{V}{R_{AC} + R_{CB}} = \frac{V}{[R_i R_m / R_i + R_m] + R_t - R_i}$$

V_m can now be calculated as:

$$V_m = I R_{AC} = \frac{V R_i R_m}{\{[R_i R_m / (R_i + R_m)] + R_t - R_i\} \{R_i + R_m\}}$$

If we express the voltage that exists across AC in the absence of the measuring instrument as V_0, then we can express the error due to the loading effect of the measuring instrument as Error $= V_0 - V_m$. From equation (19.1), $V_0 = (R_i V)/R_t$. Thus,

$$\text{Error} = V_0 - V_m = V \left(\frac{R_i}{R_t}\right) \left(\frac{R_i R_m}{\{[R_i R_m / R_i + R_m] + R_t - R_i\}\{R_i + R_m\}}\right)$$

$$\times \left(\frac{R_i^2 (R_i - R_t)}{R_t [R_i R_t + R_m R_t - R_i^2]}\right) \quad (19.2)$$

Substituting $R_i = 2R_t/3$ into equation (19.2) to find the maximum error:

$$\text{Maximum error} = \frac{2R_t}{2R_t + 9R_m}$$

For a maximum error of 1%:

$$\frac{2R_t}{2R_t + 9R_m} = 0.01 \quad (19.3)$$

Substituting $R_m = 10\,000\,\Omega$ into the above expression (19.3) gives $R_t = 454\,\Omega$. The nearest resistance values in the range of potentiometers available are $400\,\Omega$ and $500\,\Omega$. The value of $400\,\Omega$ has to be selected, as this is the only one that gives a maximum measurement error of less than 1%. The thermal rating of the potentiometers is quoted as 1 watt/0.01 m, i.e. 10 watts for the total length of 0.1 m. By Ohm's law, maximum supply voltage $= \sqrt{\text{power} \times \text{resistance}} = \sqrt{10 \times 400} = 63.25$ Volts.

Thus, the measurement sensitivity $= 63.25/0.1$ V/m $= 632.5$ V/m

19.1.2 Linear variable differential transformer (LVDT)

The linear variable differential transformer, which is commonly known by the abbreviation LVDT, consists of a transformer with a single primary winding and two secondary windings connected in the series opposing manner shown in Figure 19.2. The object whose translational displacement is to be measured is physically attached to the central iron core of the transformer, so that all motions of the body are transferred to the core. For an excitation voltage V_s given by $V_s = V_p \sin(\omega t)$, the e.m.f.s induced in the secondary windings V_a and V_b are given by:

$$V_a = K_a \sin(\omega t - \phi); \quad V_b = K_b \sin(\omega t - \phi)$$

The parameters K_a and K_b depend on the amount of coupling between the respective secondary and primary windings and hence on the position of the iron core. With the core in the central position, $K_a = K_b$, and we have:

$$V_a = V_b = K \sin(\omega t - \phi)$$

Because of the series opposition mode of connection of the secondary windings, $V_0 = V_a - V_b$, and hence with the core in the central position, $V_0 = 0$. Suppose now that the core is displaced upwards (i.e. towards winding A) by a distance x. If then $K_a = K_1$ and $K_b = K_2$, we have:

$$V_0 = (K_1 - K_2) \sin(\omega t - \phi)$$

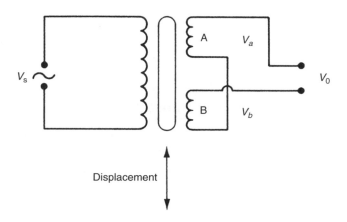

Fig. 19.2 The linear variable differential transformer (LVDT).

If, alternatively, the core were displaced downwards from the null position (i.e. towards winding B) by a distance x, the values of K_a and K_b would then be $K_a = K_2$ and $K_b = K_1$, and we would have:

$$V_0 = (K_2 - K_1)\sin(\omega t - \phi) = (K_1 - K_2)\sin(\omega t + [\pi - \phi])$$

Thus for equal magnitude displacements $+x$ and $-x$ of the core away from the central (null) position, the magnitude of the output voltage V_0 is the same in both cases. The only information about the direction of movement of the core is contained in the phase of the output voltage, which differs between the two cases by 180°. If, therefore, measurements of core position on both sides of the null position are required, it is necessary to measure the phase as well as the magnitude of the output voltage. The relationship between the magnitude of the output voltage and the core position is approximately linear over a reasonable range of movement of the core on either side of the null position, and is expressed using a constant of proportionality C as $V_0 = Cx$. The only moving part in an LVDT is the central iron core. As the core is only moving in the air gap between the windings, there is no friction or wear during operation. For this reason, the instrument is a very popular one for measuring linear displacements and has a quoted life expectancy of 200 years. The typical inaccuracy is ±0.5% of full-scale reading and measurement resolution is almost infinite. Instruments are available to measure a wide span of measurements from ±100 µm to ±100 mm. The instrument can be made suitable for operation in corrosive environments by enclosing the windings within a non-metallic barrier, which leaves the magnetic flux paths between the core and windings undisturbed. An epoxy resin is commonly used to encapsulate the coils for this purpose. One further operational advantage of the instrument is its insensitivity to mechanical shock and vibration.

Some problems that affect the accuracy of the LVDT are the presence of harmonics in the excitation voltage and stray capacitances, both of which cause a non-zero output of low magnitude when the core is in the null position. It is also impossible in practice to produce two identical secondary windings, and the small asymmetry that invariably exists between the secondary windings adds to this non-zero null output. The magnitude of this is always less than 1% of the full-scale output and in many measurement situations is of little consequence. Where necessary, the magnitude of these effects can be measured by applying known displacements to the instrument. Following this, appropriate compensation can be applied to subsequent measurements.

19.1.3 Variable capacitance transducers

Like variable inductance, the principle of variable capacitance is used in displacement measuring transducers in various ways. The three most common forms of variable capacitance transducer are shown in Figure 19.3. In Figure 19.3(a), the capacitor

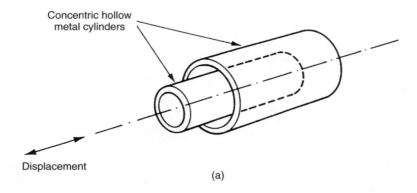

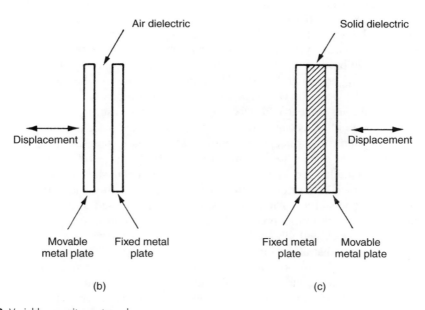

Fig. 19.3 Variable capacitance transducer.

plates are formed by two concentric, hollow, metal cylinders. The displacement to be measured is applied to the inner cylinder, which alters the capacitance. The second form, Figure 19.3(b), consists of two flat, parallel, metal plates, one of which is fixed and one of which is movable. Displacements to be measured are applied to the movable plate, and the capacitance changes as this moves. Both of these first two forms use air as the dielectric medium between the plates. The final form, Figure 19.3(c), has two flat, parallel, metal plates with a sheet of solid dielectric material between them. The displacement to be measured causes a capacitance change by moving the dielectric sheet.

Inaccuracies as low as ±0.01% are possible with these instruments, with measurement resolutions of 1 micron. Individual devices can be selected from manufacturers' ranges that measure displacements as small as 10^{-11} m or as large as 1 m. The fact that such instruments consist only of two simple conducting plates means that it is possible to fabricate devices that are tolerant to a wide range of environmental hazards such as extreme temperatures, radiation and corrosive atmospheres. As there are no contacting moving parts, there is no friction or wear in operation and the life expectancy quoted is 200 years. The major problem with variable capacitance transducers is their high impedance. This makes them very susceptible to noise and means that the length and position of connecting cables need to be chosen very carefully. In addition, very high impedance instruments need to be used to measure the value of the capacitance. Because of these difficulties, use of these devices tends to be limited to those few applications where the high accuracy and measurement resolution of the instrument are required.

19.1.4 Variable inductance transducers

One simple type of variable inductance transducer was shown earlier in Figure 13.4. This has a typical measurement range of 0–10 mm. An alternative form of variable inductance transducer shown in Figure 19.4(a) has a very similar size and physical appearance to the LVDT, but has a centre-tapped single winding. The two halves of the winding are connected, as shown in Figure 19.4(b), to form two arms of a bridge circuit that is excited with an alternating voltage. With the core in the central position, the output from the bridge is zero. Displacements of the core either side of the null position cause a net output voltage that is approximately proportional to the displacement for small movements of the core. Instruments in this second form are available to cover a wide span of displacement measurements. At the lower end of this span, instruments with a range of 0–2 mm are available, whilst at the top end, instruments with a range of 0–5 m can be obtained.

19.1.5 Strain gauges

The principles of strain gauges were covered earlier in Chapter 13. Because of their very small range of measurement (typically 0–50 µm), strain gauges are normally only used to measure displacements within devices like diaphragm-based pressure sensors rather than as a primary sensor in their own right for direct displacement measurement. However, strain gauges can be used to measure larger displacements if the range of

372 Translational motion transducers

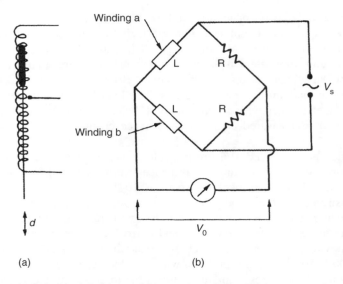

Fig. 19.4 (a) Variable inductance transducers; (b) connection in bridge circuit.

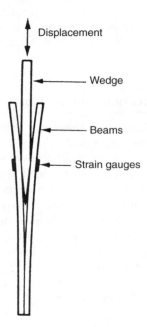

Fig. 19.5 Strain gauges measuring large displacements.

displacement measurement is extended by the scheme illustrated in Figure 19.5. In this, the displacement to be measured is applied to a wedge fixed between two beams carrying strain gauges. As the wedge is displaced downwards, the beams are forced apart and strained, causing an output reading on the strain gauges. Using this method, displacements up to about 50 mm can be measured.

19.1.6 Piezoelectric transducers

The piezoelectric transducer is effectively a force-measuring device that is used in many instruments measuring force, or the force-related quantities of pressure and acceleration. It is included within this discussion of linear displacement transducers because its mode of operation is to generate an e.m.f. that is proportional to the distance by which it is compressed. The device is manufactured from a crystal, which can be either a natural material such as quartz or a synthetic material such as lithium sulphate. The crystal is mechanically stiff (i.e. a large force is required to compress it), and consequently piezoelectric transducers can only be used to measure the displacement of mechanical systems that are stiff enough themselves to be unaffected by the stiffness of the crystal. When the crystal is compressed, a charge is generated on the surface that is measured as the output voltage. As is normal with any induced charge, the charge leaks away over a period of time. Consequently, the output voltage–time characteristic is as shown in Figure 19.6. Because of this characteristic, piezoelectric transducers are not suitable for measuring static or slowly varying displacements, even though the time constant of the charge–decay process can be lengthened by adding a shunt capacitor across the device.

As a displacement-measuring device, the piezoelectric transducer has a very high sensitivity, about one thousand times better than the strain gauge. Its typical inaccuracy is $\pm 1\%$ of full-scale reading and its life expectancy is three million reversals.

19.1.7 Nozzle flapper

The nozzle flapper is a displacement transducer that translates displacements into a pressure change. A secondary pressure-measuring device is therefore required within the instrument. The general form of a nozzle flapper is shown schematically in Figure 19.7. Fluid at a known supply pressure, P_s, flows through a fixed restriction and then through

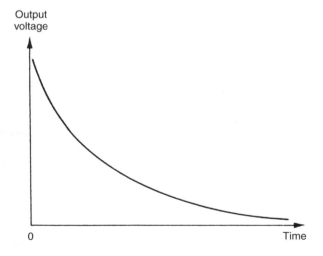

Fig. 19.6 Voltage–time characteristic of piezoelectric transducer following step displacement.

374 Translational motion transducers

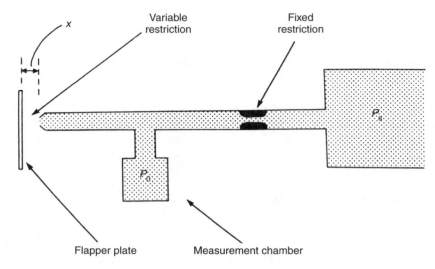

Fig. 19.7 Nozzle flapper.

a variable restriction formed by the gap, x, between the end of the main vessel and the flapper plate. The body whose displacement is being measured is connected physically to the flapper plate. The output measurement of the instrument is the pressure P_o in the chamber shown in Figure 19.7, and this is almost proportional to x over a limited range of movement of the flapper plate. The instrument typically has a first order response characteristic. Air is very commonly used as the working fluid and this gives the instrument a time constant of about 0.1 seconds. The instrument has extremely high sensitivity but its range of measurement is quite small. A typical measurement range is ± 0.05 mm with a measurement resolution of ± 0.01 μm. One very common application of nozzle flappers is measuring the displacements within a load cell, which are typically very small.

19.1.8 Other methods of measuring small displacements

Apart from the methods outlined above, several other techniques for measuring small translational displacements exist, as discussed below. Some of these involve special instruments that have a very limited sphere of application, for instance in measuring machine tool displacements. Others are very recent developments that may potentially gain wide use in the future but have few applications at present.

Linear inductosyn
The linear inductosyn is an extremely accurate instrument that is widely used for axis measurement and control within machine tools. Typical measurement resolution is 2.5 microns. The instrument consists of two magnetically coupled parts that are separated by an air gap, typically 0.125 mm wide, as shown in Figure 19.8. One part, the track, is attached to the axis along which displacements are to be measured. This would

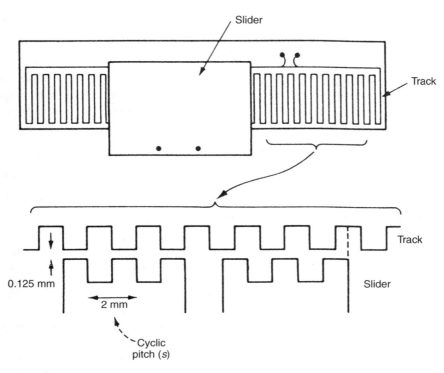

Fig. 19.8 Linear inductosyn.

generally be the bed of a machine tool. The other part, the slider, is attached to the body that is to be measured or positioned. This would usually be a cutting tool.

The track, which may be several metres long, consists of a fine metal wire formed into the pattern of a continuous rectangular waveform and deposited onto a glass base. The typical pitch (cycle length), s, of the pattern is 2 mm, and this extends over the full length of the track. The slider is usually about 50 mm wide and carries two separate wires formed into continuous rectangular waveforms that are displaced with respect to each other by one-quarter of the cycle pitch, i.e. by 90 electrical degrees. The wire waveform on the track is excited by an applied voltage given by:

$$V_s = V \sin(\omega t)$$

This excitation causes induced voltages in the slider windings. When the slider is positioned in the null position such that its first winding is aligned with the winding on the track, the output voltages on the two slider windings are given by:

$$V_1 = 0; \quad V_2 = V \sin(\omega t)$$

For any other position, the slider winding voltages are given by:

$$V_1 = V \sin(\omega t) \sin(2\pi x/s); \quad V_2 = V \sin(\omega t) \cos(2\pi x/s)$$

where x is the displacement of the slider away from the null position.

Consideration of these equations for the slider winding outputs shows that the pattern of output voltages repeats every cycle pitch. Therefore, the instrument can only discriminate displacements of the slider within one cycle pitch of the windings. This means that the typical measurement range of an inductosyn is only 2 mm. This is of no use in normal applications, and therefore an additional displacement transducer with coarser resolution but larger measurement range has to be used as well. This coarser measurement is commonly made by translating the linear displacements by suitable gearing into rotary motion, which is then measured by a rotational displacement transducer such as a synchro or resolver.

One slight problem with the inductosyn is the relatively low level of coupling between the track and slider windings. Compensation for this is made by using a high-frequency excitation voltage (5–10 kHz is common).

Translation of linear displacements into rotary motion
In some applications, it is inconvenient to measure linear displacements directly, either because there is insufficient space to mount a suitable transducer or because it is inconvenient for other reasons. A suitable solution in such cases is to translate the translational motion into rotational motion by suitable gearing. Any of the rotational displacement transducers discussed in Chapter 20 can then be applied.

Integration of output from velocity transducers and accelerometers
If velocity transducers or accelerometers already exist in a system, displacement measurements can be obtained by integration of the output from these instruments. This, however, only gives information about the relative position with respect to some arbitrary starting point. It does not yield a measurement of the absolute position of a body in space unless all motions away from a fixed starting point are recorded.

Laser interferometer
This recently developed instrument is shown in Figure 19.9. In this particular design, a dual-frequency helium–neon (He–Ne) laser is used that gives an output pair of light waves at a nominal frequency of 5×10^{14} Hz. The two waves differ in frequency by 2×10^6 Hz and have opposite polarization. This dual-frequency output waveform is split into a measurement beam and a reference beam by the first beam splitter.

The reference beam is sensed by the polarizer and photodetector, A, which converts both waves in the light to the same polarization. The two waves interfere constructively and destructively alternately, producing light–dark flicker at a frequency of 2×10^6 Hz. This excites a 2 MHz electrical signal in the photodetector.

The measurement beam is separated into the two component frequencies by a polarizing beam splitter. Light of the first frequency, f_1, is reflected by a fixed reflecting cube into a photodetector and polarizer, B. Light of the second frequency, f_2, is reflected by a movable reflecting cube and also enters B. The displacement to be measured is applied to the movable cube. With the movable cube in the null position, the light waves entering B produce an electrical signal output at a frequency of 2 MHz, which is the same frequency as the reference signal output from A. Any displacement of the movable cube causes a Doppler shift in the frequency f_2 and changes the output from B. The frequency of the output signal from B varies between 0.5 MHz and 3.5 MHz according to the speed and direction of movement of the movable cube.

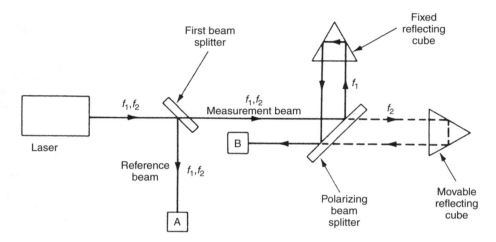

Fig. 19.9 Laser interferometer.

The outputs from A and B are amplified and subtracted. The resultant signal is fed to a counter whose output indicates the magnitude of the displacement in the movable cube and whose rate of change indicates the velocity of motion.

This technique is used in applications requiring high-accuracy measurement, such as machine tool control. Such systems can measure displacements over ranges of up to 2 m with an inaccuracy of only a few parts per million. They are therefore an attractive alternative to the inductosyn, in having both high measurement resolution and a large measurement range within one instrument.

Fotonic sensor

The Fotonic sensor is one of many instruments developed recently that make use of fibre-optic techniques. It consists of a light source, a light detector, a fibre-optic light transmission system and a plate that moves with the body whose displacement is being measured, as shown in Figure 19.10. Light from the outward fibre-optic cable travels across the air gap to the plate and some of it is reflected back into the return fibre-optic

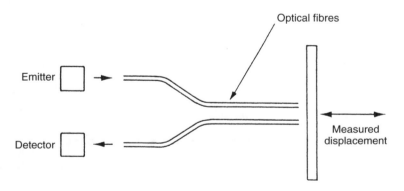

Fig. 19.10 Fotonic sensor.

cable. The amount of light reflected back from the plate is a function of the air gap length, x, and hence of the plate displacement. Measurement of the intensity of the light carried back along the return cable to the light detector allows the displacement of the plate to be calculated. Common applications of Fotonic sensors are measuring diaphragm displacements in pressure sensors and measuring the movement of bimetallic temperature sensors.

Evanescent-field fibre-optic sensors

This sensor consists of a prism and a light source/detector system, as shown in Figure 19.11. The amount of light reflected into the detector depends on the proximity of a movable silver surface to the prism. Reflection varies from 96% when the surface is touching the prism to zero when it is 1 µm away. This provides a means of measuring very tiny displacements over the range between 0 and 1 µm (1 micron).

Non-contacting optical sensor

Figure 19.12 shows an optical technique that is used to measure small displacements. The motion to be measured is applied to a vane, whose displacement progressively shades one of a pair of monolithic photodiodes that are exposed to infrared radiation. A displacement measurement is obtained by comparing the output of the reference (unshaded) photodiode with that of the shaded one. The typical range of measurement is ±0.5 mm with an inaccuracy of ±0.1% of full scale. Such sensors are used in some intelligent pressure-measuring instruments based on Bourdon tubes or diaphragms as described in Chapter 15.

19.1.9 Measurement of large displacements (range sensors)

One final class of instruments that has not been mentioned so far consists of those designed to measure relatively large translational displacements. Most of these are known as range sensors and measure the motion of a body with respect to some fixed datum point.

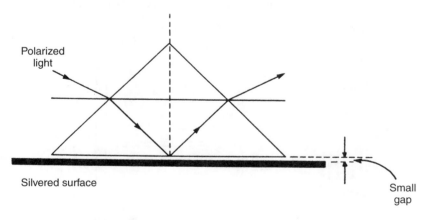

Fig. 19.11 Evanescent field fibre-optic displacement sensor.

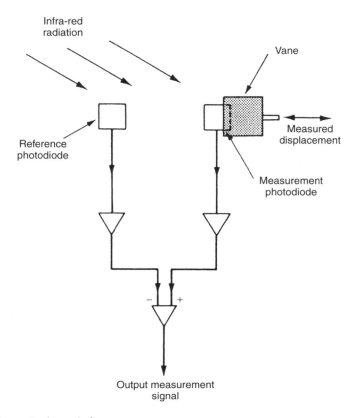

Fig. 19.12 Non-contacting optical sensor.

Rotary potentiometer and spring-loaded drum

One scheme for measuring large displacements that are beyond the measurement range of common displacement transducers is shown in Figure 19.13. This consists of a steel wire attached to the body whose displacement is being measured: the wire passes round a pulley and on to a spring-loaded drum whose rotation is measured by a rotary potentiometer. A multi-turn potentiometer is usually required for this to give an adequate measurement resolution. With this measurement system, it is possible to reduce measurement uncertainty to as little as ±0.01% of full-scale reading.

Range sensors

Range sensors provide a well-used technique of measuring the translational displacement of a body with respect to some fixed boundary. The common feature of all range sensing systems is an energy source, an energy detector and an electronic means of timing the time of flight of the energy between the source and detector. The form of energy used is either ultrasonic or light. In some systems, both energy source and detector are fixed on the moving body and operation depends on the energy being reflected back from the fixed boundary as in Figure 19.14(a). In other systems, the energy source is attached to the moving body and the energy detector is located within the fixed boundary, as shown in Figure 19.14(b).

Translational motion transducers

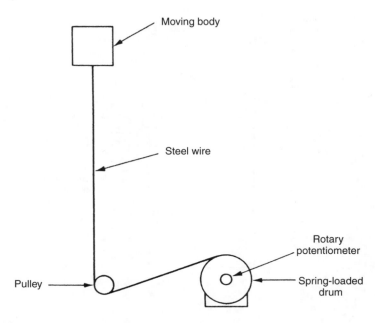

Fig. 19.13 System for measuring large displacements.

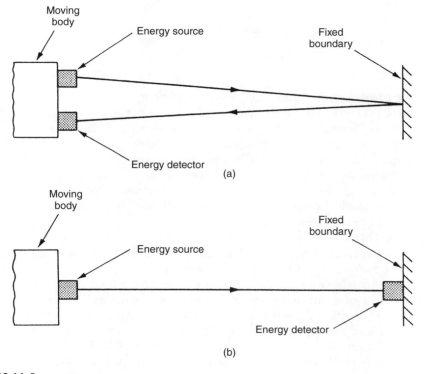

Fig. 19.14 Range sensors.

In ultrasonic systems, the energy is transmitted from the source in high-frequency bursts. A frequency of at least 20 kHz is usual, and 40 kHz is common for measuring distances up to 5 m. By measuring the time of flight of the energy, the distance of the body from the fixed boundary can be calculated, using the fact that the speed of sound in air is 340 m/s. Because of difficulties in measuring the time of flight with sufficient accuracy, ultrasonic systems are not suitable for measuring distances of less than about 300 mm. Measurement resolution is limited by the wavelength of the ultrasonic energy and can be improved by operating at higher frequencies. At higher frequencies, however, attenuation of the magnitude of the ultrasonic wave as it passes through air becomes significant. Therefore, only low frequencies are suitable if large distances are to be measured. The typical inaccuracy of ultrasonic range finding systems is $\pm 0.5\%$ of full scale.

Optical range finding systems generally use a laser light source. The speed of light in air is about 3×10^8 m/s, so that light takes only a few nanoseconds to travel a metre. In consequence, such systems are only suitable for measuring very large displacements where the time of flight is long enough to be measured with reasonable accuracy.

19.1.10 Proximity sensors

For the sake of completeness, it is proper to conclude this chapter on translational displacement transducers with consideration of proximity sensors. Proximity detectors provide information on the displacement of a body with respect to some boundary, but only insofar as to say whether the body is less than or greater than a certain distance away from the boundary. The output of a proximity sensor is thus binary in nature: the body is or is not close to the boundary.

Like range sensors, proximity detectors make use of an energy source and detector. The detector is a device whose output changes between two states when the magnitude of the incident reflected energy exceeds a certain threshold level. A common form of proximity sensor uses an infrared light-emitting diode (LED) source and a phototransistor. Light triggers the transistor into a conducting state when the LED is within a certain distance from a reflective boundary and the reflected light exceeds a threshold level. This system is physically small, occupying a volume of only a few cubic centimetres. If even this small volume is obtrusive, then fibre-optic cables can be used to transmit light from a remotely mounted LED and phototransistor. The threshold displacement detected by optical proximity sensors can be varied between 0 and 2 m.

Another form of proximity sensor uses the principle of varying inductance. Such devices are particularly suitable for operation in aggressive environmental conditions and they can be made vibration and shock resistant by vacuum encapsulation techniques. The sensor contains a high-frequency oscillator whose output is demodulated and fed via a trigger circuit to an amplifier output stage. The oscillator output radiates through the surface of the sensor and, when the sensor surface becomes close to an electrically or magnetically conductive boundary, the output voltage is reduced because of the interference with the flux paths. At a certain point, the output voltage is reduced sufficiently for the trigger circuit to change state and reduce the amplifier output to zero. Inductive sensors can be adjusted to change state at displacements in the range of 1 to 20 mm.

A third form of proximity sensor uses the capacitive principle. These can operate in similar conditions to inductive types. The threshold level of displacement detected can be varied between 5 and 40 mm.

Fibre-optic proximity sensors also exist where the amount of reflected light varies with the proximity of the fibre ends to a boundary, as shown in Figure 13.2(c).

19.1.11 Selection of translational measurement transducers

Choice between the various translational motion transducers available for any particular application depends mainly on the magnitude of the displacement to be measured, although the operating environment is also relevant. Displacements larger than five metres can only be measured by a range sensor, or possibly by the method of using a wire described in section 19.1.9 (Figure 19.14). Such methods are also used for displacements in the range between 2 and 5 metres, except where the expense of a variable inductance transducer can be justified.

For measurements within the range of 2 mm to 2 m, the number of suitable instruments grows. Both the relatively cheap potentiometer and the LVDT, which is somewhat more expensive, are commonly used for such measurements. Variable-inductance and variable-capacitance transducers are also used in some applications. Additionally, strain gauges measuring the strain in two beams forced apart by a wedge (see section 19.1.5) can measure displacements up to 50 mm. If very high measurement resolution is required, either the linear inductosyn or the laser interferometer is used.

The requirement to measure displacements of less than 2 mm usually occurs as part of an instrument that is measuring some other physical quantity such as pressure, and several types of device have evolved to fulfil this task. The LVDT, strain gauges, the Fotonic sensor, variable-capacitance transducers and the non-contacting optical transducer all find application in measuring diaphragm or Bourdon-tube displacements within pressure transducers. Load cell displacements are also very small, and these are commonly measured by nozzle flapper devices.

If the environmental operating conditions are severe (for example, hot, radioactive or corrosive atmospheres), devices that can be easily protected from these conditions must be chosen, such as the LVDT, variable inductance and variable capacitance instruments.

19.2 Velocity

Translational velocity cannot be measured directly and therefore must be calculated indirectly by other means as set out below.

19.2.1 Differentiation of displacement measurements

Differentiation of position measurements obtained from any of the translational displacement transducers described in section 19.1 can be used to produce a translational velocity signal. Unfortunately, the process of differentiation always amplifies noise in a measurement system. Therefore, if this method has to be used, a low-noise instrument such as a d.c. excited carbon film potentiometer or laser interferometer should be chosen. In the case of potentiometers, a.c. excitation must be avoided because of the problem that harmonics in the power supply would cause.

19.2.2 Integration of the output of an accelerometer

Where an accelerometer is already included within a system, integration of its output can be performed to yield a velocity signal. The process of integration attenuates rather than amplifies measurement noise and this is therefore an acceptable technique.

19.2.3 Conversion to rotational velocity

Conversion from translational to rotational velocity is the final measurement technique open to the system designer and is the one most commonly used. This enables any of the rotational velocity measuring instruments described in Chapter 20 to be applied.

19.3 Acceleration

The only class of device available for measuring acceleration is the accelerometer. These are available in a wide variety of types and ranges designed to meet particular measurement requirements. They have a frequency response between zero and a high value, and have a form of output that can be readily integrated to give displacement and velocity measurements. The frequency response of accelerometers can be improved by altering the level of damping in the instrument. Such adjustment must be done carefully, however, because frequency response improvements are only achieved at the expense of degrading the measurement sensitivity. Besides their use for general-purpose motion measurement, accelerometers are widely used to measure mechanical shocks and vibrations.

Most forms of accelerometer consist of a mass suspended by a spring and damper inside a housing, as shown in Figure 19.15. The accelerometer is rigidly fastened to the body undergoing acceleration. Any acceleration of the body causes a force, F_a, on the mass, M, given by:

$$F_a = M\ddot{x}$$

This force is opposed by the restraining effect, F_s, of a spring with spring constant K, and the net result is that the mass is displaced by a distance x from its starting position such that:

$$F_s = Kx$$

In steady state, when the mass inside is accelerating at the same rate as the case of the accelerometer, $F_a = F_s$ and so:

$$Kx = M\ddot{x} \quad \text{or} \quad \ddot{x} = (Kx)/M \qquad (19.4)$$

384 Translational motion transducers

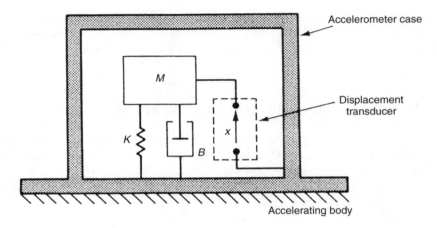

Fig. 19.15 Structure of an accelerometer.

This is the equation of motion of a second order system, and, in the absence of damping, the output of the accelerometer would consist of non-decaying oscillations. A damper is therefore included within the instrument, which produces a damping force, F_d, proportional to the velocity of the mass M given by:

$$F_d = B\dot{x}$$

This modifies the previous equation of motion (19.4) to the following:

$$Kx + B\dot{x} = M\ddot{x} \qquad (19.5)$$

One important characteristic of accelerometers is their sensitivity to accelerations at right angles to the sensing axis (the direction along which the instrument is designed to measure acceleration). This is defined as the *cross-sensitivity* and is specified in terms of the output, expressed as a percentage of the full-scale output, when an acceleration of some specified magnitude (e.g. $30g$) is applied at $90°$ to the sensing axis.

The acceleration reading is obtained from the instrument by measurement of the displacement of the mass within the accelerometer. Many different displacement-measuring techniques are used in the various types of accelerometer that are commercially available. Different types of accelerometer also vary in terms of the type of spring element and form of damping used.

Resistive potentiometers are one such displacement-measuring instrument used in accelerometers. These are used mainly for measuring slowly varying accelerations and low-frequency vibrations in the range $0-50g$. The measurement resolution obtainable is about 1 in 400 and typical values of cross-sensitivity are $\pm 1\%$. Inaccuracy is about $\pm 1\%$ and life expectancy is quoted at two million reversals. A typical size and weight are $125\,\text{cm}^3$ and 500 grams.

Strain gauges and piezoresistive sensors are also used in accelerometers for measuring accelerations up to $200g$. These serve as the spring element as well as measuring mass displacement, thus simplifying the instrument's construction. Their typical characteristics are a resolution of 1 in 1000, inaccuracy of $\pm 1\%$ and cross-sensitivity of 2%. They have a major advantage over potentiometer-based accelerometers in terms of their much smaller size and weight (3 cm^3 and 25 grams).

Another displacement transducer found in accelerometers is the LVDT. This device can measure accelerations up to 700g with a typical inaccuracy of ±1% of full scale. They are of a similar physical size to potentiometer-based instruments but are lighter in weight (100 grams).

Accelerometers based on variable inductance displacement measuring devices have extremely good characteristics and are suitable for measuring accelerations up to 40g. Typical specifications of such instruments are inaccuracy ±0.25% of full scale, resolution 1 in 10 000 and cross-sensitivity 0.5%. Their physical size and weight are similar to potentiometer-based devices. Instruments with an output in the form of a varying capacitance also have similar characteristics.

The other common displacement transducer used in accelerometers is the piezoelectric type. The major advantage of using piezoelectric crystals is that they also act as the spring and damper within the instrument. In consequence, the device is quite small (15 cm^3) and very low mass (50 grams), but because of the nature of piezoelectric crystal operation, such instruments are not suitable for measuring constant or slowly time-varying accelerations. As the electrical impedance of a piezoelectric crystal is itself high, the output voltage must be measured with a very high-impedance instrument to avoid loading effects. Many recent piezoelectric crystal-based accelerometers incorporate a high impedance charge amplifier within the body of the instrument. This simplifies the signal conditioning requirements external to the accelerometer but can lead to problems in certain operational environments because these internal electronics are exposed to the same environmental hazards as the rest of the accelerometer. Typical measurement resolution of this class of accelerometer is 0.1% of full scale with an inaccuracy of ±1%. Individual instruments are available to cover a wide range of measurements from 0.03g full scale up to 1000g full scale. *Intelligent accelerometers* are also now available that give even better performance through inclusion of processing power to compensate for environmentally induced errors.

Recently, very small microsensors have become available for measuring acceleration. These consist of a small mass subject to acceleration that is mounted on a thin silicon membrane. Displacements are measured either by piezoresistors deposited on the membrane or by etching a variable capacitor plate into the membrane.

Two forms of fibre-optic-based accelerometer also exist. One form measures the effect on light transmission intensity caused by a mass subject to acceleration resting on a multimode fibre. The other form measures the change in phase of light transmitted through a monomode fibre that has a mass subject to acceleration resting on it.

19.3.1 Selection of accelerometers

In choosing between the different types of accelerometer for a particular application, the mass of the instrument is particularly important. This should be very much less than that of the body whose motion is being measured, in order to avoid loading effects that affect the accuracy of the readings obtained. In this respect, instruments based on strain gauges are best.

19.4 Vibration

19.4.1 Nature of vibration

Vibrations are very commonly encountered in machinery operation, and therefore measurement of the accelerations associated with such vibrations is extremely important in industrial environments. The peak accelerations involved in such vibrations can be of 100g or greater in magnitude, whilst both the frequency of oscillation and the magnitude of displacements from the equilibrium position in vibrations have a tendency to vary randomly. Vibrations normally consist of linear harmonic motion that can be expressed mathematically as:

$$X = X_0 \sin(\omega t) \tag{19.6}$$

where X is the displacement from the equilibrium position at any general point in time, X_0 is the peak displacement from the equilibrium position, and ω is the angular frequency of the oscillations. By differentiating equation (19.6) with respect to time, an expression for the velocity v of the vibrating body at any general point in time is obtained as:

$$v = -\omega X_0 \cos(\omega t) \tag{19.7}$$

Differentiating equation (19.7) again with respect to time, we obtain an expression for the acceleration, α, of the body at any general point in time as:

$$\alpha = -\omega^2 X_0 \sin(\omega t) \tag{19.8}$$

Inspection of equation (19.8) shows that the peak acceleration is given by:

$$\alpha_{\text{peak}} = \omega^2 X_0 \tag{19.9}$$

This square law relationship between peak acceleration and oscillation frequency is the reason why high values of acceleration occur during relatively low-frequency oscillations. For example, an oscillation at 10 Hz produces peak accelerations of $2g$.

Example
A pipe carrying a fluid vibrates at a frequency of 50 Hz with displacements of 8 mm from the equilibrium position. Calculate the peak acceleration.

Solution
From equation (19.9),

$$\alpha_{\text{peak}} = \omega^2 X_0 = (2\pi 50)^2 \times (0.008) = 789.6 \text{ m/s}^2$$

Using the fact that the acceleration due to gravity, g, is 9.81 m/s², this answer can be expressed alternatively as:

$$\alpha_{\text{peak}} = 789.6/9.81 = 80.5g$$

19.4.2 Vibration measurement

It is apparent that the intensity of vibration can be measured in terms of either displacement, velocity or acceleration. Acceleration is clearly the best parameter to measure at high frequencies. However, because displacements are large at low frequencies according to equation (19.9), it would seem that measuring either displacement or velocity would be best at low frequencies. The amplitude of vibrations can be measured by various forms of displacement transducer. Fibre-optic-based devices are particularly attractive and can give measurement resolution as high as 1 µm. Unfortunately, there are considerable practical difficulties in mounting and calibrating displacement and velocity transducers and therefore they are rarely used. Thus, vibration is usually measured by accelerometers at all frequencies. The most common type of transducer used is the piezoaccelerometer, which has typical inaccuracy levels of ±2%.

The frequency response of accelerometers is particularly important in vibration measurement in view of the inherently high-frequency characteristics of the measurement situation. The bandwidth of both potentiometer-based accelerometers and accelerometers using variable-inductance type displacement transducers goes up to 25 Hz only. Accelerometers including either the LVDT or strain gauges can measure frequencies up to 150 Hz and the latest instruments using piezoresistive strain gauges have bandwidths up to 2 kHz. Finally, inclusion of piezoelectric crystal displacement transducers yields an instrument with a bandwidth that can be as high as 7 kHz.

When measuring vibration, consideration must be given to the fact that attaching an accelerometer to the vibrating body will significantly affect the vibration characteristics if the body has a small mass. The effect of such 'loading' of the measured system can be quantified by the following equation:

$$a_1 = a_b \left(\frac{m_b}{m_b + m_a} \right)$$

where a_1 is the acceleration of the body with accelerometer attached, a_b is the acceleration of the body without the accelerometer, m_a is the mass of the accelerometer and m_b is the mass of the body. Such considerations emphasize the advantage of piezoaccelerometers, as these have a lower mass than other forms of accelerometer and so contribute least to this system-loading effect.

As well as an accelerometer, a vibration measurement system requires other elements, as shown in Figure 19.16, to translate the accelerometer output into a recorded signal. The three other necessary elements are a signal-conditioning element, a signal analyser and a signal recorder. The signal-conditioning element amplifies the relatively weak output signal from the accelerometer and also transforms the high output impedance of the accelerometer to a lower impedance value. The signal analyser then converts the signal into the form required for output. The output parameter may be either displacement, velocity or acceleration and this may be expressed as either the peak value, r.m.s. value or average absolute value. The final element of the measurement system is the signal recorder. All elements of the measurement system, and especially the signal recorder, must be chosen very carefully to avoid distortion of the vibration waveform. The bandwidth should be such that it is at least a factor of ten better than the bandwidth of the vibration frequency components at both ends. Thus its lowest frequency limit should be less than or equal to 0.1 times the fundamental frequency of vibration and its upper frequency limit should be greater than or equal to ten times the highest significant vibration frequency component.

388 Translational motion transducers

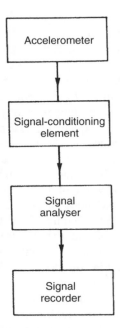

Fig. 19.16 Vibration measurement system.

If the frequency of vibration has to be known, the stroboscope is a suitable instrument to measure this. If the stroboscope is made to direct light pulses at the body at the same frequency as the vibration, the body will apparently stop vibrating.

19.5 Shock

Shock describes a type of motion where a moving body is brought suddenly to rest, often because of a collision. This is very common in industrial situations and usually involves a body being dropped and hitting the floor. Shocks characteristically involve large-magnitude deceleration (e.g. $500g$) that last for a very short time (e.g. 5 ms). An instrument having a very high-frequency response is required for shock measurement, and for this reason, piezoelectric crystal-based accelerometers are commonly used. Again, other elements for analysing and recording the signal are required as shown in Figure 19.16 and described in the last section. A storage oscilloscope is a suitable instrument for recording the output signal, as this allows the time duration as well as the acceleration levels in the shock to be measured. Alternatively, if a permanent record is required, the screen of a standard oscilloscope can be photographed. A further option is to record the output on magnetic tape, which facilitates computerized signal analysis.

Example
A body is dropped from a height of 10 m and suffers a shock when it hits the ground. If the duration of the shock is 5 ms, calculate the magnitude of the shock in terms of g.

Solution

The equation of motion for a body falling under gravity gives the following expression for the terminal velocity, v:

$$v = \sqrt{2gx}$$

where x is the height through which the body falls. Having calculated v, the average deceleration during the collision can be calculated as:

$$\alpha = v/t$$

where t is the time duration of the shock. Substituting the appropriate numerical values into these expressions:

$$v = \sqrt{(2 \times 9.81 \times 10)} = 14.0 \text{ m/s}; \quad \alpha = 14.0/0.005 = 2801 \text{ m/s} = 286g.$$

20

Rotational motion transducers

20.1 Rotational displacement

Rotational displacement transducers measure the angular motion of a body about some rotation axis. They are important not only for measuring the rotation of bodies such as shafts, but also as part of systems that measure translational displacement by converting the translational motion to a rotary form. The various devices available for measuring rotational displacements are presented below, and the arguments for choosing a particular form in any given measurement situation are considered at the end of the chapter.

20.1.1 Circular and helical potentiometers

The circular potentiometer is the cheapest device available for measuring rotational displacements. It works on almost exactly the same principles as the translational motion potentiometer, except that the track is bent round into a circular shape. The measurement range of individual devices varies from 0–10° to 0–360° depending on whether the track forms a full circle or only part of a circle. Where greater measurement range than 0–360° is required, a helical potentiometer is used, with some devices being able to measure up to 60 full turns. The helical potentiometer accommodates multiple turns of the track by forming the track into a helix shape. However, its greater mechanical complexity makes the device significantly more expensive than a circular potentiometer. The two forms of device are shown in Figure 20.1.

Both kinds of device give a linear relationship between the measured quantity and the output reading because the output voltage measured at the sliding contact is proportional to the angular displacement of the slider from its starting position. However, as with linear track potentiometers, all rotational potentiometers can give performance problems due to dirt on the track causing loss of contact. They also have a limited life because of wear between the sliding surfaces. The typical inaccuracy of this class of devices varies from ±1% of full scale for circular potentiometers down to ±0.002% of full scale for the best helical potentiometers.

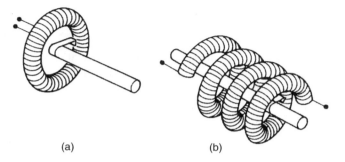

Fig. 20.1 Rotary motion potentiometers: (a) circular; (b) helical.

20.1.2 Rotational differential transformer

This is a special form of differential transformer that measures rotational rather than translational motion. The method of construction and connection of the windings is exactly the same as for the linear variable differential transformer (LVDT), except that a specially shaped core is used that varies the mutual inductance between the windings as it rotates, as shown in Figure 20.2. Like its linear equivalent, the instrument suffers no wear in operation and therefore has a very long life with almost no maintenance requirements. It can also be modified for operation in harsh environments by enclosing the windings inside a protective enclosure. However, apart from the difficulty of avoiding some asymmetry between the secondary windings, great care has to be taken in these instruments to machine the core to exactly the right shape. In consequence, the inaccuracy cannot be reduced below ±1%, and even this level of accuracy is only obtained for limited excursions of the core of ±40° away from the null position. For angular displacements of ±60°, the typical inaccuracy rises to ±3%, and the instrument is unsuitable for measuring displacements greater than this.

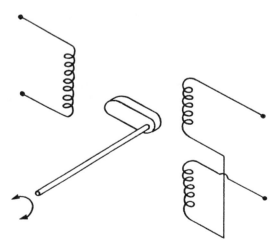

Fig. 20.2 Rotary differential transformer.

20.1.3 Incremental shaft encoders

Incremental shaft encoders are one of a class of encoder devices that give an output in digital form. They measure the instantaneous angular position of a shaft relative to some arbitrary datum point, but are unable to give any indication about the absolute position of a shaft. The principle of operation is to generate pulses as the shaft whose displacement is being measured rotates. These pulses are counted and the total angular rotation inferred from the pulse count. The pulses are generated either by optical or by magnetic means and are detected by suitable sensors. Of the two, the optical system is considerably cheaper and therefore much more common. Such instruments are very convenient for computer control applications, as the measurement is already in the required digital form and therefore the usual analogue to digital signal conversion process is avoided.

An example of an optical incremental shaft encoder is shown in Figure 20.3. It can be seen that the instrument consists of a pair of discs, one of which is fixed and one of which rotates with the body whose angular displacement is being measured. Each disc is basically opaque but has a pattern of windows cut into it. The fixed disc has only one window and the light source is aligned with this so that the light shines through all the time. The second disc has two tracks of windows cut into it that are equidistantly spaced around the disc, as shown in Figure 20.4. Two light detectors are positioned beyond the second disc so that one is aligned with each track of windows. As the second disc rotates, light alternately enters and does not enter the detectors, as windows and then opaque regions of the disc pass in front of them. These pulses of light are fed to a counter, with the final count after motion has ceased corresponding to the angular position of the moving body relative to the starting position. The primary information about the magnitude of rotation is obtained by the detector aligned with the

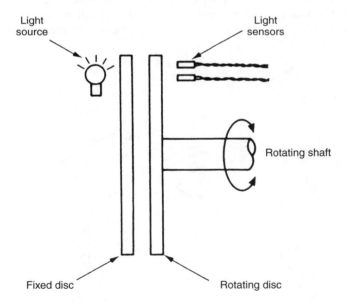

Fig. 20.3 Optical incremental shaft encoder.

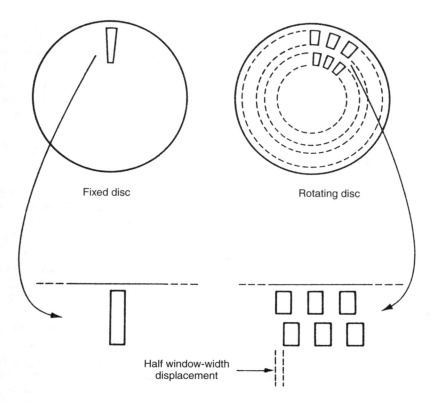

Fig. 20.4 Window arrangement in incremental shaft encoder.

outer track of windows. The pulse count obtained from this gives no information about the direction of rotation, however. Direction information is provided by the second, inner track of windows, which have an angular displacement with respect to the outer set of windows of half a window width. The pulses from the detector aligned with the inner track of windows therefore lag or lead the primary set of pulses according to the direction of rotation.

The maximum measurement resolution obtainable is limited by the number of windows that can be machined onto a disc. The maximum number of windows per track for a 150 mm-diameter disc is 5000, which gives a basic angular measurement resolution of 1 in 5000. By using more sophisticated circuits that increment the count on both the rising and falling edges of the pulses through the outer track of windows, it is possible to double the resolution to a maximum of 1 in 10 000. At the expense of even greater complexity in the counting circuit, it is possible also to include the pulses from the inner track of windows in the count, so giving a maximum measurement resolution of 1 in 20 000.

Optical incremental shaft encoders are a popular instrument for measuring relative angular displacements and are very reliable. Problems of noise in the system giving false counts can sometimes cause difficulties, although this can usually be eliminated by squaring the output from the light detectors. Such instruments are found in many applications where rotational motion has to be measured. Incremental shaft encoders are also commonly used in circumstances where a translational displacement has been transformed to a rotational one by suitable gearing. One example of this practice is in measuring the translational motions in numerically controlled (NC) drilling machines. Typical gearing used for this would give one revolution per mm of translational displacement. By using an incremental shaft encoder with 1000 windows per track in such an arrangement, a measurement resolution of 1 micron is obtained.

20.1.4 Coded-disc shaft encoders

Unlike the incremental shaft encoder that gives a digital output in the form of pulses that have to be counted, the digital shaft encoder has an output in the form of a binary number of several digits that provides an absolute measurement of shaft position. Digital encoders provide high accuracy and reliability. They are particularly useful for computer control applications, but they have a significantly higher cost than incremental encoders. Three different forms exist, using optical, electrical and magnetic energy systems respectively.

Optical digital shaft encoder
The optical digital shaft encoder is the cheapest form of encoder available and is the one used most commonly. It is found in a variety of applications, and one where it is particularly popular is in measuring the position of rotational joints in robot manipulators. The instrument is similar in physical appearance to the incremental shaft encoder. It has a pair of discs (one movable and one fixed) with a light source on one side and light detectors on the other side, as shown in Figure 20.5. The fixed

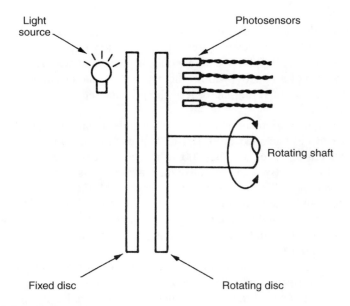

Fig. 20.5 Coded disc shaft encoder.

disc has a single window, and the principal way in which the device differs from the incremental shaft encoder is in the design of the windows on the movable disc, as shown in Figure 20.6. These are cut in four or more tracks instead of two and are arranged in sectors as well as tracks. An energy detector is aligned with each track, and these give an output of '1' when energy is detected and an output of '0' otherwise. The measurement resolution obtainable depends on the number of tracks used. For a four-track version, the resolution is 1 in 16, with progressively higher measurement resolution being attained as the number of tracks is increased. These binary outputs from the detectors are combined together to give a binary number of several digits. The number of digits corresponds to the number of tracks on the disc, which in the example shown in Figure 20.6 is four. The pattern of windows in each sector is cut such that, as that particular sector passes across the window in the fixed disc, the four energy detector outputs combine to give a unique binary number. In the binary-coded example shown in Figure 20.6, the binary number output increments by one as each sector in the rotating disc passes in turn across the window in the fixed disc. Thus the output from sector 1 is 0001, from sector 2 is 0010, from sector 3 is 0011, etc.

Whilst this arrangement is perfectly adequate in theory, serious problems can arise in practice due to the manufacturing difficulty involved in machining the windows of the movable disc such that the edges of the windows in each track are exactly aligned with each other. Any misalignment means that, as the disc turns across the boundary between one sector and the next, the outputs from each track will switch at slightly different instants of time, and therefore the binary number output will be incorrect over small angular ranges corresponding to the sector boundaries. The worst error can occur at the boundary between sectors seven and eight, where the output is switching from 0111 to 1000. If the energy sensor corresponding to the first digit switches before the others, then the output will be 1111 for a very small angular range of movement, indicating that sector 15 is aligned with the fixed disc rather than sector seven or eight. This represents an error of 100% in the indicated angular position.

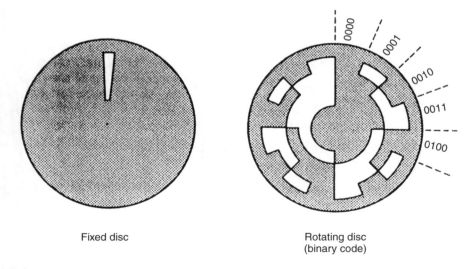

Fixed disc

Rotating disc (binary code)

Fig. 20.6 Window arrangement for coded disc shaft encoder.

396 Rotational motion transducers

There are two ways used in practice to overcome this difficulty, which both involve an alteration to the manner in which windows are machined on the movable disc, as shown in Figure 20.7. The first of these methods adds an extra outer track on the disc, known as an *anti-ambiguity track*, which consists of small windows that span a small angular range on either side of each sector boundary of the main track system. When energy sensors associated with this extra track sense energy, this is used to signify that the disc is aligned on a sector boundary and the output is unreliable.

The second method is somewhat simpler and cheaper, because it avoids the expense of machining the extra anti-ambiguity track. It does this by using a special code, known as the Gray code, to cut the tracks in each sector on the movable disc. The Gray code is a special binary representation, where only one binary digit changes in moving from one decimal number representation to the next, i.e. from one sector to the next in the digital shaft encoder. The code is illustrated in Table 20.1.

It is possible to manufacture optical digital shaft encoders with up to 21 tracks, which gives a measurement resolution of 1 part in 10^6 (about one second of arc). Unfortunately, there is a high cost involved in the special photolithography techniques used to cut the windows in order to achieve such a measurement resolution, and very high-quality mounts and bearings are needed. Hence, such devices are very expensive.

Contacting (electrical) digital shaft encoder

The contacting digital shaft encoder consists of only one disc that rotates with the body whose displacement is being measured. The disc has conducting and non-conducting segments rather than the transparent and opaque areas found on the movable disc of the optical form of instrument, but these are arranged in an identical pattern of sectors and tracks. The disc is charged to a low potential by an electrical brush in contact with one side of the disc, and a set of brushes on the other side of the disc measures the potential in each track. The output of each detector brush is interpreted as a binary value of '1' or '0' according to whether the track in that particular segment is conducting or not and hence whether a voltage is sensed or not. As for the case of the optical

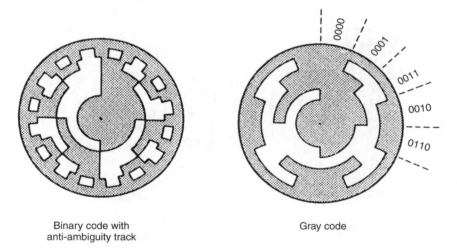

Binary code with anti-ambiguity track

Gray code

Fig. 20.7 Modified window arrangements for the rotating disc.

Table 20.1 The Gray code

Decimal	Binary	Gray
0	0000	0000
1	0001	0001
2	0010	0011
3	0011	0010
4	0100	0110
5	0101	0111
6	0110	0101
7	0111	0100
8	1000	1100
9	1001	1101
10	1010	1111
11	1011	1110
12	1100	1010
13	1101	1011
14	1110	1001
15	1111	1000

form of instrument, these outputs are combined together to give a multi-bit binary number. Contacting digital shaft encoders have a similar cost to the equivalent optical instruments and have operational advantages in severe environmental conditions of high temperature or mechanical shock. They suffer from the usual problem of output ambiguity at the sector boundaries but this problem is overcome by the same methods as used in optical instruments.

A serious problem in the application of contacting digital shaft encoders arises from their use of brushes. These introduce friction into the measurement system, and the combination of dirt and brush wear causes contact problems. Consequently, problems of intermittent output can occur, and such instruments generally have limited reliability and a high maintenance cost. Measurement resolution is also limited because of the lower limit on the minimum physical size of the contact brushes. The maximum number of tracks possible is ten, which limits the resolution to 1 part in 1000. Thus, contacting digital shaft encoders are only used where the environmental conditions are too severe for optical instruments.

Magnetic digital shaft encoder

Magnetic digital shaft encoders consist of a single rotatable disc, as in the contacting form of encoder discussed in the previous section. The pattern of sectors and tracks consists of magnetically conducting and non-conducting segments, and the sensors aligned with each track consist of small toroidal magnets. Each of these sensors has a coil wound on it that has a high or low voltage induced in it according to the magnetic field close to it. This field is dependent on the magnetic conductivity of that segment of the disc that is closest to the toroid.

These instruments have no moving parts in contact and therefore have a similar reliability to optical devices. Their major advantage over optical equivalents is an ability to operate in very harsh environmental conditions. Unfortunately, the process of manufacturing and accurately aligning the toroidal magnet sensors required makes such instruments very expensive. Their use is therefore limited to a few applications where both high measurement resolution and also operation in harsh environments are required.

20.1.5 The resolver

The resolver, also known as a *synchro-resolver*, is an electromechanical device that gives an analogue output by transformer action. Physically, resolvers resemble a small a.c. motor and have a diameter ranging from 10 mm to 100 mm. They are frictionless and reliable in operation because they have no contacting moving surfaces, and consequently they have a long life. The best devices give measurement resolutions of 0.1%.

Resolvers have two stator windings, which are mounted at right angles to one another, and a rotor, which can have either one or two windings. As the angular position of the rotor changes, the output voltage changes. The simpler configuration of a resolver with only one winding on the rotor is illustrated in Figure 20.8. This exists in two separate forms that are distinguished according to whether the output voltage changes in amplitude or changes in phase as the rotor rotates relative to the stator winding.

Varying amplitude output resolver
The stator of this type of resolver is excited with a single-phase sinusoidal voltage of frequency ω, where the amplitudes in the two windings are given by:

$$V_1 = V \sin(\beta); \quad V_2 = V \cos(\beta)$$

where $V = V_s \sin(\omega t)$

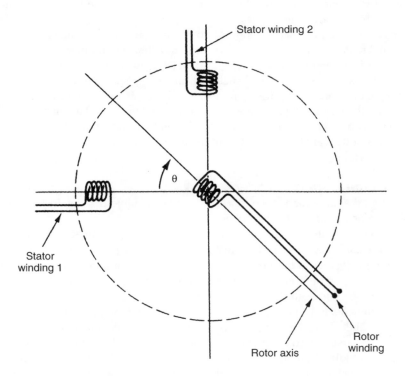

Fig. 20.8 Schematic representation of resolver windings.

The effect of this is to give a field at an angle of $(\beta + \pi/2)$ relative to stator winding 1. (A full proof of this can be found in Healey, (1975).)

Suppose that the angle of the rotor winding relative to that of the stator winding is given by θ. Then the magnetic coupling between the windings is a maximum for $\theta = (\beta + \pi/2)$ and a minimum for $(\theta = \beta)$. The rotor output voltage (see Healey (1975) for proof) is of fixed frequency and varying amplitude given by:

$$V_0 = KV_s \sin(\beta - \theta) \sin(\omega t)$$

This relationship between shaft angle position and output voltage is non-linear, but approximate linearity is obtained for small angular motions where $|\beta - \theta| < 15°$.

An intelligent version of this type of resolver is now available that uses a microprocessor to process the sine and cosine outputs, giving a measurement resolution of 2 minutes of arc (Analogue Devices, 1988).

Varying phase output resolver

This is a less common form of resolver but it is used in a few applications. The stator windings are excited with a two-phase sinusoidal voltage of frequency ω, and the instantaneous voltage amplitudes in the two windings are given by:

$$V_1 = V_s \sin(\omega t); \quad V_2 = V_s \sin(\omega t + \pi/2) = V_s \cos(\omega t)$$

The net output voltage in the rotor winding is the sum of the voltages induced due to each stator winding. This is given by:

$$\begin{aligned} V_0 &= KV_s \sin(\omega t) \cos(\theta) + KV_s \cos(\omega t) \cos(\pi/2 - \theta) \\ &= KV_s[\sin(\omega t) \cos(\theta) + \cos(\omega t) \sin(\theta)] \\ &= KV_s \sin(\omega t + \theta) \end{aligned}$$

This represents a linear relationship between shaft angle and the phase shift of the rotor output relative to the stator excitation voltage. The accuracy of shaft rotation measurement depends on the accuracy with which the phase shift can be measured. This can be improved by increasing the excitation frequency, ω, and it is possible to reduce inaccuracy to $\pm 0.1\%$. However, increasing the excitation frequency also increases magnetizing losses. Consequently, a compromise excitation frequency of about 400 Hz is used.

20.1.6 The synchro

Like the resolver, the synchro is a motor-like, electromechanical device with an analogue output. Apart from having three stator windings instead of two, the instrument is similar in appearance and operation to the resolver and has the same range of physical dimensions. The rotor usually has a dumb-bell shape and, like the resolver, can have either one or two windings.

Synchros have been in use for many years for the measurement of angular positions, especially in military applications, and achieve similar levels of accuracy and

Rotational motion transducers

measurement resolution to digital encoders. One common application is axis measurement in machine tools, where the translational motion of the tool is translated into a rotational displacement by suitable gearing. Synchros are tolerant to high temperatures, high humidity, shock and vibration and are therefore suitable for operation in such harsh environmental conditions. Some maintenance problems are associated with the slip ring and brush system used to supply power to the rotor. However, the only major source of error in the instrument is asymmetry in the windings, and reduction of measurement inaccuracy down to $\pm 0.5\%$ is easily achievable.

Figure 20.9 shows the simpler form of synchro with a single rotor winding. If an a.c. excitation voltage is applied to the rotor via slip rings and brushes, this sets up a certain pattern of fluxes and induced voltages in the stator windings by transformer action. For a rotor excitation voltage, V_r, given by:

$$V_r = V \sin(\omega t)$$

the voltages induced in the three stator windings are:

$$V_1 = V \sin(\omega t) \sin(\beta); \quad V_2 = V \sin(\omega t) \sin(\beta + 2\pi/3); \quad V_3 = V \sin(\omega t) \sin(\beta - 2\pi/3)$$

where β is the angle between the rotor and stator windings.

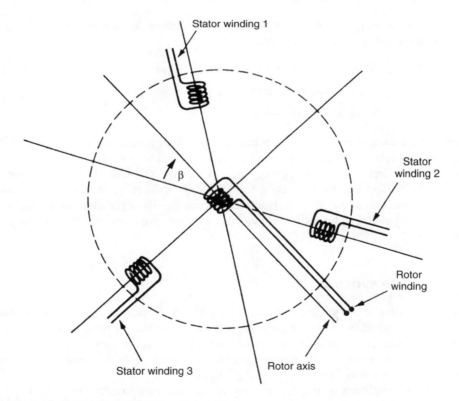

Fig. 20.9 Schematic representation of synchro windings.

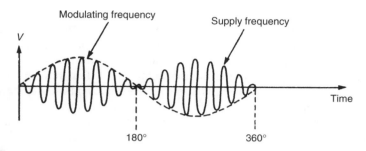

Fig. 20.10 Synchro stator voltage waveform.

If the rotor is turned at constant velocity through one full revolution, the voltage waveform induced in each stator winding is as shown in Figure 20.10. This has the form of a carrier-modulated waveform, in which the carrier frequency corresponds to the excitation frequency, ω. It follows that if the rotor is stopped at any particular angle, β', the peak-to-peak amplitude of the stator voltage is a function of β'. If therefore the stator winding voltage is measured, generally as its root-mean-squared (r.m.s.) value, this indicates the magnitude of the rotor rotation away from the null position. The direction of rotation is determined by the phase difference between the stator voltages, which is indicated by their relative instantaneous magnitudes.

Although a single synchro thus provides a means of measuring angular displacements, it is much more common to find a pair of them used for this purpose. When used in pairs, one member of the pair is known as the synchro transmitter and the other as the synchro transformer, and the two sets of stator windings are connected together, as shown in Figure 20.11. Each synchro is of the form shown in Figure 20.9, but the rotor of the transformer is fixed for displacement-measuring applications. A sinusoidal excitation voltage is applied to the rotor of the transmitter, setting up a pattern of fluxes and induced voltages in the transmitter stator windings. These voltages are transmitted to the transformer stator windings where a similar flux pattern is established. This in turn causes a sinusoidal voltage to be induced in the fixed transformer rotor winding. For an excitation voltage, $V\sin(\omega t)$, applied to the transmitter rotor, the voltage measured in the transformer rotor is given by:

$$V_0 = V\sin(\omega t)\sin(\theta)$$

where θ is the relative angle between the two rotor windings.

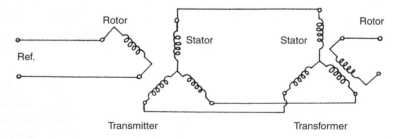

Fig. 20.11 Synchro transmitter-transformer pair.

Apart from their use as a displacement transducer, such synchro pairs are commonly used to transmit angular displacement information over some distance, for instance to transmit gyro compass measurements in an aircraft to remote meters. They are also used for load positioning, allowing a load connected to the transformer rotor shaft to be controlled remotely by turning the transmitter rotor. For these applications, the transformer rotor is free to rotate and is also damped to prevent oscillatory motions. In the simplest arrangement, a common sinusoidal excitation voltage is applied to both rotors. If the transmitter rotor is turned, this causes an imbalance in the magnetic flux patterns and results in a torque on the transformer rotor that tends to bring it into line with the transmitter rotor. This torque is typically small for small displacements, and so this technique is only useful if the load torque on the transformer shaft is very small. In other circumstances, it is necessary to incorporate the synchro pair within a servomechanism, where the output voltage induced in the transformer rotor winding is amplified and applied to a servomotor that drives the transformer rotor shaft until it is aligned with the transmitter shaft.

20.1.7 The induction potentiometer

These instruments belong to the same class as resolvers and synchros but have only one rotor winding and one stator winding. They are of a similar size and appearance to other devices in the class. A single-phase sinusoidal excitation is applied to the rotor winding and this causes an output voltage in the stator winding through the mutual inductance linking the two windings. The magnitude of this induced stator voltage varies with the rotation of the rotor. The variation of the output with rotation is naturally sinusoidal if the coils are wound such that their field is concentrated at one point, and only small excursions can be made away from the null position if the output relationship is to remain approximately linear. However, if the rotor and stator windings are distributed around the circumference in a special way, an approximately linear relationship for angular displacements of up to ±90° can be obtained.

20.1.8 The rotary inductosyn

This instrument is similar in operation to the linear inductosyn, except that it measures rotary displacements and has tracks that are arranged radially on two circular discs, as shown in Figure 20.12. Typical diameters of the instrument vary between 75 mm and 300 mm. The larger versions give a measurement resolution of up to 0.05 seconds of arc. Like its linear equivalent, however, the rotary inductosyn has a very small measurement range, and a lower-resolution, rotary displacement transducer with a larger measurement range must be used in conjunction with it.

20.1.9 Gyroscopes

Gyroscopes measure both absolute angular displacement and absolute angular velocity. The predominance of mechanical, spinning-wheel gyroscopes in the market place is now being challenged by recently introduced optical gyroscopes.

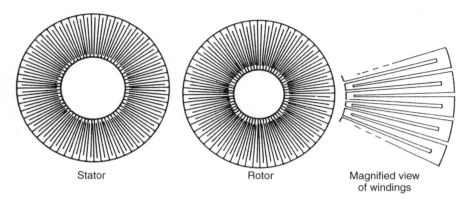

Fig. 20.12 Rotary inductosyn.

Mechanical gyroscopes

Mechanical gyroscopes consist essentially of a large, motor driven wheel whose angular momentum is such that the axis of rotation tends to remain fixed in space, thus acting as a reference point. The gyro frame is attached to the body whose motion is to be measured. The output is measured in terms of the angle between the frame and the axis of the spinning wheel. Two different forms of mechanical gyroscope are used for measuring angular displacement, the free gyro and the rate-integrating gyro. A third type of mechanical gyroscope, the rate gyro, measures angular velocity and is described in section 20.2.

Free gyroscope

The free gyroscope is illustrated in Figure 20.13. This measures the absolute angular rotation of the body to which its frame is attached about two perpendicular axes. Two alternative methods of driving the wheel are used in different versions of the instrument. One of these is to enclose the wheel in stator-like coils that are excited with a sinusoidal voltage. A voltage is applied to the wheel via slip rings at both ends of the spindle carrying the wheel. The wheel behaves as a rotor and motion is produced by motor action. The other, less common, method is to fix vanes on the wheel that is then driven by directing a jet of air onto the vanes.

The instrument can measure angular displacements of up to $10°$ with a high accuracy. For greater angular displacements, interaction between the measurements on the two perpendicular axes starts to cause a serious loss of accuracy. The physical size of the coils in the motor-action driven system also limits the measurement range to $10°$. For these reasons, this type of gyroscope is only suitable for measuring rotational displacements of up to $10°$. A further operational problem of free gyroscopes is the presence of angular drift (precession) due to bearing friction torque. This has a typical magnitude of $0.5°$ per minute and means that the instrument can only be used over short time intervals of say, 5 minutes. This time duration can be extended if the angular momentum of the spinning wheel is increased.

A major application of gyroscopes is in inertial navigation systems. Only two free gyros mounted along orthogonal axes are needed to monitor motions in three dimensions, because each instrument measures displacement about two axes. The limited

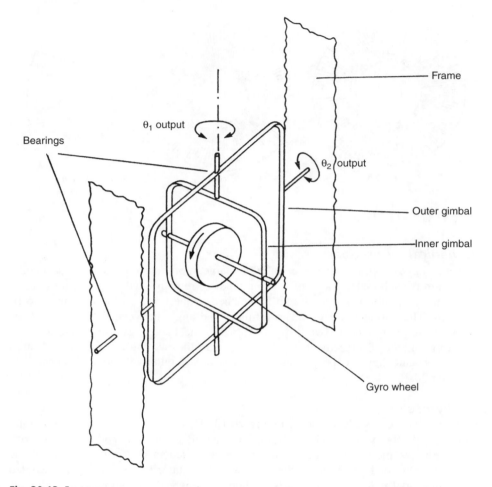

Fig. 20.13 Free gyroscope.

angular range of measurement is not usually a problem in such applications, as control action prevents the error in the direction of motion about any axis ever exceeding one or two degrees. Precession is a much greater problem, however, and for this reason, the rate-integrating gyro is used much more commonly.

Rate integrating gyroscope

The rate-integrating gyroscope, or *integrating gyro* as it is commonly known, is illustrated in Figure 20.14. It measures angular displacements about a single axis only, and therefore three instruments are required in a typical inertial navigation system. The major advantage of the instrument over the free gyro is the almost total absence of precession, with typical specifications quoting drifts of only 0.01°/hour. The instrument has a first order type of response given by:

$$\frac{\theta_o}{\theta_i}(D) = \frac{K}{\tau D + 1} \qquad (20.1)$$

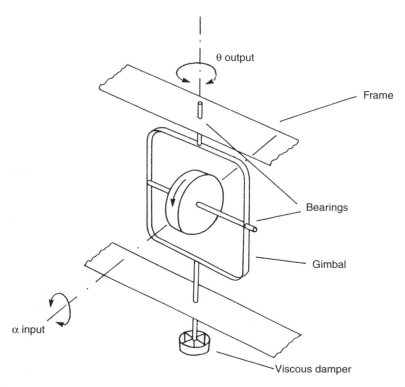

Fig. 20.14 Rate-integrating gyroscope.

where $K = H/\beta$, $\tau = M/\beta$, θ_i is the input angle, θ_o is the output angle, D is the D-operator, H is the angular momentum, M is the moment of inertia of the system about the measurement axis and β is the damping coefficient.

Inspection of equation (20.1) shows that to obtain a high value of measurement sensitivity, K, a high value of H and low value of β are required. A large H is normally obtained by driving the wheel with a hysteresis-type motor revolving at high speeds of up to 24 000 rpm. The damping coefficient β can only be reduced so far, however, because a small value of β results in a large value for the system time constant, τ, and an unacceptably low speed of system response. Therefore, the value of β has to be chosen as a compromise between these constraints.

Besides their use as a fixed reference in inertial guidance systems, integrating gyros are also commonly used within aircraft autopilot systems and in military applications such as stabilizing weapon systems in tanks.

Optical gyroscopes

Optical gyroscopes have been developed only recently and come in two forms, the ring laser gyroscope and the fibre-optic gyroscope.

The *ring laser gyroscope* consists of a glass ceramic chamber containing a helium–neon gas mixture in which two laser beams are generated by a single anode/twin cathode system, as shown in Figure 20.15. Three mirrors, supported by the

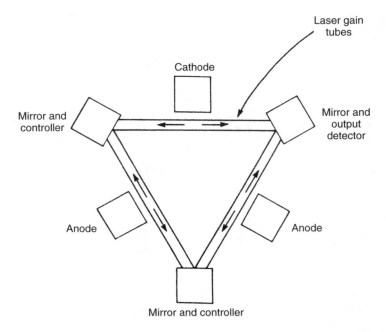

Fig. 20.15 Ring laser gyroscope.

ceramic block and mounted in a triangular arrangement, direct the pair of laser beams around the cavity in opposite directions. Any rotation of the ring affects the coherence of the two beams, raising one in frequency and lowering the other. The clockwise and anticlockwise beams are directed into a photodetector that measures the beat frequency according to the frequency difference, which is proportional to the angle of rotation. A more detailed description of the mode of operation can be found elsewhere (Nuttall, 1987). The advantages of the ring laser gyroscope are considerable. The measurement accuracy obtained is substantially better than that afforded by mechanical gyros in a similar price range. The device is also considerably smaller physically, which is of considerable benefit in many applications.

The *fibre-optic gyroscope* measures angular velocity and is described in section 20.2.

20.1.10 Choice between rotational displacement transducers

Choice between the various rotational displacement transducers that might be used in any particular measurement situation depends first of all upon whether absolute measurement of angular position is required or whether the measurement of rotation relative to some arbitrary starting point is acceptable. Other factors affecting the choice between instruments are the required measurement range, the resolution of the transducer and the measurement accuracy afforded.

Where only measurement of relative angular position is required, the incremental encoder is a very suitable instrument. The best commercial instruments of this type can measure rotations to a resolution of 1 part in 20 000 of a full revolution, and the measurement range is an infinite number of revolutions. Instruments with such a high measurement resolution are very expensive, but much cheaper versions are available according to what lower level of measurement resolution is acceptable.

All the other instruments presented in this chapter provide an absolute measurement of angular position. The required measurement range is a dominant factor in the choice between these. If this exceeds one full revolution, then the only instrument available is the helical potentiometer. Such devices can measure rotations of up to 60 full turns, but they are expensive because the procedure involved in manufacturing a helical resistance element to a reasonable standard of accuracy is difficult.

For measurements of less than one full revolution, the range of available instruments widens. The cheapest one available is the circular potentiometer, but much better measurement accuracy and resolution is obtained from coded-disc encoders. The cheapest of these is the optical form, but certain operating environments necessitate the use of the alternative contacting (electrical) and magnetic versions. All types of coded-disc encoder are very reliable and are particularly attractive in computer control schemes, as the output is in digital form. A varying phase output resolver is yet another instrument that can measure angular displacements up to one full revolution in magnitude. Unfortunately, this instrument is expensive because of the complicated electronics incorporated to measure the phase variation and convert it to a varying-amplitude output signal, and hence use is not common.

An even greater range of instruments becomes available as the required measurement range is reduced further. These include the synchro ($\pm 90°$), the varying amplitude output resolver ($\pm 90°$), the induction potentiometer ($\pm 90°$) and the differential transformer ($\pm 40°$). All these instruments have a high reliability and a long service life.

Finally, two further instruments are available for satisfying special measurement requirements, the rotary inductosyn and the gyroscope. The rotary inductosyn is used in applications where very high measurement resolution is required, although the measurement range afforded is extremely small and a coarser-resolution instrument must be used in parallel to extend the measurement range. Gyroscopes, in both mechanical and optical forms, are used to measure small angular displacements up to $\pm 10°$ in magnitude in inertial navigation systems and similar applications.

20.2 Rotational velocity

The main application of rotational velocity transducers is in speed control systems. They also provide the usual means of measuring translational velocities, which are transformed into rotational motions for measurement purposes by suitable gearing. Many different instruments and techniques are available for measuring rotational velocity as presented below.

20.2.1 Digital tachometers

Digital tachometers, or to give them their proper title, digital *tachometric generators*, are usually non-contact instruments that sense the passage of equally spaced marks on the surface of a rotating disc or shaft. Measurement resolution is governed by the

408 Rotational motion transducers

number of marks around the circumference. Various types of sensor are used, such as optical, inductive and magnetic ones. As each mark is sensed, a pulse is generated and input to an electronic pulse counter. Usually, velocity is calculated in terms of the pulse count in unit time, which of course only yields information about the mean velocity. If the velocity is changing, instantaneous velocity can be calculated at each instant of time that an output pulse occurs, using the scheme shown in Figure 20.16. In this circuit, the pulses from the transducer gate the train of pulses from a 1 MHz clock into a counter. Control logic resets the counter and updates the digital output value after receipt of each pulse from the transducer. The measurement resolution of this system is highest when the speed of rotation is low.

Optical sensing

Digital tachometers with optical sensors are often known as *optical tachometers*. Optical pulses can be generated by one of the two alternative photoelectric techniques illustrated in Figure 20.17. In Figure 20.17(a), the pulses are produced as the windows in a slotted disc pass in sequence between a light source and a detector. The alternative form, Figure 20.17(b), has both light source and detector mounted on the same side of a reflective disc which has black sectors painted onto it at regular angular intervals. Light sources are normally either lasers or LEDs, with photodiodes and phototransistors being used as detectors. Optical tachometers yield better accuracy than other forms of digital tachometer but are not as reliable because dust and dirt can block light paths.

Inductive sensing

Variable reluctance velocity transducers, also known as *induction tachometers*, are a form of digital tachometer that use inductive sensing. They are widely used in the automotive industry within anti-skid devices, anti-lock braking systems (ABS) and traction control. One relatively simple and cheap form of this type of device was

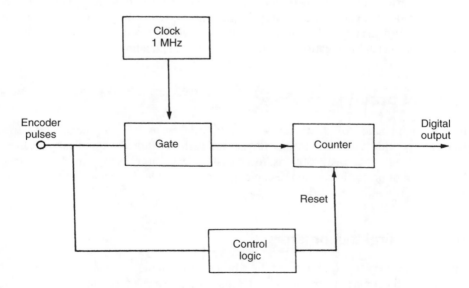

Fig. 20.16 Scheme to measure instantaneous angular velocities.

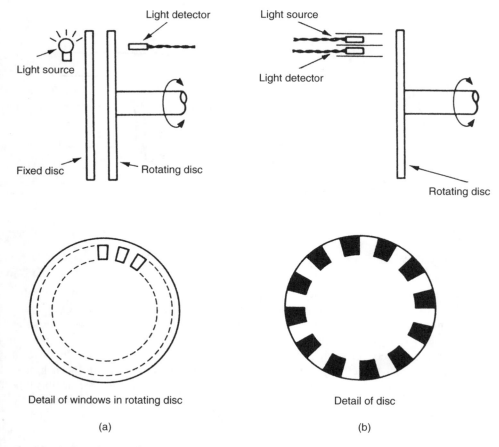

Fig. 20.17 Photoelectric pulse generation techniques.

described earlier in section 13.2 (Figure 13.2). A more sophisticated version shown in Figure 20.18 has a rotating disc that is constructed from a bonded-fibre material into which soft iron poles are inserted at regular intervals around its periphery. The sensor consists of a permanent magnet with a shaped pole piece, which carries a wound coil. The distance between the pick-up and the outer perimeter of the disc is around 0.5 mm. As the disc rotates, the soft iron inserts on the disc move in turn past the pick-up unit. As each iron insert moves towards the pole piece, the reluctance of the magnetic circuit increases and hence the flux in the pole piece also increases. Similarly, the flux in the pole piece decreases as each iron insert moves away from the sensor. The changing magnetic flux inside the pick-up coil causes a voltage to be induced in the coil whose magnitude is proportional to the rate of change of flux. This voltage is positive whilst the flux is increasing and negative whilst it is decreasing. Thus, the output is a sequence of positive and negative pulses whose frequency is proportional to the rotational velocity of the disc. The maximum angular velocity that the instrument can measure is limited to about 10 000 rpm because of the finite width of the induced pulses. As the velocity increases, the distance between the pulses is

Rotational motion transducers

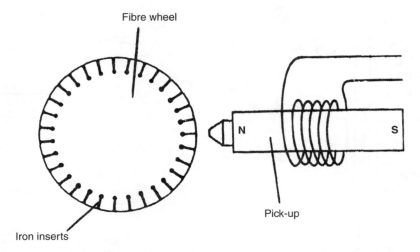

Fig. 20.18 Variable reluctance transducer.

reduced, and at a certain velocity, the pulses start to overlap. At this point, the pulse counter ceases to be able to distinguish the separate pulses. The optical tachometer has significant advantages in this respect, since the pulse width is much narrower, allowing measurement of higher velocities.

A simpler and cheaper form of variable reluctance transducer also exists that uses a ferromagnetic gear wheel in place of a fibre disc. The motion of the tip of each gear tooth towards and away from the pick-up unit causes a similar variation in the flux pattern to that produced by the iron inserts in the fibre disc. The pulses produced by these means are less sharp, however, and consequently the maximum angular velocity measurable is lower.

Magnetic (Hall-effect) sensing

The rotating element in *Hall-effect* or *magnetostrictive tachometers* has a very simple design in the form of a toothed metal gearwheel. The sensor is a solid-state, Hall-effect device that is placed between the gear wheel and a permanent magnet. When an inter-tooth gap on the gear wheel is adjacent to the sensor, the full magnetic field from the magnet passes through it. Later, as a tooth approaches the sensor, the tooth diverts some of the magnetic field, and so the field through the sensor is reduced. This causes the sensor to produce an output voltage that is proportional to the rotational speed of the gear wheel.

20.2.2 Stroboscopic methods

The stroboscopic technique of rotational velocity measurement operates on a similar physical principle to digital tachometers except that the pulses involved consist of flashes of light generated electronically and whose frequency is adjustable so that it can be matched with the frequency of occurrence of some feature on the rotating body being measured. This feature can either be some naturally occurring one such as gear teeth or the spokes of a wheel, or it can be an artificially created pattern of black and white stripes. In either case, the rotating body appears stationary when the frequencies of the light pulses and body features are in synchronism. Flashing rates available in commercial stroboscopes vary from 110 up to 150 000 per minute according to the range of velocity measurement required, and typical measurement inaccuracy is ±1% of the reading. The instrument is usually in the form of a hand-held device that is pointed towards the rotating body.

It must be noted that measurement of the flashing rate at which the rotating body appears stationary does not automatically indicate the rotational velocity, because synchronism also occurs when the flashing rate is some integral sub-multiple of the rotational speed. The practical procedure followed is therefore to adjust the flashing rate until synchronism is obtained at the largest flashing rate possible, R_1. The flashing rate is then carefully decreased until synchronism is again achieved at the next lower flashing rate, R_2. The rotational velocity is then given by:

$$V = \frac{R_1 R_2}{R_1 - R_2}$$

20.2.3 Analogue tachometers

Analogue tachometers are less accurate than digital tachometers but are nevertheless still used successfully in many applications. Various forms exist.

The *d.c. tachometer* has an output that is approximately proportional to its speed of rotation. Its basic structure is identical to that found in a standard d.c. generator used for producing power, and is shown in Figure 20.19. Both permanent-magnet types and separately excited field types are used. However, certain aspects of the design are optimized to improve its accuracy as a speed-measuring instrument. One significant design modification is to reduce the weight of the rotor by constructing the windings on a hollow fibreglass shell. The effect of this is to minimize any loading effect of the instrument on the system being measured. The d.c. output voltage from the instrument is of a relatively high magnitude, giving a high measurement sensitivity that is typically 5 volts per 1000 rpm. The direction of rotation is determined by the polarity of the output voltage. A common range of measurement is 0–6000 rpm. Maximum non-linearity is usually about ±1% of the full-scale reading. One problem with these devices that can cause difficulties under some circumstances is the presence of an a.c. ripple in the output signal. The magnitude of this can be up to 2% of the output d.c. level.

The *a.c. tachometer* has an output approximately proportional to rotational speed like the d.c. tachogenerator. Its mechanical structure takes the form of a two-phase induction motor, with two stator windings and (usually) a drag-cup rotor, as shown in Figure 20.20. One of the stator windings is excited with an a.c. voltage and the measurement signal is taken from the output voltage induced in the second winding. The magnitude of this output voltage is zero when the rotor is stationary, and otherwise proportional to the angular velocity of the rotor. The direction of rotation is determined by the phase of the output voltage, which switches by 180° as the direction reverses. Therefore, both the phase and magnitude of the output voltage have to be measured. A

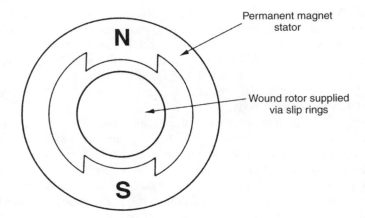

Fig. 20.19 D.c. tachometer.

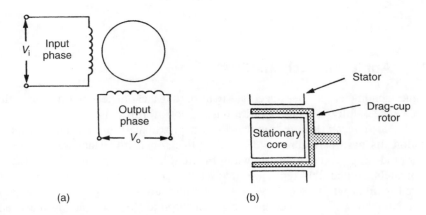

Fig. 20.20 A.c. tachometer.

typical range of measurement is 0–4000 rpm, with an inaccuracy of ±0.05% of full-scale reading. Cheaper versions with a squirrel-cage rotor also exist, but measurement inaccuracy in these is typically ±0.25%.

The *drag-cup tachometer*, also known as an *eddy-current tachometer*, has a central spindle carrying a permanent magnet that rotates inside a non-magnetic drag-cup consisting of a cylindrical sleeve of electrically conductive material, as shown in Figure 20.21. As the spindle and magnet rotate, a voltage is induced which causes circulating eddy currents in the cup. These currents interact with the magnetic field from the permanent magnet and produce a torque. In response, the drag-cup turns until the induced torque is balanced by the torque due to the restraining springs connected to the cup. When equilibrium is reached, the angular displacement of the cup is proportional to the rotational velocity of the central spindle. The instrument has a typical measurement inaccuracy of ±0.5% and is commonly used in the speedometers of motor vehicles and as a speed indicator for aero-engines. It is capable of measuring velocities up to 15 000 rpm.

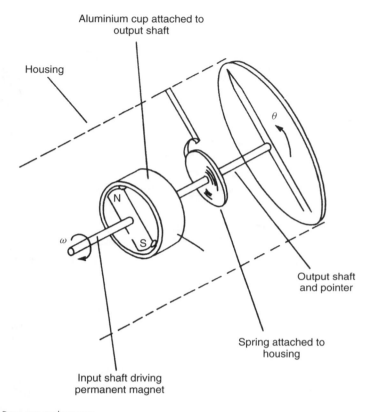

Fig. 20.21 Drag-cup tachometer.

Analogue-output forms of the *variable reluctance velocity transducer* (see section 20.2.1) also exist in which the output voltage pulses are converted into an analogue, varying-amplitude, d.c. voltage by means of a frequency-to-voltage converter circuit. However, the measurement accuracy is inferior to digital output forms.

20.2.4 Mechanical flyball

The mechanical flyball (alternatively known as a *centrifugal tachometer*) is a velocity-measuring instrument that was invented in 1817 and so might now be regarded as being old-fashioned. However, because it can act as a control actuator as well as a measuring instrument, it still finds substantial use in speed-governing systems for engines and turbines in which the measurement output is connected via a system of mechanical links to the throttle. The output is linear, typical measurement inaccuracy is ±1%, and velocities up to 40 000 rpm can be measured. As shown in Figure 20.22, the device consists of a pair of spherical balls pivoted on the rotating shaft. These balls move outwards under the influence of centrifugal forces as the rotational velocity of the shaft increases and lift a pointer against the resistance of a spring. The pointer can be arranged to give a visual indication of speed by causing it to move in front

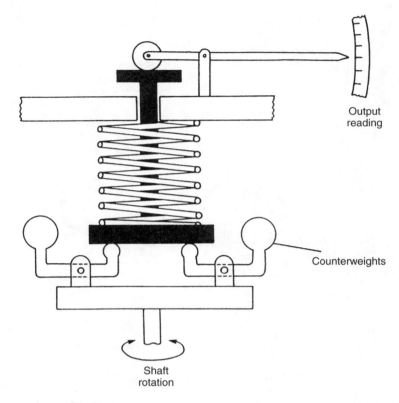

Fig. 20.22 Mechanical flyball.

of a calibrated scale, or its motion can be converted by a translational displacement transducer into an electrical signal.

In equilibrium, the centrifugal force, F_c, is balanced by the spring force, F_s, where:

$$F_c = K_c \omega^2; \quad F_s = K_s x$$

and K_c and K_s are constants, ω is the rotational velocity and x is the displacement of the pointer.

Thus:

$$K_c \omega^2 = K_s x \quad \text{or} \quad \omega = \sqrt{\left(\frac{K_s x}{K_c}\right)}$$

This is inconvenient because it involves a non-linear relationship between the pointer displacement and the rotational velocity. If this is not acceptable, a linear relationship can be obtained by using a spring with a non-linear characteristic such that $F_s = K'_s x^2$.

Then, equating expressions for F_c and F_s as before gives:

$$\omega = x\sqrt{\left(\frac{K'_s}{K_c}\right)}$$

20.2.5 The rate gyroscope

The rate gyro, illustrated in Figure 20.23, has an almost identical construction to the rate integrating gyro (Figure 20.14), and differs only by including a spring system which acts as an additional restraint on the rotational motion of the frame. The instrument measures the absolute angular velocity of a body, and is widely used in generating stabilizing signals within vehicle navigation systems. The typical measurement resolution given by the instrument is 0.01°/s and rotation rates up to 50°/s can be measured. The angular velocity, α, of the body is related to the angular deflection of the gyroscope, θ, by the equation:

$$\frac{\theta}{\alpha}(D) = \frac{H}{MD^2 + \beta D + K} \tag{20.2}$$

where H is the angular momentum of the spinning wheel, M is the moment of inertia of the system, β is the viscous damping coefficient, K is the spring constant, and D is the D-operator.

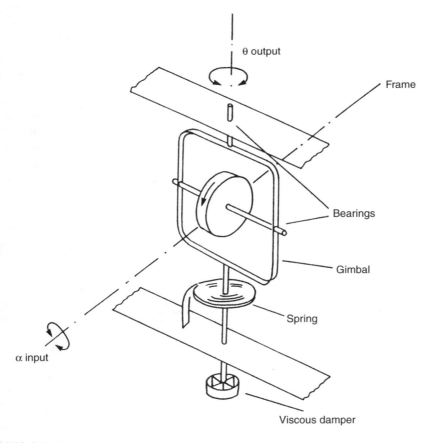

Fig. 20.23 Rate gyroscope.

This relationship (20.2) is a second order differential equation and therefore we must expect the device to have a response typical of second order instruments, as discussed in Chapter 2. The instrument must therefore be designed carefully so that the output response is neither oscillatory nor too slow in reaching a final reading. To assist in the design process, it is useful to re-express equation (20.2) in the following form:

$$\frac{\theta}{\alpha}(D) = \frac{K'}{D^2/\omega^2 + 2\xi D/\omega + 1} \quad (20.3)$$

where

$$K' = H/K, \omega = \sqrt{K/M} \quad \text{and} \quad \xi = \frac{\beta}{2\sqrt{KM}}$$

The static sensitivity of the instrument, K', is made as large as possible by using a high-speed motor to spin the wheel and so make H high. Reducing the spring constant K further improves the sensitivity but this cannot be reduced too far as it makes the resonant frequency ω of the instrument too small. The value of β is chosen such that the damping ratio ξ is close to 0.7.

20.2.6 Fibre-optic gyroscope

This is a relatively new instrument that makes use of fibre-optic technology. Incident light from a source is separated by a beam splitter into a pair of beams a and b, as shown in Figure 20.24. These travel in opposite directions around an optic-fibre coil (which may be several hundred metres long) and emerge from the coil as the beams marked a' and b'. The beams a' and b' are directed by the beam splitter into an interferometer. Any motion of the coil causes a phase shift between a' and b' which is detected by the interferometer. Further details can be found in Nuttall (1987).

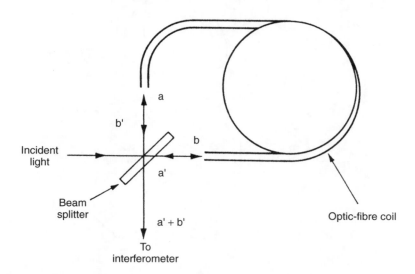

Fig. 20.24 Fibre-optic gyroscope.

20.2.7 Differentiation of angular displacement measurements

Angular velocity measurements can be obtained by differentiating the output signal from angular displacement transducers. Unfortunately, the process of differentiation amplifies any noise in the measurement signal, and therefore this technique has only rarely been used in the past. The technique has become more feasible with the advent of intelligent instruments, and one such instrument which processes the output of a resolver claims a maximum velocity measurement inaccuracy of $\pm 1\%$ (Analogue Devices, 1988).

20.2.8 Integration of the output from an accelerometer

In measurement systems that already contain an angular acceleration transducer, it is possible to obtain a velocity measurement by integrating the acceleration measurement signal. This produces a signal of acceptable quality, as the process of integration attenuates any measurement noise. However, the method is of limited value in many measurement situations because the measurement obtained is the average velocity over a period of time, rather than a profile of the instantaneous velocities as motion takes place along a particular path.

20.2.9 Choice between rotational velocity transducers

Choice between different rotational velocity transducers is influenced strongly by whether an analogue or digital form of output is required. Digital output instruments are now widely used and a choice has to be made between the variable reluctance transducer, devices using electronic light pulse counting methods, and the stroboscope. The first two of these are used to measure angular speeds up to about 10 000 rpm and the last one can measure speeds up to 25 000 rpm.

Probably the most common form of analogue output device used is the d.c. tachometer. This is a relatively simple device that measures speeds up to about 5000 rpm with a maximum inaccuracy of $\pm 1\%$. Where better accuracy is required within a similar range of speed measurement, a.c. tachometers are used. The squirrel-cage rotor type has an inaccuracy of only $\pm 0.25\%$ and drag-cup rotor types can have inaccuracies as low as $\pm 0.05\%$.

Other devices with an analogue output that are also sometimes used are the drag-cup tachometer and the mechanical flyball. The drag-cup tachometer has a typical inaccuracy of $\pm 5\%$ but it is cheap and therefore very suitable for use in vehicle speedometers. The Mechanical flyball has a better accuracy of $\pm 1\%$ and is widely used in speed governors, as noted earlier.

20.3 Measurement of rotational acceleration

Rotational accelerometers work on very similar principles to translational motion accelerometers. They consist of a rotatable mass mounted inside a housing that is attached to the accelerating, rotating body. Rotation of the mass is opposed by a torsional spring and damping. Any acceleration of the housing causes a torque $J\ddot{\theta}$ on the mass. This torque is opposed by a backward torque due to the torsional spring and in equilibrium:

$$J\ddot{\theta} = K\theta \quad \text{and hence:} \quad \ddot{\theta} = k\theta/J$$

A damper is usually included in the system to avoid undying oscillations in the instrument. This adds an additional backward torque $B\dot{\theta}$ to the system and the equation of motion becomes:

$$J\ddot{\theta} = B\dot{\theta} + K\theta$$

References and further reading

Analogue Devices (1988) Resolver to digital converter, *Measurement and Control*, **21**(10), p. 291.

Healey, M. (1975) *Principles of Automatic Control*, Hodder and Stoughton, London.

Nuttall, J.D. (1987) Optical gyroscopes, *Electronics and Power*, **33**(11), pp. 703–707.

21

Summary of other measurements

21.1 Dimension measurement

Dimension measurement includes measurement of the length, width and height of components and also the depth of holes and slots. Tapes and rules are commonly used to give approximate measurements, and various forms of calliper and micrometer are used where more accurate measurements are required. Gauge blocks and length bars are also used when very high accuracy is required, although these are primarily intended for calibration duties.

A flat and level *reference plane*, on which components being measured are placed, is often essential in dimension measurement. Such reference planes are available in a range of standard sizes, and a means of adjusting the feet is always provided to ensure that the surface is exactly level. Smaller sizes exist as a *surface plate* resting on a supporting table, whereas larger sizes take the form of free standing tables that usually have a projection at the edge to facilitate the clamping of components. They are normally used in conjunction with box cubes and vee blocks (see Figure 21.1) that locate components in a fixed position. In modern tables, granite has tended to supersede iron as the preferred material for the plate, although iron plates are still available. Granite is ideal for this purpose as it does not corrode, is dimensionally very stable and does not form burrs when damaged. Iron plates, on the other hand, are prone to rusting and susceptible to damage: this results in burrs on the surface that interfere with measurement procedures.

21.1.1 Rules and tapes

Rules and tapes are the simplest way of measuring larger dimensions. Steel rules are generally only available to measure dimensions up to 1 metre. Beyond this, steel tapes (measuring to 30 m) or an ultrasonic rule (measuring to 10 m) are used.

The *steel rule* is undoubtedly the simplest instrument available for measuring length. Measurement accuracy is only modest using standard rules, which typically have rulings at 0.5 mm intervals, but the best rules have rulings at 0.05 mm intervals and a measurement resolution of 0.02 mm. When used by placing the rule against an object,

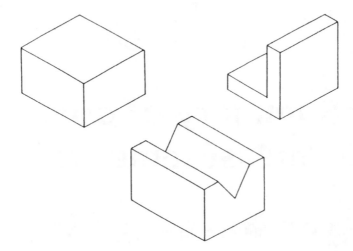

Fig. 21.1 Box cubes and vee blocks.

the measurement accuracy is much dependent upon the skill of the human measurer and, at best, the inaccuracy is likely to be at least ±0.5%.

The retractable *steel tape* is another well-known instrument. The end of the tape is usually provided with a flat hook that is loosely fitted so as to allow for automatic compensation of the hook thickness when the rule is used for internal measurements. Again, measurement accuracy is governed by human skill, but, with care, the measurement inaccuracy can be made to be as low as ±0.01% of full-scale reading.

The *ultrasonic rule* consists of an ultrasonic energy source, an ultrasonic energy detector and battery-powered, electronic circuitry housed within a hand-held box, as shown in Figure 21.2. Both source and detector often consist of the same type of piezoelectric crystal excited at a typical frequency of 40 kHz. Energy travels from the source to a target object and is then reflected back into the detector. The time of flight of this energy is measured and this is converted into a distance reading by the enclosed electronics. Maximum measurement inaccuracy of ±1% of the full-scale reading is claimed. This is only a modest level of accuracy, but it is sufficient for such purposes as measuring rooms by estate agents prior to producing sales literature, where the ease and speed of making measurements is of great value.

A fundamental problem in the use of ultrasonic energy of this type is the limited measurement resolution (7 mm) imposed by the 7 mm wavelength of sound at this frequency. Further problems are caused by the variation in the speed of sound with humidity (variations of ±0.5% possible) and the temperature-induced variation of 0.2% per °C. Therefore, the conditions of use must be carefully controlled if the claimed accuracy figure is to be met.

21.1.2 Callipers

Callipers are generally used in situations where measurement of dimensions using a rule or tape is not accurate enough. Two types exist, the standard calliper and the vernier calliper.

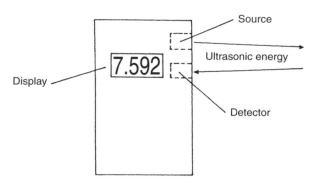

Fig. 21.2 Ultrasonic rule.

Figure 21.3 shows two types of *standard calliper*. The range of measurement, according to the version used, is up to 600 mm. These are used to transfer the measured dimension from the workpiece to a steel rule. This avoids the necessity to align the end of the rule exactly with the edge of the workpiece and reduces the measurement inaccuracy by a factor of two. In the basic calliper, careless use can allow the setting of the calliper to be changed during transfer from the workpiece to the rule. Hence, the spring-loaded type, which prevents this happening, is preferable.

The *vernier calliper*, shown in Figure 21.4(a), is a combination of a standard calliper and a steel rule. The main body of the instrument includes the main scale with a fixed anvil at one end. This carries a sliding anvil that is provided with a second, vernier scale. This second scale is shorter than the main scale and is divided into units that are slightly smaller than the main scale units but related to them by a fixed factor. Determination of the point where the two scales coincide enables very accurate measurements to be made, with typical inaccuracy levels down to $\pm 0.01\%$.

Figure 21.4(b) shows details of a typical combination of main and vernier scales. The main scale is ruled in 1 mm units. The vernier scale is 49 mm long and divided into 50 units, thereby making each unit 0.02 mm smaller than the main scale units. Each group of five units on the vernier scale thus differs from the main scale by 0.1 mm and the numbers marked on the scale thus refer to these larger units of 0.1 mm. In

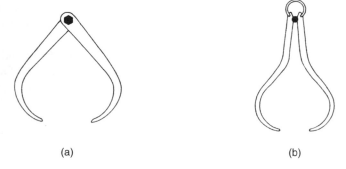

Fig. 21.3 (a) Standard calliper; (b) spring-loaded calliper.

422 Summary of other measurements

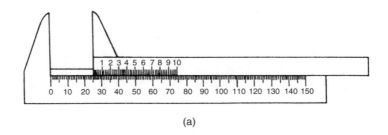

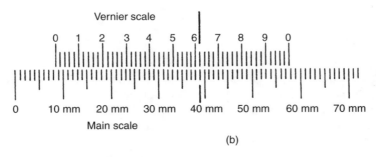

Fig. 21.4 Vernier calliper: (a) basic instrument; (b) details of scale.

the particular position shown in the figure, the zero on the vernier scale is indicating a measurement between 8 and 9 mm. Both scales coincide at a position of 6.2 (large units). This defines the interval between 8 and 9 mm to be $6.2 \times 0.1 = 0.62$ mm, i.e. the measurement is 8.62 mm.

Intelligent digital callipers are now available that give a measurement resolution of 0.01 mm and a low inaccuracy of ±0.03 mm. These have automatic compensation for wear, and hence calibration checks have to be very infrequent. In some versions, the digital display can be directly interfaced to an external computer monitoring system.

21.1.3 Micrometers

Micrometers provide a means of measuring dimensions to high accuracy. Different forms provide measurement of both internal and external dimensions of components, and of holes, slots etc. within components. In the *standard micrometer*, shown in Figure 21.5(a), measurement is made between two anvils, one fixed and one that is moved along by the rotation of an accurately machined screw thread. One complete rotation of the screw typically moves the anvil by a distance of 0.5 mm. Such movements of the anvil are measured using a scale marked with divisions every 0.5 mm along the barrel of the instrument. A scale marked with 50 divisions is etched around the circumference of the spindle holder: each division therefore corresponds to an axial movement of 0.01 mm. Assuming that the user is able to judge the position of the spindle on this circular scale against the datum mark to within one-fifth of a division, a measurement resolution of 0.002 mm is possible.

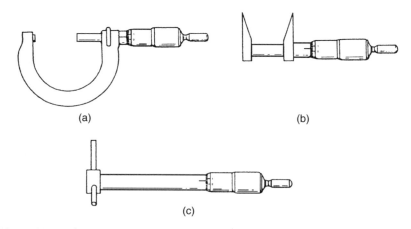

Fig. 21.5 Micrometers: (a) standard (external) micrometer; (b) internal micrometer; (c) bore micrometer.

The most common measurement ranges are either 0–25 mm or 25–50 mm, with inaccuracy levels down to ±0.003%. However, a whole family of micrometers is available, where each has a measurement span of 25 mm, but with the minimum distance measured varying from 0 mm up to 575 mm. Thus, the last instrument in this family measures the range from 575 to 600 mm. Some manufacturers also provide micrometers with two or more interchangeable anvils, which extend the span measurable with one instrument to between 50 mm and 100 mm according to the number of anvils supplied. Therefore, an instrument with four anvils might for instance measure the range from 300 mm to 400 mm, by making appropriate changes to the anvils.

The *internal micrometer* (see Figure 21.5(b)) is able to measure internal dimensions such as the diameters of holes. In the case of measuring holes, micrometers are inaccurate if there is any ovality in the hole, unless the diameter is measured at several points. An alternative solution to this problem is to use a special type of instrument known as a *bore micrometer* (Figure 21.5(c)). In this, three probes move out radially from the body of the instrument as the spindle is turned. These probes make contact with the sides of the hole at three equidistant points, thus averaging out any ovality.

Intelligent micrometers in the form of the electronic *digital micrometer* are now available. These have a self-calibration capability and a digital readout, with a measurement resolution of 0.001 mm (1 micron).

21.1.4 Gauge blocks (slip gauges) and length bars

Gauge blocks, also known as *slip gauges* (see Figure 21.6(a)), consist of rectangular blocks of hardened steel that have flat and parallel end faces. These faces are machined to very high standards of accuracy in terms of their surface finish and flatness. The purpose of gauge blocks is to provide a means of checking whether a particular dimension in a component is within the allowable tolerance rather than actually measuring what the dimension is. To do this, a number of gauge blocks are joined together to make up the required dimension to be checked. Gauge blocks are available in five grades of accuracy known as calibration, 00, 0, 1 and 2. Grades 1 and 2 are used for

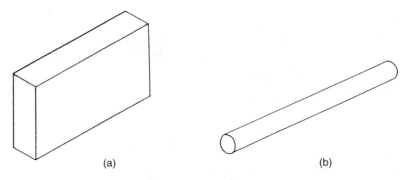

Fig. 21.6 (a) Gauge block; (b) length bar.

normal production and inspection measurements, with the other grades being intended only for calibration procedures at various levels.

Gauge blocks are available in boxed sets containing a range of block sizes, which allows any dimension up to 200 mm to be constructed by joining together an appropriate number of blocks. Whilst 200 mm is the maximum dimension that should be set up with gauge blocks alone, they can be used in conjunction with length bars to set up much greater standard dimensions. Blocks are joined by 'wringing', a procedure in which the two end faces are rotated slowly against each other. This removes the air film and allows adhesion to develop by intermolecular attraction. Adhesion is so good in fact that, if groups of blocks were not separated within a few hours, the molecular diffusion process would continue to the point where the blocks would be permanently welded together. The typical interblock gap resulting from wringing has been measured as 0.001 µm, which is effectively zero. Thus, any number of blocks can be joined without creating any significant measurement error. It is fairly common practice with blocks of grades 0, 1 and 2 to include an extra pair of 2 mm thick blocks in the set that are made from wear-resisting tungsten carbide. These are marked with a letter P and are designed to protect the other blocks from wear during use. Where such protector blocks are used, due allowance has to be made for their thickness (4 mm) in calculating the sizes of block needed to make up the required length.

A necessary precaution when using gauge blocks is to avoid handling them more than is necessary. The length of a bar that was 100 mm long at 20°C would increase to 100.02 mm at 37°C (body temperature). Hence, after wringing bars together, they should be left to stabilize back to the ambient room temperature before use. This wait might need to be several hours if the blocks have been handled to any significant extent.

Where a greater dimension than 200 mm is required, gauge blocks are used in conjunction with *length bars* (Figure 21.6(b)). Length bars consist of straight, hardened, high-quality steel bars of a uniform 22 mm diameter and in a range of lengths between 100 mm and 1200 mm. They are available in four grades of accuracy, *reference, calibration, grade 1* and *grade 2*. Reference and calibration grades have accurately flat end faces, which allows a number of bars to be wrung together to obtain the required standard length. Bars of grades 1 and 2 have threaded ends that allow them to be screwed together. Grade 2 bars are used for general measurement duties, with grade 1 bars being reserved for inspection duties. By combining length bars with gauge blocks, any dimension up to about 2 m can be set up with a resolution of 0.0005 mm.

21.1.5 Height and depth measurement

The height of objects and the depth of holes, slots etc. are measured by the height gauge and depth gauge respectively. A dial gauge is often used in conjunction with these instruments to improve measurement accuracy. The *height gauge*, shown in Figure 21.7(a), effectively consists of a vernier calliper mounted on a flat base. Measurement inaccuracy levels down to ±0.015% are possible. The *depth gauge* (Figure 21.7(b)) is a further variation on the standard vernier calliper principle that has the same measurement accuracy capabilities as the height gauge.

In practice, certain difficulties can arise in the use of these instruments where either the base of the instrument is not properly located on the measuring table or where the point of contact between the moving anvil and the workpiece is unclear. In such cases, a dial gauge, which has a clearly defined point of contact with the measured object, is used in conjunction with the height or depth gauge to avoid these possible sources of error. These instruments can also be obtained in intelligent versions that give a digital display and have self-calibration capabilities.

The *dial gauge*, shown in Figure 21.8(a), consists of a spring-loaded probe that drives a pointer around a circular scale via rack and pinion gearing. Typical measurement resolution is 0.01 mm. When used to measure the height of objects, it is clamped in a retort stand and a measurement taken of the height of the unknown component. Then it is put in contact with a height gauge (Figure 21.8(b)) that is adjusted until the reading on the dial gauge is the same. At this stage, the height gauge is set to the height

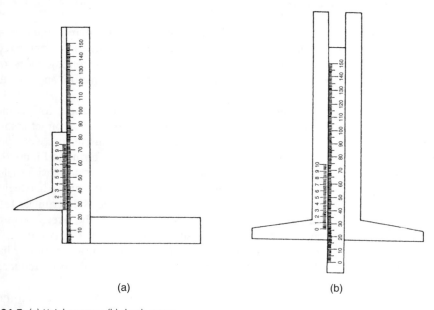

(a)　　　　　　　　　　(b)

Fig. 21.7 (a) Height gauge; (b) depth gauge.

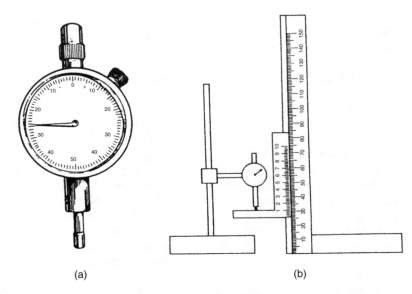

Fig. 21.8 Dial gauge: (a) basic instrument; (b) use in conjunction with height gauge.

of the object. The dial gauge is also used in conjunction with the depth gauge in an identical manner. (Gauge blocks can be used instead of height/depth gauges in such measurement procedures if greater accuracy is required.)

21.2 Angle measurement

Measurement of angles is one of the less common measurement requirements that instrumentation technologists are likely to meet. However, angle measurement is required in some circumstances, such as when the angle between adjoining faces on a component must be checked. The main instruments used are protractors and a form of angle-measuring spirit level.

In some circumstances, a simple protractor of the sort used in school for geometry exercises can be used. However, the more sophisticated form of angle protractor shown in Figure 21.9 provides better measurement accuracy. This consists of two straight edges, one of which is able to rotate with respect to the other. Referring to Figure 21.9, the graduated circular scale A attached to the straight edge C rotates inside a fixed circular housing attached to the other straight edge B. The relative angle between the two straight edges in contact with the component being measured is determined by the position of the moving scale with respect to a reference mark on the fixed housing B. With this type of instrument, measurement inaccuracy is at least ±1%. An alternative form, the *bevel protractor*, is similar to this form of angle protractor, but it has a vernier scale on the fixed housing. This allows the inaccuracy level to be reduced to ±10 minutes of arc.

The *spirit level* shown in Figure 21.10 is an alternative angle-measuring instrument. It consists of a standard spirit level attached to a rotatable circular scale that is mounted

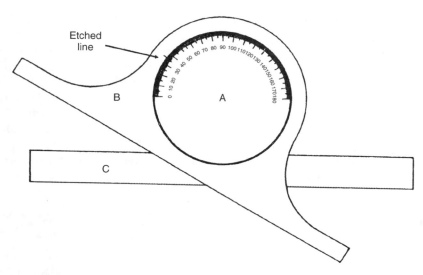

Fig. 21.9 Angle protractor.

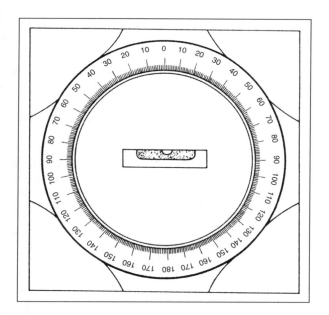

Fig. 21.10 Angle-measuring spirit level.

inside an accurately machined square frame. When placed on the sloping surfaces of components, rotation of the scale to centralize the bubble in the spirit level allows the angle of slope to be measured. Again, measuring inaccuracies down to ±10 minutes of arc are possible if a vernier scale is incorporated in the instrument.

The *electronic spirit level* contains a pendulum whose position is sensed electrically. Measurement resolution as good as 0.2 seconds of arc is possible.

21.3 Flatness measurement

The only dimensional parameter not so far discussed where a measurement requirement sometimes exists is the flatness of the surface of a component. This is measured by a *variation gauge*. As shown in Figure 21.11, this has four feet, three of which are fixed and one of which floats in a vertical direction. Motion of the floating foot is measured by a dial gauge that is calibrated such that its reading is zero when the floating foot is exactly level with the fixed feet. Thus, any non-zero reading on the dial gauge indicates non-flatness at the point of contact of the floating foot. By moving the variation gauge over the surface of a component and taking readings at various points, a contour map of the flatness of the surface can be obtained.

21.4 Volume measurement

Volume measurement is required in its own right as well as being required as a necessary component in some techniques for the measurement and calibration of other quantities such as volume flow rate and viscosity. The volume of vessels of a regular shape, where the cross-section is circular or oblong in shape, can be readily calculated from the dimensions of the vessel. Otherwise, for vessels of irregular shape, it is necessary to use either gravimetric techniques or a set of calibrated volumetric measures.

In the gravimetric technique, the dry vessel is weighed and then is completely filled with water and weighed again. The volume is then simply calculated from this weight difference and the density of water.

The alternative technique involves transferring the liquid from the vessel into an appropriate number of volumetric measures taken from a standard-capacity, calibrated set. Each vessel in the set has a mark that shows the volume of liquid contained when the vessel is filled up to the mark. Special care is needed to ensure that the meniscus of the water is in the correct position with respect to the reference mark on the vessel when it is deemed to be full. Normal practice is to set the water level such that the

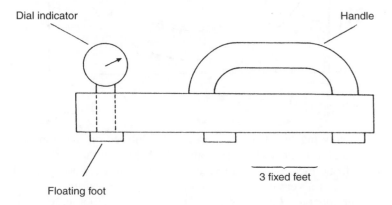

Fig. 21.11 Variation gauge.

Table 21.1 Typical measurement uncertainty of calibrated volumetric measures

Capacity	Volumetric uncertainty
1 ml	±4%
10 ml	±0.8%
100 ml	±0.2%
1 l	±0.1%
10 l	±0.05%
100 l	±0.02%
1000 l	±0.02%

reference mark forms a smooth tangent with the convex side of the meniscus. This is made easier to achieve if the meniscus is viewed against a white background and the vessel is shaded from stray illumination.

The measurement uncertainty using calibrated volumetric measures depends on the number of measures used for any particular measurement. The total error is a multiple of the individual error of each measure, typical values of which are shown in Table 21.1.

21.5 Viscosity measurement

Viscosity measurement is important in many process industries. In the food industry, the viscosity of raw materials such as dough, batter and ice cream has a direct effect on the quality of the product. Similarly, in other industries such as the ceramic one, the quality of raw materials affects the final product quality. Viscosity control is also very important in assembly operations that involve the application of mastics and glue flowing through tubes. Clearly, successful assembly requires such materials to flow through tubes at the correct rate and therefore it is essential that their viscosity is correct.

Viscosity describes the way in which a fluid flows when it is subject to an applied force. Consider an elemental cubic volume of fluid and a shear force F applied to one of its faces of area A. If this face moves a distance L and at a velocity V relative to the opposite face of the cube under the action of F, the shear stress (s) and shear rate (r) are given by:

$$s = F/A; \quad r = V/L$$

The *coefficient of viscosity* (C_V) is the ratio of shear stress to shear rate, i.e.

$$C_V = s/r.$$

C_V is often described simply as the 'viscosity'. A further term, *kinematic viscosity*, is also sometimes used, given by $K_V = C_V/\rho$, where K_V is the kinematic viscosity and ρ is the fluid density. To avoid confusion, C_V is often known as the *dynamic viscosity*, to distinguish it from K_V. C_V is measured in units of poise or Ns/m^2 and K_V is measured in units of stokes or m^2/s.

Viscosity was originally defined by Newton, who assumed that it was constant with respect to shear rate. However, it has since been shown that the viscosity of many

430 Summary of other measurements

fluids varies significantly at high shear rates and the viscosity of some varies even at low shear rates. The worst non-Newtonian characteristics tend to occur with emulsions, pastes and slurries. For non-Newtonian fluids, subdivision into further classes can also be made according to the manner in which the viscosity varies with shear rate, as shown in Figure 21.12.

The relationship between the input variables and output measurement for instruments that measure viscosity normally assumes that the measured fluid has Newtonian characteristics. For non-Newtonian fluids, a correction must be made for shear rate variations (see Miller, 1975a). If such a correction is not made, the measurement obtained is known as the *apparent viscosity*, and this can differ from the true viscosity by a large factor. The true viscosity is often called the *absolute viscosity* to avoid ambiguity. Viscosity also varies with fluid temperature and density.

Instruments for measuring viscosity work on one of three physical principles:

- Rate of flow of the liquid through a tube
- Rate of fall of a body through the liquid
- Viscous friction force exerted on a rotating body.

21.5.1 Capillary and tube viscometers

These are the most accurate types of viscometer, with typical measurement inaccuracy levels down to ±0.3%. Liquid is allowed to flow, under gravity from a reservoir, through a tube of known cross-section. In different instruments, the tube can vary from capillary-sized to a large diameter. The pressure difference across the ends of the tube and the time for a given quantity of liquid to flow are measured, and then the

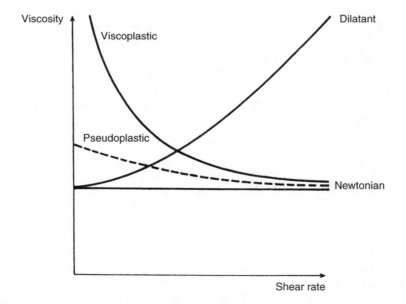

Fig. 21.12 Different viscosity/shear-rate relationships.

liquid viscosity for Newtonian fluids can be calculated as (in units of poise):

$$C_V = \frac{1.25\pi R^4 PT}{LV}$$

where R is the radius (m) of the tube, L is its length (m), P is the pressure difference (N/m^2) across the ends and V is the volume of liquid flowing in time T (m^3/s).

For non-Newtonian fluids, corrections must be made for shear rate variations (Miller, 1975a). For any given viscometer, R, L and V are constant and equation (21.1) can be written as:

$$C_V = KPT$$

where K is known as the viscometer constant.

21.5.2 Falling body viscometer

The falling body viscometer is particularly recommended for the measurement of high-viscosity fluids. It can give measurement uncertainty levels down to $\pm 1\%$. It involves measuring the time taken for a spherical body to fall a given distance through the liquid. The viscosity for Newtonian fluids is then given by Stoke's formula as (in units of poise):

$$C_V = \frac{R^2 g(\rho_s - \rho_l)}{450\, V}$$

where R is the radius (m) of the sphere, g is the acceleration due to gravity (m/s^2), ρ_s and ρ_l are the specific gravities (g/m^3) of the sphere and liquid respectively and V is the velocity (m/s) of the sphere.

For non-Newtonian fluids, correction for the variation in shear rate is very difficult.

21.5.3 Rotational viscometers

Rotational viscometers are relatively easy to use but their measurement inaccuracy is at least $\pm 10\%$. All types have some form of element rotating inside the liquid at a constant rate. One common version has two coaxial cylinders with the fluid to be measured contained between them. One cylinder is driven at a constant angular velocity by a motor and the other is suspended by torsion wire. After the driven cylinder starts from rest, the suspended cylinder rotates until an equilibrium position is reached where the force due to the torsion wire is just balanced by the viscous force transmitted through the liquid. The viscosity (in poise) for Newtonian fluids is then given by:

$$C_V = 2.5G \left(\frac{1/R_1^2 - 1/R_2^2}{\pi h \omega} \right)$$

where G is the couple (Nm) formed by the force exerted by the torsion wire and its deflection, R_1 and R_2 are the radii (m) of the inner and outer cylinders, h is the length of the cylinder (m) and ω is the angular velocity (rad/s) of the rotating cylinder. Again, corrections have to be made for non-Newtonian fluids.

21.6 Moisture measurement

There are many industrial requirements for the measurement of the moisture content. This can be required in solids, liquids or gases. The physical properties and storage stability of most solid materials is affected by their water content. There is also a statutory requirement to limit the moisture content in the case of many materials sold by weight. In consequence, the requirement for moisture measurement pervades a large number of industries involved in the manufacture of foodstuffs, pharmaceuticals, cement, plastics, textiles and paper.

Measurement of the water content in liquids is commonly needed for fiscal purposes, but is also often necessary to satisfy statutory requirements. The petrochemical industry has wide-ranging needs for moisture measurement in oil etc. The food industry also needs to measure the water content of products such as beer and milk.

In the case of moisture in gases, the most common measurement is the amount of moisture in air. This is usually known as the humidity level. Humidity measurement and control is an essential requirement in many buildings, greenhouses and vehicles.

As there are several ways in which humidity can be defined, three separate terms have evolved so that ambiguity can be avoided. *Absolute humidity* is the mass of water in a unit volume of moist air; *specific humidity* is the mass of water in a unit mass of moist air; *relative humidity* is the ratio of the actual water vapour pressure in air to the saturation vapour pressure, usually expressed as a percentage.

21.6.1 Industrial moisture measurement techniques

Industrial methods for measuring moisture are based on the variation of some physical property of the material with moisture content. Many different properties can be used and therefore the range of available techniques, as listed below, is large.

Electrical methods

Measuring the amount of absorption of *microwave energy* beamed through the material is the most common technique for measuring moisture content and is described in detail in Anderson (1989), and Thompson (1989). Microwaves at wavelengths between 1 mm and 1 m are absorbed to a much greater extent by water than most other materials. Wavelengths of 30 mm or 100 mm are commonly used because 'off-the-shelf' equipment to produce these is readily available from instrument suppliers. The technique is suitable for moisture measurement in solids, liquids and gases at moisture-content levels up to 45% and measurement uncertainties down to $\pm 0.3\%$ are possible.

The *capacitance moisture meter* uses the principle that the dielectric constant of materials varies according to their water content. Capacitance measurement is therefore related to moisture content. The instrument is useful for measuring moisture-content levels up to 30% in both solids and liquids, and measurement uncertainty down to $\pm 0.3\%$ has been claimed for the technique (Slight, 1989). Drawbacks of the technique include (a) limited measurement resolution owing to the difficulty in measuring small changes in a relatively large standing capacitance value and (b) difficulty when the sample has a high electrical conductivity. An alternative capacitance charge transfer technique has been reported (Gimson, 1989) that overcomes these problems by measuring the charge carrying capacity of the material. In this technique, wet and dry samples of the material are charged to a fixed voltage and then simultaneously discharged into charge-measuring circuits.

The *electrical conductivity* of most materials varies with moisture content and this therefore provides another means of measurement. Techniques using electrical conductivity variation are cheap and can measure moisture levels up to 25%. However, the presence of other conductive substances in the material such as salts or acids affects the measurement.

A further technique is to measure the frequency change in a *quartz crystal* that occurs as it takes in moisture.

Neutron moderation

Neutron moderation measures moisture content using a radioactive source and a neutron counter. Fast neutrons emitted from the source are slowed down by hydrogen nuclei in the water, forming a cloud whose density is related to the moisture content. Measurements take a long time because the output density reading may take up to a few minutes to reach steady state, according to the nature of the materials involved. Also, the method cannot be used with any materials that contain hydrogen molecules, such as oils and fats, as these slow down neutrons as well. Specific humidities up to 15% ($\pm 1\%$ error) can be measured.

Low resolution nuclear magnetic resonance (NMR)

Low-resolution nuclear magnetic resonance involves subjecting the sample to both an unidirectional and an alternating radio-frequency (RF) magnetic field. The amplitude of the unidirectional field is varied cyclically, which causes resonance once per cycle in the coil producing the RF field. Under resonance conditions, protons are released from the hydrogen content of the water in the sample. These protons cause a measurable moderation of the amplitude of the RF oscillator waveform that is related to the moisture content of the sample. The technique is described more fully in Young (1989).

Materials that naturally have a hydrogen content cannot normally be measured. However, pulsed NMR techniques have been developed that overcome this problem by taking advantage of the different relaxation times of hydrogen nuclei in water and oil. In such pulsed techniques, the dependence on the relaxation time limits the maximum fluid flow rate for which moisture can be measured.

Optical methods

The *refractometer* is a well-established instrument that is used for measuring the water content of liquids. It measures the refractive index of the liquid, which changes according to the moisture content.

Moisture-related *energy absorption* of near-infra-red light can be used for measuring the moisture content of solids, liquids and gases. At a wavelength of 1.94 µm, energy absorption due to moisture is high, whereas at 1.7 µm, absorption due to moisture is zero. Therefore, measuring absorption at both 1.94 µm and 1.7 µm allows absorption due to components in the material other than water to be compensated for, and the resulting measurement is directly related to energy content. The latest instruments use multiple-frequency infra-red energy and have an even greater capability for eliminating the effect of components in the material other than water that absorb energy. Such multi-frequency instruments also cope much better with variations in particle size in the measured material.

In alternative versions of this technique, energy is either transmitted through the material or reflected from its surface. In either case, materials that are either very dark or highly reflective give poor results. The technique is particularly attractive, where applicable, because it is a non-contact method that can be used to monitor moisture content continuously at moisture levels up to 50%, with inaccuracy as low as $\pm 0.1\%$ in the measured moisture level. A deeper treatment can be found in Benson (1989).

Ultrasonic methods

The presence of water changes the speed of propagation of ultrasonic waves through liquids. The moisture content of liquids can therefore be determined by measuring the transmission speed of ultrasound. This has the inherent advantage of being a non-invasive technique but temperature compensation is essential because the velocity of ultrasound is particularly affected by temperature changes. The method is best suited to measurement of high moisture levels in liquids that are not aerated or of high viscosity. Typical measurement uncertainty is $\pm 1\%$ but measurement resolution is very high, with changes in moisture level as small as 0.05% being detectable. Further details can be found in Wiltshire (1989).

Mechanical properties

Density changes in many liquids and slurries can be measured and related to moisture content, with good measurement resolution up to 0.2% moisture. Moisture content can also be estimated by measuring the moisture level-dependent viscosity of liquids, pastes and slurries.

21.6.2 Laboratory techniques for moisture measurement

Laboratory techniques for measuring moisture content generally take much longer to obtain a measurement than the industrial techniques described above. However, the measurement accuracy obtained is usually much better.

Water separation

Various laboratory techniques are available that enable the moisture content of liquids to be measured accurately by separating the water from a sample of the host liquid. Separation is effected either by titration (Karl Fischer technique), distillation (Dean and Stark technique) or a centrifuge. Any of these methods can measure water content in a liquid with measurement uncertainty levels down to $\pm 0.03\%$.

Gravimetric methods

Moisture content in solids can be measured accurately by weighing the moist sample, drying it and then weighing again. Great care must be taken in applying this procedure, as many samples rapidly take up moisture again if they are removed from the drier and exposed to the atmosphere before being weighed. Normal procedure is to put the sample in an open container, dry it in an oven and then screw an airtight top onto the container before it is removed from the oven.

Phase-change methods

The boiling and freezing point of materials is altered by the presence of moisture, and therefore the moisture level can be determined by measuring the phase-change temperature. This technique is used for measuring the moisture content in many food products and in some oil and alcohol products.

Equilibrium relative humidity measurement

This technique involves placing a humidity sensor in close proximity to the sample in an airtight container. The water vapour pressure close to the sample is related to the moisture content of the sample. The moisture level can therefore be determined from the humidity measurement.

21.6.3 Humidity measurement

The three major instruments used for measuring humidity in industry are the electrical hygrometer, the psychrometer and the dew point meter. The dew point meter is the most accurate of these and is commonly used as a calibration standard. The various types of hygrometer are described more fully in Miller (1975b).

The electrical hygrometer

The electrical hygrometer measures the change in capacitance or conductivity of a hygroscopic material as its moisture level changes. Conductivity types use two noble metal electrodes either side of an insulator coated in a hygroscopic salt such as calcium chloride. Capacitance types have two plates either side of a hygroscopic dielectric such as aluminium oxide.

These instruments are suitable for measuring moisture levels between 15% and 95%, with typical measurement uncertainty of ±3%. Atmospheric contaminants and operation in saturation conditions both cause characteristics drift, and therefore the recalibration frequency has to be determined according to the conditions of use.

The psychrometer (wet and dry bulb hygrometer)

The psychrometer, also known as the wet and dry bulb hygrometer, has two temperature sensors, one exposed to the atmosphere and one enclosed in a wet wick. Air is blown across the sensors, which causes evaporation and a reduction in temperature in the wet sensor. The temperature difference between the sensors is related to the humidity level. The lowest measurement uncertainty attainable is ±4%.

Dew point meter

The elements of the dew point meter, also known as the dew point hygrometer, are shown in Figure 21.13. The sample is introduced into a vessel with an electrically cooled mirror surface. The mirror surface is cooled until a light source-light detector system detects the formation of dew on the mirror, and the condensation temperature is measured by a sensor bonded to the mirror surface. The dew point is the temperature at which the sample becomes saturated with water. Therefore, this temperature is related to the moisture level in the sample. A microscope is also provided in the instrument

Summary of other measurements

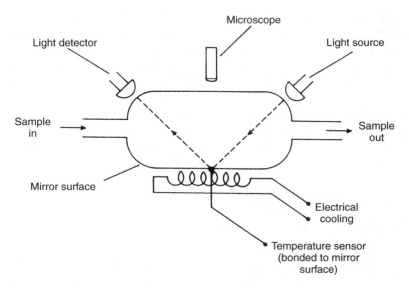

Fig. 21.13 Dew point meter.

so that the thickness and nature of the condensate can be observed. The instrument is described in greater detail in Pragnell (1989).

Even small levels of contaminants on the mirror surface can cause large changes in the dew point and therefore the instrument must be kept very clean. When necessary, the mirror should be cleaned with deionized or distilled water applied with a lint-free swab. Any contamination can be detected by a skilled operator, as this makes the condensate look 'blotchy' when viewed through the microscope. The microscope also shows up other potential problems such as large ice crystals in the condensate that cause temperature gradients between the condensate and the temperature sensor. When used carefully, the instrument is very accurate and is often used as a reference standard.

21.7 Sound measurement

Noise can arise from many sources in both industrial and non-industrial environments. Even low levels of noise can cause great annoyance to the people subjected to it and high levels of noise can actually cause hearing damage. Apart from annoyance and possible hearing loss, noise in the workplace also causes loss of output where the persons subjected to it are involved in tasks requiring high concentration. Extreme noise can even cause material failures through fatigue stresses set up by noise-induced vibration.

Various items of legislation exist to control the creation of noise. Court orders can be made against houses or factories in a neighbourhood that create noise exceeding a certain acceptable level. In extreme cases, where hearing damage may be possible, health and safety legislation comes into effect. Such legislation clearly requires the existence of accurate methods of quantifying sound levels. Sound is measured in terms

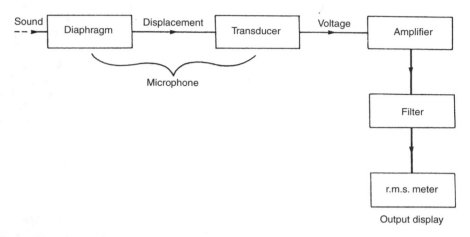

Fig. 21.14 Sound meter.

of the *sound pressure level*, S_P, which is defined as:

$$S_P = 20 \log_{10}\left(\frac{P}{0.0002}\right) \text{ decibels (dB)}$$

where P is the r.m.s. sound pressure in µbar.

The quietest sound that the average human ear can detect is a tone at a frequency of 1 kHz and sound pressure level of 0 dB (2×10^{-4} µbar). At the upper end, sound pressure levels of 144 dB (3.45 mbar) cause physical pain.

Sound is usually measured with a sound meter. This essentially processes the signal collected by a microphone, as shown in Figure 21.14. The microphone is a diaphragm-type pressure-measuring device that converts sound pressure into a displacement. The displacement is applied to a displacement transducer (normally capacitive, inductive or piezoelectric type) which produces a low magnitude voltage output. This is amplified, filtered and finally gives an output display on an r.m.s. meter. The filtering process has a frequency response approximating that of the human ear so that the sound meter 'hears' sounds in the same way as a human ear. In other words, the meter selectively attenuates frequencies according to the sensitivity of the human ear at each frequency, so that the sound level measurement output accurately reflects the sound level heard by humans. If sound level meters are being used to measure sound to predict vibration levels in machinery, then they are used without filters so that the actual rather than the human-perceived sound level is measured.

21.8 pH measurement

pH is a parameter that quantifies the level of acidity or akalinity in a chemical solution. It defines the concentration of hydrogen atoms in the solution in grams/litre and is expressed as:

$$\text{pH} = \log_{10}[1/\text{H}^+]$$

where H^+ is the hydrogen ion concentration in the solution.

The value of pH can range from 0, which describes extreme acidity, to 14, which describes extreme akalinity. Pure water has a pH of 7. pH measurement is required in many process industries, and especially those involving food and drink production. The most universally known method of measuring pH is to use litmus paper or some similar chemical indicator that changes colour according to the pH value. Unfortunately, this method gives only a very approximate indication of pH unless used under highly controlled laboratory conditions. Much research is ongoing into on-line pH sensors and the various activities are described later. However, at the present time, the device known as the glass electrode is by far the most common on-line sensor used.

21.8.1 The glass electrode

The glass electrode consists of a glass probe containing two electrodes, a measuring one and a reference one, separated by a solid glass partition. Neither of the electrodes is in fact glass. The reference electrode is a screened electrode, immersed in a buffer solution, which provides a stable reference e.m.f. that is usually 0 V. The tip of the measuring electrode is surrounded by a pH-sensitive glass membrane at the end of the probe, which permits the diffusion of ions according to the hydrogen ion concentration in the fluid outside the probe. The measuring electrode therefore generates an e.m.f. proportional to pH that is amplified and fed to a display meter. The characteristics of the glass electrode are very dependent on ambient temperature, with both zero drift and sensitivity drift occurring. Thus, temperature compensation is essential. This is normally achieved through calibrating the system output before use by immersing the probe in solutions at reference pH values. Whilst being theoretically capable of measuring the full range of pH values between 0 and 14, the upper limit in practice is generally a pH value of about 12 because electrode contamination at very high alkaline concentrations becomes a serious problem and also glass starts to dissolve at such high pH values. Glass also dissolves in acid solutions containing fluoride, and this represents a further limitation in use. If required, the latter problems can be overcome to some extent by using special types of glass.

Great care is necessary in the use of the glass electrode type of pH probe. Firstly, the measuring probe has a very high resistance (typically $10^8\ \Omega$) and a very low output. Hence, the output signal from the probes must be electrically screened to prevent any stray pick-up and electrical insulation of the assembly must be very high. The assembly must also be very efficiently sealed to prevent the ingress of moisture.

A second problem with the glass electrode is the deterioration in accuracy that occurs as the glass membrane becomes coated with various substances it is exposed to in the measured solution. Cleaning at prescribed intervals is therefore necessary and this must be carried out carefully, using the correct procedures, to avoid damaging the delicate glass membrane at the end of the probe. The best cleaning procedure varies according to the nature of the contamination. In some cases, careful brushing or wiping is adequate, whereas in other cases spraying with chemical solvents is necessary. Ultrasonic cleaning is often a useful technique, though it tends to be expensive. Steam cleaning should not be attempted, as this damages the pH-sensitive membrane. Mention must also be made about storage. The glass electrode must not be allowed to dry out during storage, as this would cause serious damage to the pH-sensitive layer.

Finally, caution must be taken about the response time of the instrument. The glass electrode has a relatively large time constant of one to two minutes, and so it must be left to settle for a long time before the reading is taken. If this causes serious difficulties, special forms of low-resistivity glass electrode are now available that have smaller time constants.

21.8.2 Other methods of pH measurement

Whilst the glass electrode predominates at present in pH measurement, several other devices and techniques exist. Whilst most of these are still under development and unproven in long-term use, a few are in practical use, especially for special measurement situations.

One alternative, which is in current use, is the antimony electrode. This is of a similar construction to the glass electrode but uses antimony instead of glass. The device is more robust than the glass electrode and can be cleaned by rubbing it with emery cloth. However, its time constant is very large and its output response is grossly non-linear, limiting its application to environments where the glass electrode is unsuitable. Such applications include acidic environments containing fluoride and environments containing very abrasive particles. The normal measurement range is pH 1 to 11.

A fibre-optic pH sensor is another available device, as described earlier in Chapter 13, in which the pH level is indicated by the intensity of light reflected from the tip of a probe coated in a chemical indicator whose colour changes with pH. Unfortunately, this device only has the capability to measure over a very small range of pH (typically 2 pH) and it has a short life.

21.9 Gas sensing and analysis

Gas sensing and analysis is required in many applications. A primary role of gas sensing is in hazard monitoring to predict the onset of conditions where flammable gases are reaching dangerous concentrations. Danger is quantified in terms of the *lower explosive level*, which is usually reached when the concentration of gas in air is in the range of between 1% and 5%.

Gas sensing also provides a fire detection and prevention function. When materials burn, a variety of gaseous products result. Most sensors that are used for fire detection measure carbon monoxide concentration, as this is the most common combustion product. Early fire detection enables fire extinguishing systems to be triggered, preventing serious damage from occurring in most cases. However, fire prevention is even better than early fire detection, and solid-state sensors, based on a sintered mass of polycrystalline tin oxide, can now detect the gaseous products (generally various types of hydrocarbon) that are generated when materials become hot but before they actually burn.

Health and safety legislation creates a further requirement for gas sensors. Certain gases, such as carbon monoxide, hydrogen sulphide, chlorine and nitrous oxide, cause fatalities above a certain concentration and sensors must provide warning of impending danger. For other gases, health problems are caused by prolonged exposure and so the sensors in this case must integrate gas concentration over time to determine whether the allowable exposure limit over a given period of time has been exceeded. Again, solid-state sensors are now available to fulfil this function.

Concern about general environmental pollution is also making the development of gas sensors necessary in many new areas. Legislation is growing rapidly to control the emission of everything that is proven or suspected to cause health problems or environmental damage. The present list of controlled emissions includes nitrous oxide, oxides of sulphur, carbon monoxide and dioxide, CFCs, ammonia and hydrocarbons. Sensors are required both at the source of these pollutants, where concentrations are high, and also to monitor the much lower concentrations in the general environment. Oxygen concentration measurement is often of great importance also in pollution control, as the products of combustion processes are greatly affected by the air/fuel ratio.

Sensors associated with pollution monitoring and control often have to satisfy quite stringent specifications, particularly where the sensors are located at the pollutant source. Robustness is usually essential, as such sensors are subjected to bombardment from a variety of particulate matters, and they must also endure conditions of high humidity and temperature. They are also frequently located in inaccessible locations, such as in chimneys and flues, which means that they must have stable characteristics over long periods of time without calibration checks being necessary. The need for such high-specification sensors makes such pollutant-monitoring potentially very expensive if there are several problem gases involved. However, because the concentration of all output gases tends to vary to a similar extent according to the condition of filters etc., it is frequently only necessary to measure the concentration of one gas, from which the concentration of other gases can be predicted reliably. This greatly reduces the cost involved in such monitoring.

A number of devices that sense, measure the concentration of or analyse gases exist. In terms of frequency of usage, they vary from those that have been in use for a number of years, to those that have appeared recently, and finally to those that are still under research and development. In the following list of devices, their status in terms of current usage will be indicated. Fuller information can be found in Jones (1989).

21.9.1 Catalytic (calorimetric) sensors

Catalytic sensors, otherwise known as calorimetric sensors, have widespread use for measuring the concentration of flammable gases. Their principle of operation is to measure the heat evolved during the catalytic oxidation of reducing gases. They are cheap and robust but are unsuitable for measuring either very low or very high gas concentrations. The catalysts that have been commonly used in these devices in the past are adversely affected by many common industrial substances such as lead, phosphorus, silicon and sulphur, and this catalyst poisoning has previously prevented this type of device being used in many applications. However, new types of poison-resistant catalyst are now becoming available that are greatly extending the applicability of this type of device.

21.9.2 Paper tape sensors

By moving a paper tape impregnated with a reagent sensitive to a specific gas (e.g. lead acetate tape to detect hydrogen sulphide) through an air stream, the time history of the concentration of gas is indicated by the degree of colour change in the tape. This is used as a low accuracy but reliable and cheap means of detecting the presence of hydrogen sulphide and ammonia.

21.9.3 Liquid electrolyte electrochemical cells

These consist of two electrodes separated by electrolyte, to which the measured air supply is directed through a permeable membrane, as shown in Figure 21.15. The gas in the air to which the cell is sensitive reacts at the electrodes to form ions in the solution. This produces a voltage output from the cell.

Electrochemical cells have stable characteristics and give good measurement sensitivity. However, they are expensive and their durability is relatively poor, with life being generally limited to about one or two years at most. A further restriction is that they cannot be used above temperatures of about 50°C, as their performance deteriorates rapidly at high temperatures because of interference from other atmospheric substances.

The main use of such cells is in measuring toxic gases in satisfaction of health and safety legislation. Versions of the cell for this purpose are currently available to measure carbon monoxide, chlorine, nitrous oxide, hydrogen sulphide and ammonia. Cells to measure other gases are currently under development.

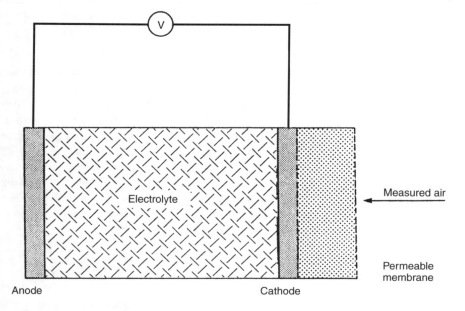

Fig. 21.15 Liquid electrolyte electrochemical cell.

In addition, electrochemical cells are also used to a limited extent to monitor carbon monoxide emissions in flue gases for environmental control purposes. Pre-cooling of the emitted gases is a necessary condition for this application.

21.9.4 Solid-state electrochemical cells (zirconia sensor)

At present, these cells are used only for measuring oxygen concentration, but ways of extending their use to other gases are currently in progress. The oxygen-measurement cell consists of two chambers separated by a zirconia wall. One chamber contains gas with a known oxygen concentration and the other contains the air being measured. Ions are conducted across the zirconia wall according to the difference in oxygen concentration across it and this produces an output e.m.f. The device is rugged but requires high temperatures to operate efficiently. It is, however, well proven and a standard choice for oxygen measurement. In industrial uses, it is often located in chimneystacks, where quite expensive mounting and protection systems are needed. However, very low cost versions (around £200) are now used in some vehicle exhaust systems as part of the engine management system.

21.9.5 Catalytic gate FETs

These consist of field effect transistors with a catalytic, palladium gate that is sensitive to hydrogen ions in the environment. The gate voltage, and hence characteristics of the device, change according to the hydrogen concentration. They can be made sensitive to gases such as hydrogen sulphide, ammonia and hydrocarbons as well as hydrogen. They are cheap and find application in workplace monitoring, in satisfaction of health and safety legislation, and in fire detection (mainly detecting hydrocarbon products).

21.9.6 Semiconductor (metal oxide) sensors

In these devices use is made of the fact that the surface conductivity of semiconductor metal oxides (generally tin or zinc oxides) changes according to the concentration of certain gases with which they are in contact. Unfortunately, they have a similar response for the range of gases to which they are sensitive. Hence, they show that a gas is present but not which one. Such sensors are cheap, robust, very durable and sensitive to very low gas concentrations. However, because their discrimination between gases is low and their accuracy in quantitative measurement is poor, they are mainly used only for qualitative indication of gas presence. In this role, they are particularly useful for fire prevention in detecting the presence of the combustion products that occur in low concentrations when the temperature starts to rise due to a fault.

21.9.7 Organic sensors

These work on similar principles to metal oxide semiconductors but use an organic surface layer that is designed to respond selectively to only one gas. At present, these devices are still the subjects of ongoing research, but industrial exploitation is anticipated in the near future. They promise to be cheap and have high stability and sensitivity.

21.9.8 Piezoelectric devices

In these devices, piezoelectric crystals are coated with an absorbent layer. As this layer absorbs gases, the crystal undergoes a change in resonant frequency that can be measured. There is no discrimination in this effect between different gases but the technique potentially offers a high sensitivity mechanism for detecting gas presence. At the present time, problems of finding a suitable type of coating material where absorption is reversible have not been generally solved, and the device only finds limited application at present for measuring moisture concentrations.

21.9.9 Infra-red absorption

This technique uses infra-red light at a particular wavelength that is directed across a chamber between a source and detector. The amount of light absorption is a function of the unknown gas concentration in the chamber. The instrument normally has a second chamber containing gas at a known concentration across which infra-red light at the same wavelength is directed to provide a reference. Sensitivity to carbon monoxide, carbon dioxide, ammonia or hydrocarbons can be provided according to the wavelength used. Microcomputers are now routinely incorporated in the instrument to reduce its sensitivity to gases other than the one being sensed and so improve measurement accuracy. The instrument finds widespread use in chimney/flue emission monitoring and in general process measurements.

21.9.10 Mass spectrometers

The mass spectrometer is a laboratory device for analysing gases. It first reduces a gas sample to a very low pressure. The sample is then ionized, accelerated and separated into its constituent components according to the respective charge-to-mass ratios. Almost any mixture of gases can be analysed and the individual components quantified, but the instrument is very expensive and requires a skilled user. Mass spectrometers have existed for over half a century but recent advances in electronic data processing techniques have greatly improved their performance.

21.9.11 Gas chromatography

This is also a laboratory instrument in which a gaseous sample is passed down a packed column. This separates the gas into its components, which are washed out of the column in turn and measured by a detector. Like the mass spectrometer, the instrument is versatile but expensive and it requires skilled use.

References and further reading

Dimension, angles and flatness:
Anthony, D.M. (1986) *Engineering Metrology*, Pergamon, Oxford, UK.
Hume, K.J. (1970) *Engineering Metrology*, McDonald, London.
Viscosity:
Miller, J.T. (ed.) (1975a) *The Instrument Manual*, United Trade Press, London, pp. 62–106.
Moisture and humidity:
Anderson, J.G. (1989) Paper moisture measurement using microwaves, *Measurement and Control*, **22**, pp. 82–84.
Benson, I.B. (1989) Industrial applications of near infrared reflectance for the measurement of moisture, *Measurement and Control*, **22**, pp. 45–49.
Gimson, C. (1989) Using the capacitance charge transfer principle for water content measurement, *Measurement Control*, **22**, pp. 79–81.
Miller, J.T. (ed.) (1975b) *The Instrument Manual*, United Trade Press, London, pp. 180–209.
Pragnell, R.F. (1989) The modern condensation dewpoint hygrometer, *Measurement and Control*, **22**, pp. 74–77.
Slight, H.A. (1989) Further thoughts on moisture measurement, *Measurement and Control*, **22**, pp. 85–86.
Thompson, F. (1989) Moisture measurement using microwaves, *Measurement and Control*, **22**, pp. 210–215.
Wiltshire, M.P. (1989) Ultrasonic moisture measurement, *Measurement and Control*, **22**, pp. 51–53.
Young, L. (1989) Moisture measurement using low resolution nuclear magnetic resonance, *Measurement and Control*, **22**, pp. 54–55.
Gas sensing and analysis:
Jones, T.A. (1989) Trends in the development of gas sensors, *Measurement and Control*, **22**(6), pp. 176–182.

Appendix 1
Imperial–metric–SI conversion tables

Length

SI units: mm, m, km
Imperial units: in, ft, mile

	mm	m	km	in	ft	mile
mm	1	10^{-3}	10^{-6}	0.039 3701	3.281×10^{-3}	–
m	1000	1	10^{-3}	39.3701	3.280 84	6.214×10^{-4}
km	10^6	10^3	1	39 370.1	3280.84	0.621 371
in	25.4	0.0254	–	1	0.083 333	–
ft	304.8	0.3048	3.048×10^{-4}	12	1	1.894×10^{-4}
mile	–	1609.34	1.609 34	63 360	5280	1

Area

SI units: mm^2, m^2, km^2
Imperial units: in^2, ft^2, $mile^2$

	mm^2	m^2	km^2	in^2	ft^2	$mile^2$
mm^2	1	10^{-6}	–	1.550×10^{-3}	1.076×10^{-5}	–
m^2	10^6	1	10^{-6}	1550	10.764	–
km^2	–	10^6	1	–	1076×10^7	0.3861
in^2	645.16	6.452×10^{-4}	–	1	6.944×10^{-3}	–
ft^2	92 903	0.092 90	–	144	1	–
$mile^2$	–	2.590×10^6	2.590	–	2.788×10^7	1

Second moment of area

SI units: mm^4, m^4
Imperial units: in^4, ft^4

	mm^4	m^4	in^4	ft^4
mm^4	1	10^{-12}	2.4025×10^{-6}	1.159×10^{-10}
m^4	10^{12}	1	2.4025×10^{6}	115.86
in^4	416 231	4.1623×10^{-7}	1	4.8225×10^{-5}
ft^4	8.631×10^{9}	8.631×10^{-3}	20 736	1

Volume

SI units: mm^3, m^3
Metric units: ml, l
Imperial units: in^3, ft^3, UK gallon

	mm^3	ml	l	m^3	in^3	ft^3	UK gallon
mm^3	1	10^{-3}	10^{-6}	10^{-9}	6.10×10^{-5}	–	–
ml	10^3	1	10^{-3}	10^{-6}	0.061 024	3.53×10^{-5}	2.2×10^{-4}
l	10^6	10^3	1	10^{-3}	61.024	0.035 32	0.22
m^3	10^9	10^6	10^3	1	61 024	35.31	220
in^3	16 387	16.39	0.0164	1.64×10^{-5}	1	5.79×10^{-4}	3.61×10^{-3}
ft^3	–	2.83×10^4	28.32	0.028 32	1728	1	6.229
UK gallon	–	4546	4.546	4.55×10^{-3}	277.4	0.1605	1

Note: Additional unit: 1 US gallon = 0.8327 UK gallon.

Density

SI unit: kg/m^3
Metric unit: g/cm^3
Imperial units: lb/ft^3, lb/in^3

	kg/m^3	g/cm^3	lb/ft^3	lb/in^3
kg/m^3	1	10^{-3}	0.062 428	3.605×10^{-5}
g/cm^3	1000	1	62.428	0.036 127
lb/ft^3	16.019	0.016 019	1	5.787×10^{-4}
lb/in^3	27 680	27.680	1728	1

Mass

SI units: g, kg, t
Imperial units: lb, cwt, ton

	g	kg	t	lb	cwt	ton
g	1	10^{-3}	10^{-6}	2.205×10^{-3}	1.968×10^{-5}	9.842×10^{-7}
kg	10^3	1	10^{-3}	2.204 62	0.019 684	9.842×10^{-4}
t	10^6	10^3	1	2204.62	19.6841	0.984 207
lb	453.592	0.453 59	4.536×10^{-4}	1	8.929×10^{-3}	4.464×10^{-4}
cwt	50 802.3	50.8023	0.050 802	112	1	0.05
ton	1.016×10^6	1016.05	1.016 05	2240	20	1

Force

SI units: N, kN
Metric unit: kg_f
Imperial units: pdl (poundal), lb_f, UK ton_f

	N	kg_f	kN	pdl	lb_f	UK ton_f
N	1	0.1020	10^{-3}	7.233	0.2248	1.004×10^{-4}
kg_f	9.807	1	9.807×10^{-3}	70.93	2.2046	9.842×10^{-4}
kN	1000	102.0	1	7233	224.8	0.1004
pdl	0.1383	0.0141	1.383×10^{-4}	1	0.0311	1.388×10^{-5}
lb_f	4.448	0.4536	4.448×10^{-3}	32.174	1	4.464×10^{-4}
UK ton_f	9964	1016	9.964	72 070	2240	1

Note: Additional unit: 1 dyne $= 10^{-5} N = 7.233 \times 10^{-5}$ pdl.

Torque (moment of force)

SI unit: N m
Metric unit: kg_f m
Imperial units: pdl ft, lb_f ft

	N m	kg_f m	pdl ft	lb_f ft
N m	1	0.1020	23.73	0.7376
kg_f m	9.807	1	232.7	7.233
pdl ft	0.042 14	4.297×10^{-3}	1	0.031 08
lb_f ft	1.356	0.1383	32.17	1

Inertia

SI unit: N m^2
Imperial unit: lb$_f$ ft^2

$$1 \text{ lb}_f \text{ ft}^2 = 0.4132 \text{ N m}^2$$
$$1 \text{ N m}^2 = 2.420 \text{ lb}_f \text{ ft}^2$$

Pressure

SI units: mbar, bar, N/m^2 (pascal)
Imperial units: lb/in^2, in Hg, atm

	mbar	bar	N/m^2	lb/in^2	in Hg	atm
mbar	1	10^{-3}	100	0.014 50	0.029 53	9.869 × 10^{-4}
bar	1000	1	10^5	14.50	29.53	0.9869
N/m^2	0.01	10^{-5}	1	1.450 × 10^{-4}	2.953 × 10^{-4}	9.869 × 10^{-6}
lb/in^2	68.95	0.068 95	6895	1	2.036	0.068 05
in Hg	33.86	0.033 86	3386	0.4912	1	0.033 42
atm	1013	1.013	1.013 × 10^5	14.70	29.92	1

Additional conversion factors

1 inch water = 0.073 56 in Hg = 2.491 mbar
1 torr = 1.333 mbar
1 pascal = 1 N/m^2

Energy, work, heat

SI unit: J
Metric units: kg$_f$ m, kW h
Imperial units: ft lb$_f$, cal, Btu

	J	kg$_f$ m	kW h	ft lb$_f$	cal	Btu
J	1	0.1020	2.778 × 10^{-7}	0.7376	0.2388	9.478 × 10^{-4}
kg$_f$ m	9.8066	1	2.724 × 10^{-6}	7.233	2.342	9.294 × 10^{-3}
kW h	3.600 × 10^6	367 098	1	2.655 × 10^6	859 845	3412.1
ft lb$_f$	1.3558	0.1383	3.766 × 10^{-7}	1	0.3238	1.285 × 10^{-3}
cal	4.1868	0.4270	1.163 × 10^{-6}	3.0880	1	3.968 × 10^{-3}
Btu	1055.1	107.59	2.931 × 10^{-4}	778.17	252.00	1

Additional conversion factors

1 therm = 10^5 Btu = 1.0551×10^8 J
1 thermie = 4.186×10^6 J
1 hp h = 0.7457 kW h = 2.6845×10^6 J
1 ft pdl = 0.042 14 J
1 erg = 10^{-7} J

Power

SI units W, kW
Imperial units: HP, ft lb_f/s

	W	kW	HP	ft lb_f/s
W	1	10^{-3}	1.341×10^{-3}	0.735 64
kW	10^3	1	1.341 02	735.64
HP	745.7	0.7457	1	548.57
ft lb_f/s	1.359 35	1.359×10^{-3}	1.823×10^{-3}	1

Velocity

SI units: mm/s, m/s
Metric unit: km/h
Imperial units: ft/s, mile/h

	mm/s	m/s	km/h	ft/s	mile/h
mm/s	1	10^{-3}	3.6×10^{-3}	3.281×10^{-3}	2.237×10^{-3}
m/s	1000	1	3.6	3.280 84	2.236 94
km/h	277.778	0.277 778	1	0.911 344	0.621 371
ft/s	304.8	0.3048	1.097 28	1	0.681 818
mile/h	447.04	0.447 04	1.609 344	1.466 67	1

Acceleration

SI unit: m/s^2
Other metric unit: cm/s^2
Imperial unit: ft/s^2
Other unit: g

	m/s^2	cm/s^2	ft/s^2	g
m/s^2	1	100	3.281	0.102
cm/s^2	0.01	1	0.0328	0.001 02
ft/s^2	0.3048	30.48	1	0.031 09
g	9.81	981	32.2	1

Mass flow rate

SI unit: g/s
Metric units: kg/h, tonne/d
Imperial units: lb/s, lb/h, ton/d

	g/s	kg/h	tonne/d	lb/s	lb/h	ton/d
g/s	1	3.6	0.086 40	2.205×10^{-3}	7.937	0.085 03
kg/h	0.2778	1	0.024 00	6.124×10^{-4}	2.205	0.023 62
tonne/d	11.57	41.67	1	0.025 51	91.86	0.9842
lb/s	453.6	1633	39.19	1	3600	38.57
lb/h	0.1260	0.4536	0.010 89	2.788×10^{-4}	1	0.010 71
ton/d	11.76	42.34	1.016	0.025 93	93.33	1

Volume flow rate

SI unit: m^3/s
Metric units: l/h, ml/s
Imperial units: gal/h, ft^3/s, ft^3/h

	l/h	ml/s	m^3/s	gal/h	ft^3/s	ft^3/h
l/h	1	0.2778	2.778×10^{-7}	0.2200	9.810×10^{-6}	0.035 316
ml/s	3.6	1	10^{-6}	0.7919	3.532×10^{-5}	0.127 14
m^3/s	3.6×10^6	10^6	1	7.919×10^5	35.31	1.271×10^5
gal/h	4.546	1.263	1.263×10^{-6}	1	4.460×10^{-5}	0.160 56
ft^3/s	1.019×10^5	2.832×10^4	0.028 32	2.242×10^4	1	3600
ft^3/h	28.316	7.8653	7.865×10^{-6}	6.2282	2.778×10^{-4}	1

Specific energy (heat per unit volume)

SI units: J/m^3, kJ/m^3, MJ/m^3
Imperial units: $kcal/m^3$, Btu/ft^3, therm/UK gal

	J/m^3	kJ/m^3	MJ/m^3	$kcal/m^3$	Btu/ft^3	therm/UK gal
J/m^3	1	10^{-3}	10^{-6}	1.388×10^{-4}	2.684×10^{-5}	–
kJ/m^3	1000	1	10^{-3}	0.2388	0.026 84	–
MJ/m^3	10^6	1000	1	238.8	26.84	4.309×10^{-5}
$kcal/m^3$	4187	4.187	4.187×10^{-3}	1	0.1124	1.804×10^{-7}
Btu/ft^3	3.726×10^4	37.26	0.037 26	8.899	1	1.605×10^{-6}
therm/gal	–	–	2.321×10^4	5.543×10^6	6.229×10^5	1

Dynamic viscosity

SI unit: $N\, s/m^2$
Metric unit: cP (centipoise), P (poise) [1 P = 100 g/m s]
Imperial unit: $lb_m/ft\, h$

	$lb_m/ft\, h$	P	cP	$N\, s/m^2$
$lb_m/ft\, h$	1	4.133×10^{-3}	0.4134	4.134×10^{-4}
P	241.9	1	100	0.1
cP	2.419	0.01	1	10^{-3}
$N\, s/m^2$	2419	10	1000	1

Note: Additional unit: 1 pascal second = $1\, N\, s/m^2$.

Kinematic viscosity

SI unit: m^2/s
Metric unit: cSt (centistokes), St (stokes)
Imperial unit: ft^2/s

	ft^2/s	m^2/s	cSt	St
ft^2/s	1	0.0929	9.29×10^4	929
m^2/s	10.764	1	10^6	10^4
cSt	1.0764×10^{-5}	10^{-6}	1	0.01
St	1.0764×10^{-3}	10^{-4}	100	1

Appendix 2 Thévenin's theorem

Thévenin's theorem is extremely useful in the analysis of complex electrical circuits. It states that any network which has two accessible terminals A and B can be replaced, as far as its external behaviour is concerned, by a single e.m.f. acting in series with a single resistance between A and B. The single equivalent e.m.f. is that e.m.f. which is measured across A and B when the circuit external to the network is disconnected. The single equivalent resistance is the resistance of the network when all current and voltage sources within it are reduced to zero. To calculate this internal resistance of the network, all current sources within it are treated as open circuits and all voltage sources as short circuits. The proof of Thévenin's theorem can be found in Skilling (1967).

Figure A2.1 shows part of a network consisting of a voltage source and four resistances. As far as its behaviour external to the terminals A and B is concerned, this can be regarded as a single voltage source V_t and a single resistance R_t. Applying Thévenin's theorem, R_t is found first of all by treating V_1 as a short circuit, as shown in Figure A2.2. This is simply two resistances, R_1 and $(R_2 + R_4 + R_5)$ in parallel. The

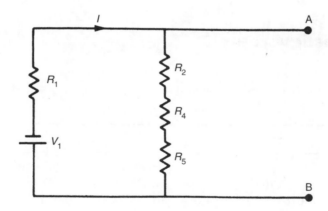

Fig. A2.1

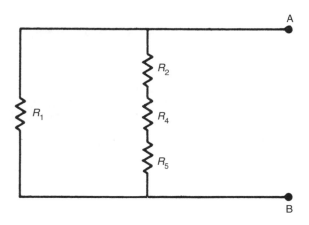

Fig. A2.2

equivalent resistance R_t is thus given by:

$$R_t = \frac{R_1(R_2 + R_4 + R_5)}{R_1 + R_2 + R_4 + R_5}$$

V_t is the voltage drop across AB. To calculate this, it is necessary to carry out an intermediate step of working out the current flowing, I. Referring to Figure A2.1, this is given by:

$$I = \frac{V_1}{R_1 + R_2 + R_4 + R_5}$$

Now, V_t can be calculated from:

$$V_t = I(R_2 + R_4 + R_5)$$
$$= \frac{V_1(R_2 + R_4 + R_5)}{R_1 + R_2 + R_4 + R_5}$$

The network of Figure A2.1 has thus been reduced to the simpler equivalent network shown in Figure A2.3.

Let us now proceed to the typical network problem of calculating the current flowing in the resistor R_3 of Figure A2.4. R_3 can be regarded as an external circuit or load on

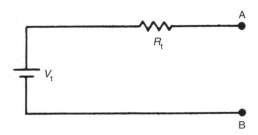

Fig. A2.3

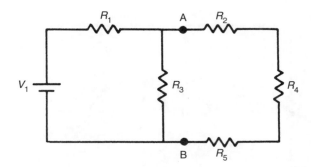

Fig. A2.4

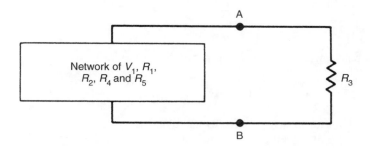

Fig. A2.5

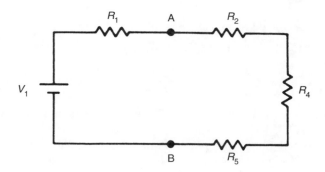

Fig. A2.6

the rest of the network consisting of V_1, R_1, R_2, R_4 and R_5, as shown in Figure A2.5. This network of V_1, R_1, R_2, R_4 and R_5 is that shown in Figure A2.6. This can be rearranged to the network shown in Figure A2.1, which is equivalent to the single voltage source and resistance, V_t and R_t, calculated above. The whole circuit is then equivalent to that shown in Figure A2.7, and the current flowing through R_3 can be written as:

$$I_{AB} = \frac{V_t}{R_t + R_3}$$

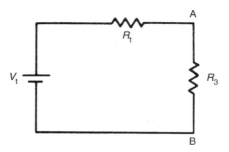

Fig. A2.7

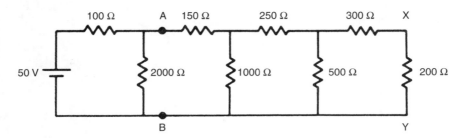

Fig. A2.8

Thévenin's theorem can be applied successively to solve ladder networks of the form shown in Figure A2.8. Suppose in this network that it is required to calculate the current flowing in branch XY.

The first step is to imagine two terminals in the circuit A and B and regard the network to the right of AB as a load on the circuit to the left of AB. The circuit to the left of AB can be reduced to a single equivalent voltage source, E_{AB}, and resistance, R_{AB}, by Thévenin's theorem. If the 50 V source is replaced by its zero internal resistance (i.e. by a short circuit), then R_{AB} is given by:

$$\frac{1}{R_{AB}} = \frac{1}{100} + \frac{1}{2000} = \frac{2000 + 100}{200\,000}$$

Hence:
$$R_{AB} = 95.24\,\Omega$$

When AB is open circuit, the current flowing round the loop to the left of AB is given by:

$$I = \frac{50}{100 + 2000}$$

Hence, E_{AB}, the open-circuit voltage across AB, is given by:

$$E_{AB} = I \times 2000 = 47.62\,\text{V}$$

We can now replace the circuit shown in Figure A2.8 by the simpler equivalent circuit shown in Figure A2.9.

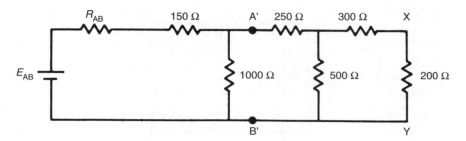

Fig. A2.9

The next stage is to apply an identical procedure to find an equivalent circuit consisting of voltage source $E_{A'B'}$ and resistance $R_{A'B'}$ for the network to the left of points A' and B' in Figure A2.9:

$$\frac{1}{R_{A'B'}} = \frac{1}{R_{AB} + 150} + \frac{1}{1000} = \frac{1}{245.24} + \frac{1}{1000} = \frac{1245.24}{245\,240}$$

Hence:

$$R_{A'B'} = 196.94\,\Omega$$

$$E_{A'B'} = \frac{1000}{R_{AB} + 150 + 1000} E_{AB} = 38.24\,\text{V}$$

The circuit can now be represented in the yet simpler form shown in Figure A2.10. Proceeding as before to find an equivalent voltage source and resistance, $E_{A''B''}$ and $R_{A''B''}$, for the circuit to the left of A''B'' in Figure A2.10:

$$\frac{1}{R_{A''B''}} = \frac{1}{R_{A'B'} + 250} + \frac{1}{500} = \frac{500 + 446.94}{223\,470}$$

Hence:

$$R_{A''B''} = 235.99\,\Omega$$

$$E_{A''B''} = \frac{500}{R_{A'B'} + 250 + 500} E_{A'B'} = 20.19\,\text{V}$$

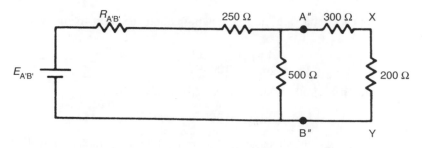

Fig. A2.10

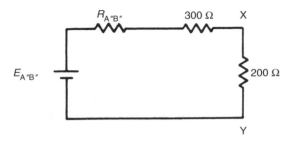

Fig. A2.11

The circuit has now been reduced to the form shown in Figure A2.11, where the current through branch XY can be calculated simply as:

$$I_{XY} = \frac{E_{A''B''}}{R_{A''B''} + 300 + 200} = \frac{20.19}{735.99} = 27.43 \, \text{mA}$$

References and further reading

Skilling, H.H. (1967) *Electrical Engineering Circuits*, Wiley: New York.

Appendix 3
Thermocouple tables

Type E: chromel–constantan
Type J: iron–constantan
Type K: chromel–alumel
Type N: nicrosil–nisil
Type S: platinum/10% rhodium–platinum
Type T: copper–constantan

Temp. (°C)	Type E	Type J	Type K	Type N	Type S	Type T
−270	−9.834		−6.458	−4.345		
−260	−9.795		−6.441	−4.336		
−250	−9.719		−6.404	−4.313		
−240	−9.604		−6.344	−4.277		−6.105
−230	−9.456		−6.262	−4.227		−6.003
−220	−9.274		−6.158	−4.162		−5.891
−210	−9.063	−8.096	−6.035	−4.083		−5.753
−200	−8.824	−7.890	−5.891	−3.990		−5.603
−190	−8.561	−7.659	−5.730	−3.884		−5.438
−180	−8.273	−7.402	−5.550	−3.766		−5.261
−170	−7.963	−7.122	−5.354	−3.634		−5.070
−160	−7.631	−6.821	−5.141	−3.491		−4.865
−150	−7.279	−6.499	−4.912	−3.336		−4.648
−140	−6.907	−6.159	−4.669	−3.170		−4.419
−130	−6.516	−5.801	−4.410	−2.994		−4.177
−120	−6.107	−5.426	−4.138	−2.807		−3.923
−110	−5.680	−5.036	−3.852	−2.612		−3.656
−100	−5.237	−4.632	−3.553	−2.407		−3.378
−90	−4.777	−4.215	−3.242	−2.193		−3.089
−80	−4.301	−3.785	−2.920	−1.972		−2.788
−70	−3.811	−3.344	−2.586	−1.744		−2.475
−60	−3.306	−2.892	−2.243	−1.509		−2.152
−50	−2.787	−2.431	−1.889	−1.268	−0.236	−1.819

Temp. (°C)	Type E	Type J	Type K	Type N	Type S	Type T
−40	−2.254	−1.960	−1.527	−1.023	−0.194	−1.475
−30	−1.709	−1.481	−1.156	−0.772	−0.150	−1.121
−20	−1.151	−0.995	−0.777	−0.518	−0.103	−0.757
−10	−0.581	−0.501	−0.392	−0.260	−0.053	−0.383
0	0.000	0.000	0.000	0.000	0.000	0.000
10	0.591	0.507	0.397	0.261	0.055	0.391
20	1.192	1.019	0.798	0.525	0.113	0.789
30	1.801	1.536	1.203	0.793	0.173	1.196
40	2.419	2.058	1.611	1.064	0.235	1.611
50	3.047	2.585	2.022	1.339	0.299	2.035
60	3.683	3.115	2.436	1.619	0.365	2.467
70	4.329	3.649	2.850	1.902	0.432	2.908
80	4.983	4.186	3.266	2.188	0.502	3.357
90	5.646	4.725	3.681	2.479	0.573	3.813
100	6.317	5.268	4.095	2.774	0.645	4.277
110	6.996	5.812	4.508	3.072	0.719	4.749
120	7.683	6.359	4.919	3.374	0.795	5.227
130	8.377	6.907	5.327	3.679	0.872	5.712
140	9.078	7.457	5.733	3.988	0.950	6.204
150	9.787	8.008	6.137	4.301	1.029	6.702
160	10.501	8.560	6.539	4.617	1.109	7.207
170	11.222	9.113	6.939	4.936	1.190	7.718
180	11.949	9.667	7.338	5.258	1.273	8.235
190	12.681	10.222	7.737	5.584	1.356	8.757
200	13.419	10.777	8.137	5.912	1.440	9.286
210	14.161	11.332	8.537	6.243	1.525	9.820
220	14.909	11.887	8.938	6.577	1.611	10.360
230	15.661	12.442	9.341	6.914	1.698	10.905
240	16.417	12.998	9.745	7.254	1.785	11.456
250	17.178	13.553	10.151	7.596	1.873	12.011
260	17.942	14.108	10.560	7.940	1.962	12.572
270	18.710	14.663	10.969	8.287	2.051	13.137
280	19.481	15.217	11.381	8.636	2.141	13.707
290	20.256	15.771	11.793	8.987	2.232	14.281
300	21.033	16.325	12.207	9.340	2.323	14.860
310	21.814	16.879	12.623	9.695	2.414	15.443
320	22.597	17.432	13.039	10.053	2.506	16.030
330	23.383	17.984	13.456	10.412	2.599	16.621
340	24.171	18.537	13.874	10.772	2.692	17.217
350	24.961	19.089	14.292	11.135	2.786	17.816
360	25.754	19.640	14.712	11.499	2.880	18.420
370	26.549	20.192	15.132	11.865	2.974	19.027
380	27.345	20.743	15.552	12.233	3.069	19.638

Appendix 3 Thermocouple tables

Temp. (°C)	Type E	Type J	Type K	Type N	Type S	Type T
390	28.143	21.295	15.974	12.602	3.164	20.252
400	28.943	21.846	16.395	12.972	3.260	20.869
410	29.744	22.397	16.818	13.344	3.356	
420	30.546	22.949	17.241	13.717	3.452	
430	31.350	23.501	17.664	14.091	3.549	
440	32.155	24.054	18.088	14.467	3.645	
450	32.960	24.607	18.513	14.844	3.743	
460	33.767	25.161	18.938	15.222	3.840	
470	34.574	25.716	19.363	15.601	3.938	
480	35.382	26.272	19.788	15.981	4.036	
490	36.190	26.829	20.214	16.362	4.135	
500	36.999	27.388	20.640	16.744	4.234	
510	37.808	27.949	21.066	17.127	4.333	
520	38.617	28.511	21.493	17.511	4.432	
530	39.426	29.075	21.919	17.896	4.532	
540	40.236	29.642	22.346	18.282	4.632	
550	41.045	30.210	22.772	18.668	4.732	
560	41.853	30.782	23.198	19.055	4.832	
570	42.662	31.356	23.624	19.443	4.933	
580	43.470	31.933	24.050	19.831	5.034	
590	44.278	32.513	24.476	20.220	5.136	
600	45.085	33.096	24.902	20.609	5.237	
610	45.891	33.683	25.327	20.999	5.339	
620	46.697	34.273	25.751	21.390	5.442	
630	47.502	34.867	26.176	21.781	5.544	
640	48.306	35.464	26.599	22.172	5.648	
650	49.109	36.066	27.022	22.564	5.751	
660	49.911	36.671	27.445	22.956	5.855	
670	50.713	37.280	27.867	23.348	5.960	
680	51.513	37.893	28.288	23.740	6.064	
690	52.312	38.510	28.709	24.133	6.169	
700	53.110	39.130	29.128	24.526	6.274	
710	53.907	39.754	29.547	24.919	6.380	
720	54.703	40.382	29.965	25.312	6.486	
730	55.498	41.013	30.383	25.705	6.592	
740	56.291	41.647	30.799	26.098	6.699	
750	57.083	42.283	31.214	26.491	6.805	
760	57.873	42.922	31.629	26.885	6.913	
770	58.663	43.563	32.042	27.278	7.020	
780	59.451	44.207	32.455	27.671	7.128	
790	60.237	44.852	32.866	28.063	7.236	
800	61.022	45.498	33.277	28.456	7.345	
810	61.806	46.144	33.686	28.849	7.454	

Temp. (°C)	Type E	Type J	Type K	Type N	Type S	Type T
820	62.588	46.790	34.095	29.241	7.563	
830	63.368	47.434	34.502	29.633	7.672	
840	64.147	48.076	34.908	30.025	7.782	
850	64.924	48.717	35.314	30.417	7.892	
860	65.700	49.354	35.718	30.808	8.003	
870	66.473	49.989	36.121	31.199	8.114	
880	67.245	50.621	36.524	31.590	8.225	
890	68.015	51.249	36.925	31.980	8.336	
900	68.783	51.875	37.325	32.370	8.448	
910	69.549	52.496	37.724	32.760	8.560	
920	70.313	53.115	38.122	33.149	8.673	
930	71.075	53.729	38.519	33.538	8.786	
940	71.835	54.341	38.915	33.926	8.899	
950	72.593	54.949	39.310	34.315	9.012	
960	73.350	55.553	39.703	34.702	9.126	
970	74.104	56.154	40.096	35.089	9.240	
980	74.857	56.753	40.488	35.476	9.355	
990	75.608	57.349	40.879	35.862	9.470	
1000	76.357	57.942	41.269	36.248	9.585	
1010		58.533	41.657	36.633	9.700	
1020		59.121	42.045	37.018	9.816	
1030		59.708	42.432	37.402	9.932	
1040		60.293	42.817	37.786	10.048	
1050		60.877	43.202	38.169	10.165	
1060		61.458	43.585	38.552	10.282	
1070		62.040	43.968	38.934	10.400	
1080		62.619	44.349	39.315	10.517	
1090		63.199	44.729	39.696	10.635	
1100		63.777	45.108	40.076	10.754	
1110		64.355	45.486	40.456	10.872	
1120		64.933	45.863	40.835	10.991	
1130		65.510	46.238	41.213	11.110	
1140		66.087	46.612	41.590	11.229	
1150		66.664	46.985	41.966	11.348	
1160		67.240	47.356	42.342	11.467	
1170		67.815	47.726	42.717	11.587	
1180		68.389	48.095	43.091	11.707	
1190		68.963	48.462	43.464	11.827	
1200		69.536	48.828	43.836	11.947	
1210			49.192	44.207	12.067	
1220			49.555	44.577	12.188	
1230			49.916	44.947	12.308	
1240			50.276	45.315	12.429	

Appendix 3 Thermocouple tables

Temp. (°C)	Type E	Type J	Type K	Type N	Type S	Type T
1250			50.633	45.682	12.550	
1260			50.990	46.048	12.671	
1270			51.344	46.413	12.792	
1280			51.697	46.777	12.913	
1290			52.049	47.140	13.034	
1300			52.398	47.502	13.155	
1310			52.747		13.276	
1320			53.093		13.397	
1330			53.438		13.519	
1340			53.782		13.640	
1350			54.125		13.761	
1360			54.467		13.883	
1370			54.807		14.004	
1380					14.125	
1390					14.247	
1400					14.368	
1410					14.489	
1420					14.610	
1430					14.731	
1440					14.852	
1450					14.973	
1460					15.094	
1470					15.215	
1480					15.336	
1490					15.456	
1500					15.576	
1510					15.697	
1520					15.817	
1530					15.937	
1540					16.057	
1550					16.176	
1560					16.296	
1570					16.415	
1580					16.534	
1590					16.653	
1600					16.771	
1610					16.890	
1620					17.008	
1630					17.125	
1640					17.243	
1650					17.360	
1660					17.477	
1670					17.594	

Temp. (°C)	Type E	Type J	Type K	Type N	Type S	Type T
1680					17.711	
1690					17.826	
1700					17.942	
1710					18.056	
1720					18.170	
1730					18.282	
1740					18.394	
1750					18.504	
1760					18.612	

Appendix 4 Solutions to self-test questions

Chapter 2

Q5. 0.0175 mV/°C

Q7. (a) 2.62 ; (b) 2.94 ; 0.32

Q8. (a) 20 μm/kg; 22 μm/kg; (b) 200 μm; 2 μm/kg; (c) 14.3 μm/°C; 0.143 μm (°C)$^{-1}$ (kg)$^{-1}$

Q9. (a)

Time	Depth	Temp. reading	Temp. error
0	0	20.0	0.0
100	50	19.716	0.216
200	100	19.245	0.245
300	150	18.749	0.249
400	200	18.250	0.250
500	250	17.750	0.250

(b) 10.25°C

Chapter 3

Q3. 3.9%

Q5. 5.0%; 24 750 Ω

Q6. 10.0%

Q9. mean 31.1; median 30.5; standard deviation 3.0

Q10. mean 1.537; standard deviation 0.021; accuracy of mean value = ±0.007 i.e. mean value = 1.537 ± 0.007; for 1000 measurements, accuracy would be improved by a factor of 10

Q11. Mean value = 21.8 mA

Measurement	21.5	22.1	21.3	21.7	22.0	22.2	21.8	21.4	21.9	22.1
Deviation from mean	−0.3	+0.3	−0.5	−0.1	+0.2	+0.4	0.0	−0.4	+0.1	+0.3
(deviations)^2	0.09	0.09	0.25	0.01	0.04	0.16	0.0	0.16	0.01	0.09

Standard deviation = 0.32

Q12. 86.6%

Q13. 97.7%

Q14. ±0.7%

Q15. ±4.7%

Q16. ±2.2%

Q17. 46.7 ohm ± 2.5%

Q18. 2.5%

Q19. (a) 0.31 m^3/min; (b) ±4.1%

Chapter 7

Q1. 81.9 mV

Q2. 378 mV

Q3. (a) 0.82 mV/°C; (b) indicated temperature 101.9°C; error 1.9°C

Q4. 24 V; 1.2 W

Q6. 85.9 mV

Q7. (a) 69.6 Ω, 930.4 Ω; (b) 110.3 Ω

Q8. (a) $R_u = R_2R_3/R_1$; $L_u = R_2R_3C$; (b) 1.57 Ω; 100 mH; (c) 20

Q9. 2.538 V r.m.s.

Q10. 50 µF

Q11. (a) At balance $\dfrac{R_1 + j\omega L}{R_3} = \dfrac{R_2}{R_4 - j/\omega C}$. Then, by taking real and imaginary parts and manipulating $L = \dfrac{R_1}{\omega^2 R_4 C}$ and $R_1 = \dfrac{R_2 R_3}{R_4(1 + 1/\omega^2 R_4^2 C^2)}$. Hence, at balance $L = \dfrac{R_2 R_3 C}{1 + \omega^2 R_4^2 C^2}$

(b) $Q = \omega L/R_1 = 1/\omega R_4 C$ using the equations developed in part (a) above. For large Q, $\omega^2 R_4^2 C^2 \ll 1$, and the equation for L above becomes $L = R_2 R_3 C$. This is independent of frequency because there is no ω term in the expression

(c) 20 mH

Chapter 9

Q2.
	One's compl.	Two's compl.
(a)	01010000	01010001
(b)	10001000	10001001
(c)	10011010	10011011
(d)	00101001	00101010
(e)	00010011	00010100

Q3. (a) 111001 71 39
    (b) 1100101 145 65
    (c) 10101111 257 AF
    (d) 100000011 403 103
    (e) 1111100111 1747 3E7
    (f) 10011010010 2322 4D2

Q4. (a) 7515; (b) F4D

Q5. (a) 130645; (b) B1A5

Q6. (a) 1214; (b) 28C; (c) 3352

Chapter 11

Q1. (b) $\omega_n = \sqrt{K_s/J}$; $\beta = \sqrt{K_I/2RJK_S}$; sensitivity $= K_I/K_S R$
    (d) 0.7; (e) typical bandwidth 100 Hz; maximum frequency 30 Hz

Q3. $a = 12.410$; $b = 40.438$

Q4. $9.8 \, \Omega$

Q5. $a = 1.12$; $b = 2.00$

Q6. (a) $C = 5.77 \times 10^{-7}$; $T_0 = 11027$; (b) $T = 428°K$

Chapter 12

Q1. 30.4 days

Q2. 40.6 days

Q3. 222 days

Q4. 0.988

Q5. 0.61 or 61%

Q6. 0.95 or 95%

Q7. 0.86

Q8. 0.9975

Q9. 0.92

Q10. 24

Q11. 39
Q12. (a) 95.2%; (b) 49 seeded errors
Q13. 97.2%
Q14. 94.2%

Chapter 14

Q1. 300°C
Q2. 147.1°C
Q3. 700°C
Q4. 610°C; 678.4°C
Q5. 15.55 mV; 228.5°C

Index

absolute pressure, 304, 305, 307, 310
a.c. carrier, 153–4
acceleration measurement, 254, 258, 269, 383–6, 417–8
accuracy, 16–17
acoustic thermometer, 298–9, 302
active filters, 85–6
active instruments, 12–13
address bus, 167
address decoding, 174–5
aggregation of measurement errors, 56–9
air-vane meter, 337
alarms, 241
aliasing, 96
amplification, 9, 87–8, 89, 101
amplifier *see* operational amplifier
amplitude modulation (AM), 153
analogue–digital conversion, 95–6, 97–8
analogue filters, 78–86
analogue instruments, 14–15
analogue meters, 104–113
angle measurement, 426–7
angular motion, 390–418
annubar, 322, 327
anti–ambiguity track, 396
antimony electrode, 439
apex bridge-circuit balancing, 129–30
ASIC (Application specific integrated circuit), 184
asynchronous transmission, 189–90
attenuation, 80, 82, 84, 88–9, 101
autocorrelation, 100–1

band pass filter, 80, 81, 83, 85
band stop filter, 80, 81, 83–4, 85
bandwidth, 114
bathtub reliability curve, 226

beam balance, 356
bell-shaped distribution, 48
bellows, 307–8, 317
bevel protractor, 426
bias, 21, 91, 101
bimetallic thermometer, 296, 302
bimetallic thermostat, 296
binary numbers, 168–73
Bourdon tube, 308–10, 317
box cube, 419
bridge circuits, 8, 119–135, 138, 144
    a.c., 130–4, 138, 144
    balancing, 129–30
    d.c., 119–30, 135
    error analysis, 128–9
British Calibration Service (BCS), 68
BS 5750, 66
bubbler unit, 342
bus network, 194–5

calibration, 21, 29–30, 41, 64–72, 179, 182
    calibration chain, 67–70
    calibration frequency, 65
    documentation, 69–72
calipers, 420–2
calorimetric sensors, 440
Cambridge ring, 195
capacitance measurement, 138–9
capacitive coupling, 74
capacitive sensors, 247, 260, 306, 343, 370–1, 432
catalytic gate FET, 442
catalytic sensors, 440
cathode ray oscilloscope, 114–8, 143–4, 147
centrifugal tachometer, 413
characteristic impedance, 82

chart recorders, 202–11
chi-squared test, 55
choice of instrument *see* instrument choice
chromatography, 443
clamp-on meter, 108, 141
coded disc shaft encoder *see* shaft encoders
coefficient of viscosity, 429
colour codes (resistors and capacitors), 137, 139
colour temperature indicators, 299, 302
common mode rejection, 88
communications, 183–4, 187–99
compensating leads, 275
compensating resistance, 39
computer data logging, 210–1
computer networks, 187–99
computing principles, 165–77
confidence tests, 216, 220
constant-k filter, 80, 83, 85
contention protocol, 195
continuous thermocouple, 282–3
control bus, 167, 175
conversion tables: imperial-SI units, 445–51
Coriolis meter, 320, 338
corona discharge, 75
correlation test, 220–1
counter–timer, 142, 145
CPLD (Complex programmable logic device), 184
crayon temperature indicators, 299
creep, 352
cross correlation flowmeter, 336
cross sensitivity, 384
cross talk sensor, 257, 349
cumulative distribution function, 48
current loop transmission, 152–3, 183
current measurement, 140–1
current to voltage conversion, 153
current transformer, 140
curve fitting, 214–21
cut off frequency, 79

Dall flow tube, 322, 325–6
damping ratio, 29, 205–6
data analysis, 43–56
data bus, 167, 175
data logging, 210–1
data presentation, 212–21
data transmission, 187–99
dead space, 23
dead weight gauge, 14, 312, 317

decibel (dB), 114
deflection instruments, 13–14
depth gauge, 425
design of instruments *see* instrument design
dew point meter, 435
dial gauge, 425
diaphragm–type pressure transducer, 305–7, 317
differential amplifier, 89–90
differential pressure, 304, 305, 307, 310, 318
differential transformers
    linear, 368
    rotational, 391
digital–analog converter, 99
digital filters, 100
digital instruments, 14–15
digital meters, 102–4, 136, 140, 141, 142
digital recorder, 210–1
digital (storage) oscilloscope, 118, 211
digital thermometer, 270, 282
digital voltmeter (DVM), 102–4, 136
dimension measurement, 419–27
diode temperature sensors, 287
dipstick, 340
discharge coefficient, 324
displacement measurement
    rotational, 390–407
    translational, 255, 365–82
display of signals, 200–1
distributed control system, 187
distributed sensors, 254, 259, 282, 298
Doppler effect, 265–7, 333
draft gauge, 311
dual diverse temperature sensor, 301
duplex communication, 188
DVM *see* digital voltmeter
dynamic characteristics of instruments, 23–9, 205–6
dynamic viscosity, 429
dynamometer, 107–8

earthing, 77
eddy current sensors, 248
electrical signals:
    measurement, 34–7, 102–18, 119–47
    recording, 202–11
electrochemical cells, 441–2
electrochemical potential, 75
electrodynamic meter, 107–8
electromagnetic balance, 359

electromagnetic flowmeter, 330–2, 339
electronic balance, 352, 354
electronic spirit level, 427
electronic voltmeter, 111
electrostatic coupling, 74
emissivity, 288
EN50170 fieldbus, 197
encoders *see* shaft encoders
environmentally-induced errors, 37
environmental pollution, 440
equal-arm balance *see* beam balance
error frequency distribution, 49
errors in measurement systems, 32–59, 91, 125–8
Ethernet, 195
evanescent field effect displacement sensor, 378
extension leads, 274

Farad, 138
fault detection, 180, 182
fibre optic principles, 156–60
   data networks, 193, 198
   recorder, 209
   sensors, 253–9, 296, 297–8, 302, 306, 349, 406, 416
   signal transmission, 155–60
Fieldbus, 196–9
filters, 78–86, 100
fire detection/prevention, 282, 293–4, 439
first order type instruments, 25–8
fixed points (temperature measurement), 271
flatness measurement, 428
float and tape gauge, 341
float systems, 340–1
floating point, 170
flow measurement, 178, 258, 319–39
   mass flow rate, 319–21
   volume flow rate, 178, 321–39
flow nozzle, 322, 325–6
force measurement, 359–61
fotonic sensor, 254, 377
FPGA (Field programmable gate array), 184
frequency attenuation, 82–5
frequency distribution, 46–56
frequency measurement, 141–5
frequency modulation (FM), 153
full duplex mode, 188

gas chromatography, 443
gas sensing and analysis, 258, 439–40

gate-type meter, 336
gateway, 195
gauge block, 423–4
gauge factor, 251
gauge pressure, 304, 305, 307, 308, 310, 311, 312
Gaussian distribution, 48–56
Gaussian tables, 50–1
glass electrode, 438
goodness of fit, 54–6
graphical data analysis, 46–56
graphical data presentation, 213–21
Gray code, 396
gyroscopes, 258, 402–6, 415–6

half duplex mode, 188
Hall-effect sensors, 249–50
HART, 195–6
Hay's bridge, 149
heat detection, 282, 293–4
heat-sensitive cable, 282
height gauge, 425
henry, 138
hertz (Hz), 141
hexadecimal numbers, 171–3
high pass filter, 80, 81, 83, 85
histogram, 46–7
hot wire element level gauge, 348
humidity measurement *see* moisture measurement
hydrostatic systems, 341–3
hygrometers, 435–6
hysteresis, 22

IEC bus (IEC625), 191
IEC61158 fieldbus
IEC61508, 237–8
IEE488 bus, 191–2
Imaging, 267, 293–4
imperial–SI units conversion tables, 445–51
imperial units, 5–6
inaccuracy, 16–17
incremental shaft encoder, 392
indicating instruments, 15–16
inductance measurement, 138
induction potentiometer, 402
induction tachometer, 408
inductive coupling, 74
inductive sensors, 247–50, 371, 408
inductosyn, 374, 402

inertial navigation systems, 403
instrument choice, 9–11
instrument design, 9–11, 39, 41, 43
instumentation amplifier, 87–8
instrumentation networks, 187–99
integrated circuit transistor sensors, 286
intelligent devices, 42, 165–85
  in acceleration measurement, 385
  in dimension measurement, 422, 423
  in displacement measurement, 399
  in flow measurement, 178, 338
  in force and mass measurement, 355
  in level measurement, 351
  in pressure measurement, 316
  in temperature measurement, 300, 302
interfacing, 167, 174–7, 187–99
international practical temperature scale (ITPS), 271
intrinsic safety, 236
ionisation gauge, 315
ISO-7 protocol, 197, 198–9
ISO 9000, 66, 69

kinematic viscosity, 429

LAN see local area network
laser Doppler flowmeter, 258, 337
laser interferometer, 376
law of intermediate metals, 275–6
law of intermediate temperatures, 277–9
least squares regression, 216–20
length bar, 424
level measurement, 12, 257, 340–51
line-type heat detector, 282
linear variable differential transformer ((LVDT), 368
linearization, 90–1
linearity, 19
liquid-in-glass thermometer, 26, 295, 302
Lisajous patterns, 143–4
load cell, 352–6
local area network, 187, 190, 192–9
lock-in amplifier, 78, 93–4
low pass filter, 80, 81, 83, 85
lower explosive level, 439
LVDT see linear variable differential transformer

magnetic sensors, 247–50, 410
magnetic tape recorder, 209–10
magnetostrictive tachometer, 410

manometers, 310–1, 317
manufacturing tolerances, 53–6
MAP (manufacturing automation protocol), 199
mass flow rate, see flow measurement
mass measurement, 342–9
mass spectrometer, 443
Maxwell bridge, 131–2
McLeod gauge, 314–5
mean, 43–4
mean-time-between-failures, 225
mean-time-to-repair, 225
measurement disturbance, 33–8, 125–8, 140
measurement system design, 8, 37–9
measurement uncertainty, 16–17
measuring units, 3–6, 445–51
mechanical flyball, 413
median, 43–4
metal oxide gas sensors, 442
meters, 102–13
metric units, 6
metropolitan area network (MAN), 198
microbend sensor, 307
micrometers, 69–70, 422–3
microprocessor, 166
microsensors, 268–70, 306
mirror galvanometer, 208
modem, 193
modifying inputs see environmentally induced errors
moisture measurement, 432–6
moving coil meters, 105, 140
moving iron meters, 106, 140
multimeters, 102, 104, 108–9, 136
multiple earths, 74
multiplexing, 155, 160
multivariable transmitter, 180

National Measurement Accreditation Service (NAMAS), 68
National Standards Organisations, 67–70
National Testing Laboratory Accreditation Scemem (NATLAS), 68
natural frequency of instruments, 28–9
networks, 192–9
neutron moderation, 433
noise, 73–8
normal distribution, 48
normal probability plot, 54–5
notch filter see band pass filter

nozzle flapper, 373
nuclear magnetic resonance (NMR), 433
nuclear sensors, 267–8, 319
null type instruments, 13–4
numerically controlled machine tools, 374, 377, 394, 400

octal numbers, 170–3
ohmmeter, 136
one-out-of-two voting, 240
one's complement, 168
open systems interconnection seven layer model, 197, 198–9
operational amplifier, 87–95
optical fibres *see* fibre optics
optical incremental shaft encoder, 392–4
optical pyrometer, 289–90
optical resonator, 258
optical sensors, 252–9
optical shaft encoder, 393–4
optical tachometer, 408
optical wireless telemetry, 160–1
organic gas sensors, 442
orifice plate, 322–5, 339
oscilloscope *see* cathode ray oscilloscope

paper-tape gas sensor, 441
paperless recorder, 211
parallax error, 102
parallel communication/interface, 187–8, 190–2
parity bit, 189, 190
pass band, 79, 82
passive filters, 81–5
passive instruments, 12–13
PCI (Peripheral component interconnect), 174, 184
PCM *see* pulse code modulation
pendulum scale, 358
pH measurement, 437
phase-locked loop, 77, 142–3
phase measurement, 145–7
phase-sensitive detector, 93–4, 147
photon detector, 290, 292
piezoelectric gas sensor, 443
piezoelectric transducers, 250–1
piezoresistive transducers, 252, 306
Pirani gauge, 313–4
Pitot tube, 322, 326–7
platinum resistance thermometer, 284
pneumatic signal transmission, 154

pollution monitoring and control, 440
positive displacement flowmeter, 328–9, 339
potentiometers, 25, 365–8, 390, 402
  induction, 402
  rotational, 390
  translational, 365–8
preamplifier, 92
precession, 403
precision, 17
presentation of data *see* data presentation
pressure measurement, 12–14, 304–18
  high pressures, 315–6
  low pressures, 312–5
pressure thermometer, 296–7, 302
primary fixed point (of temperature), 271
primary reference standard, 69
probability curve, 47
probability density function, 47
programming and program execution, 173–4
Prony brake, 361
protractors, 426–7
proximity sensors, 381
PRT *see* platinum resistance thermometer
psychrometer, 435
pulse code modulation (PCM), 163
pulsed temperature sensor, 301
pyrometers, 287–93, 301
pyrometric cone, 299

Q factor (quality factor), 131
quantization, 97
quartz thermometer, 297, 302

radiation pyrometer/thermometer, *see* pyrometers
radio telemetry, 161–3
random access memory (RAM), 166
random errors, 33, 42–56
range (of instrument), 18–19
range measurement, 263, 378–81
ratio pyrometer, 292–3
read only memory (ROM), 166–7
recorders *see* signal recorders
recording oscilloscopes, 209
redundancy, 230
reference standards, 67–70
refractive index measurement, 257
refractometer, 433
regression techniques, 215–220

relative humidity, 432
reliability, 224–35
  components in parallel, 229
  components in series, 228
  manufacturing systems, 224–31
  measurement systems, 224–31
  safety systems, 236–41
  software, 232–5
repeatability/reproducibility, 17
resistance measurement, 119–30, 134–7
resistance temperature device (RTD), 283–5, 301
resistance thermometer, 283–5, 301
resistive sensors, 247
resolution, 20
resolver, 398–9
resonant-wire pressure sensor, 311–2, 317
Reynolds number, 324
ring laser gyroscope, 405
ring network, 194–5
rise time, 114
risk analysis, 237
rogue data points, 55–6
rotameter, 327
rotary differential transformer, 391
rotary piston flowmeter, 328–9
rotational acceleration, 417–8
RS232 interface, 190
RTD *see* resistance thermometers
rules (measuring), 419–20

safety systems, 235–41
  safety integrity level (SIL), 237
sample and hold circuit, 97
sampling of signals, 95–7
scale factor drift, 21–2
second order type instruments, 28–9, 205
secondary fixed point (of temperature), 271
secondary reference standard, 67
Seger cone, 299
selected waveband pyrometer, 293
selection of instruments *see* instrument choice
self-calibration, 179, 182
self-diagnosis, 180, 182
semiconductor gas sensors, 442
semiconductor strain gauge, 251–2
semiconductor temperature sensor, 286–7, 301
sensitivity drift, 21–2, 37
sensitivity of measurement, 19–20, 25

sensitivity to disturbance, 20–22
sensor, 8
serial communication/interface, 187–90
shaft encoders, 392–7
shielding, 77, 152
shock measurement, 388–9
shot noise, 75
SI units, 6, 445–51
sigma-delta technique, 268
signal display, 200–1
signal measurement, 102–18
signal processing, 8, 78–101
  analog, 78–95
  digital, 95–101
signal recording, 202–11
signal sampling, 95–7
signal-to-noise ratio, 73
signal transmission, 9, 151–64, 188–99
simplex communication, 188
sing-around flowmeter, 336
slip gauge, 423–4
smart microsensor, 185
smart sensor, 165, 177, 179–80
smart transmitter, 165, 177, 179, 180–4
smoke detector, 254
solid-state electrochemical cells, 442
solid-state gas sensors, 442
sound measurement, 436
span, 18–19
specific humidity, 432
spirit level (angle-measuring), 426–7
spring balance, 359
standard deviation, 44–6
standard error of the mean, 52–3
standard measurement units, 4–6, 445–51
Standards Laboratories, 67–9
standby systems, 240–1
star network, 193–4
static characteristics of instruments, 16–23
static sensitivity of instruments, 26
statistical analysis of data, 42–56
steel rule/tape, 419–20
stop band, 79, 82
storage oscilloscope, 118, 211
strain gauge, 251–2, 258, 305, 371–2
stroboscopic velocity measurement, 410
student-t distribution, 56
synchro, 399–402
synchro-resolver, 398–9
synchro transformer/transmitter, 401
systematic errors, 32–42, 91

Index 475

tabular data presentation, 212–3
tachometric generators (tachometers), 407–10
tank gauge, 341
target meter, 337
telemetry, 160–3
temperature coefficient, 38, 39, 285, 366
temperature measurement, 255–9, 270, 271–303
thermal detector, 290
thermal e.m.f., 75, 272
thermal imaging, 293–4, 349
thermal mass flow meter, 320–1
thermistor, 285–6, 301
thermistor gauge, 314
thermocouple, 272–83, 300–1
thermocouple gauge, 313
thermocouple meter, 110
thermocouple tables, 276–7, 458–63
thermoelectric effect, 75, 272–83
thermography, 293–4, 349
thermometer (liquid–in–glass type), 26, 295, 302
thermopile, 282
Thevenin's theorem, 34–7, 125–8, 452–7
threshold, 20
time base circuit, 117
time constant, 27
tolerance, 17–18, 53–6
torque measurement, 361–4
total measurement error, 56–9
touch screens, 201
traceability, 67–70
transducer, 8
transmitter, 9
triggering, 117
turbine meters, 329–30, 339
twisted pair, 76
two-colour pyrometer, 292–3
two-out-of-three voting, 239
two's complement, 168

UART interface, 174
U-tube manometer, 310–1
ultrasonic flowmeters, 332–6, 339
ultrasonic imaging, 267
ultrasonic level gauge, 344
ultrasonic principles, 260–7
ultrasonic rule, 420
ultrasonic thermometer, 299
ultrasonic transducers, 259–67
ultraviolet (UV) recorder, 208–9
uncertainty, 16–17
units of measurement, 3–6, 445–51
USB (universal serial bus), 174

V24 interface, 190
vacuum pressures, 312–5
variable area flowmeter, 327–8, 339
variable reluctance sensors, 248, 408, 413
variance, 44–6
variation gauge, 428
vee block, 419
velocity measurement:
  rotational, 407–17
  translational, 382–3
venturi, 322, 323, 325
vibrating level sensor, 348
vibrating wire force sensor, 360
vibration measurement, 386–8
viscosity measurement (viscometers), 429–31
voltage comparator, 92–3
voltage follower, 92
voltage to current conversion, 152
voltage to frequency conversion, 153–4
volume flow rate measurement- see flow measurement
volume measurement, 428–9
vortex shedding flowmeter, 332, 339

weigh beam, 357
weighing see mass measurement
Wein bridge, 144–5
wet and dry bulb hygrometer, 435
Wheatstone bridge, 120–1
Wide area network, 198
wringing (gauge blocks), 424

x–y plotter, 145–6

zero drift, 21, 37, 91
zero order type instruments, 25
zirconia gas sensor, 442

MAY 3 1 2002